Algebra II

Course Workbook

2021-22 Edition

Donny Brusca

www.CourseWorkBooks.com

Algebra II
Course Workbook

2021-22 Edition

Donny Brusca

ISBN 979-8-71197-620-2

www.CourseWorkBooks.com

Table of Contents

Introduction

ABOUT THE AUTHOR

Donny Brusca founded CourseWorkBooks in 2010.

He has retired from teaching and school administration after about 30 years of employment, mostly on the high school and college levels. He has a B.S. in mathematics and M.A. in computer and information science from Brooklyn College (CUNY) and a post-graduate P.D. in educational administration and supervision from St. John's University.

His college-level instructional experiences include Brooklyn College, Pace University, and Touro College Graduate School of Technology. His high school teaching experiences include public, charter, Catholic, and Jewish schools.

His administrative experiences include serving as the Student Data Systems Manager at a charter management organization and high school and as the Academic Dean and Mathematics Chairperson at a Manhattan business college. For several years, he was responsible for student scheduling in a high school of about 1000 students.

He has taught courses in basic mathematics, logic, algebra, geometry, probability, discrete structures, computer programming, web design, data structures, switching theory, computer architecture, and application software. He has taught all three Regents level mathematics courses and AP Computer Science A and AP Computer Science Principles.

He owned and operated a successful part-time disc jockey business (Sound Sensation), performing personally at nearly 1000 weddings and private events over a 12-year span.

He has managed and coached high school baseball teams, after playing for over 40 years in organized baseball and softball leagues. He currently works part time as a baseball umpire.

He lives in Staten Island, NY, with his wife, Camille, and their cats.

ABOUT THIS BOOK

The topics in this book are aligned to the national Common Core high school curriculum (C.C.S.S.). The book is intended for use in any state, but it is specifically arranged to correspond to the scope and sequence of topics established by New York State for the successful completion of a course leading to the state Regents examination.

Every topic section begins with an explanation of the Key Terms and Concepts. I have intentionally limited the content here to the most essential ideas. The notes should supplement a fuller presentation of the concepts by the teacher through a more developmental approach.

Calculator Tips explain how to use the graphing calculator to solve problems or check solutions. Keystrokes include button names in rectangles, [STO▸], alternate button features in brackets, [SIN⁻¹], and on-screen text in larger rectangles, [NUM]. Directions for selecting on-screen text (arrow keys and [ENTER]) are usually omitted. Screenshots from a TI-84 Plus CE are shown by default, but differences from the TI-83 or other TI-84 models are noted.

Topic sections include one or more Model Problems, each with a solution and an explanation of steps needed to solve the problem. Steps lettered (A), (B), etc., in the explanations refer to the corresponding lettered steps shown in the solutions. General wording is used in the explanations so that students may apply the steps directly to new but similar problems. However, for clarity, the text often refers to the specific model problem by using *[italicized text in brackets]*. To make the most sense of this writing style, insert the words "in this case" before reading any *[italicized text in brackets]*.

After the Model Problem are a number of Practice Problems in boxed work spaces. These numbered problems are generally arranged in order of increasing level of difficulty.

A supplemental text, Algebra II Regents Questions, containing all Common Core exam questions related to each chapter of this book, is available at CourseWorkBooks.com. Both books are available in digital format as eWorkBooks, allowing students to not only read the books online, but also write directly onto the pages and save their work as PNG files.

Answer Keys are also available at CourseWorkBooks.com.

My goal is to have an error-free book and answer key and I am willing to offer a monetary reward to the first person who contacts me about any error in mathematical content. Simply email me at donny@courseworkbooks.com and be as specific as possible about the error. Corrections are posted under "Errata" at CourseWorkBooks.com.

NEXT GENERATION LEARNING STANDARDS

Several years ago, the New York State Education Department (NYSED) proposed changes to the state mathematics learning standards, which they called **Next Generation**. Despite the name change, there was no major overhaul -- most of the Common Core based standards from 2011 remain intact or were only slightly modified. The introduction of the updated standards has been delayed multiple times, most recently due to the virus pandemic. Here is the most recent timeline for implementation:

Timetable for Next Generation Regents Exam Administration in New York State

Regents Exam	First Administration of Exam Aligned to Next Generation Standards	Last Administration of Exam Aligned to 2011 Standards
Algebra I	June, 2023	June, 2024
Geometry	June, 2024	June, 2025
Algebra II	June, 2025	June, 2026

The following topics will be **added to** the Algebra I curriculum when the Next Generation standards are implemented. They are marked by a [NG] in the section titles.

- Factor a Trinomial by Grouping [NG] (Section 3.1)
- Solve Quadratics with a≠1 [NG] (Section 3.2)
- Inequality of Two Functions [NG] (Section 13.6)

The first two topics are being moved from the Algebra I curriculum.

The following topics will be **eliminated from** the Algebra II curriculum when the Next Generation standards are implemented. They are marked by a [CC] in the section titles.

- Linear Systems in Three Variables (Section 1.1)
- Operations with Square Roots [CC] (Section 2.1)
- Rationalize Monomial Denominators [CC] (Section 2.2)
- Focus and Directrix [CC] (Section 3.5)
- Polynomial Identities [CC] (Section 6.6)
- Algebraically Determine Even or Odd (Section 13.2)
- Series of Events [CC] (Section 15.4)
- Mean Difference [CC] (Section 16.6)

The two topics on Radicals are being moved to the Algebra I curriculum.

Officially, the use of sigma notation is a new standard; however, it is used in the sections on finding the sums of series since it helps students to better understand these concepts.

Chapter 1. Linear Functions

1.1 Linear Systems in Three Variables [CC]

Key Terms and Concepts

In Algebra I, we learned two methods that can be used to solve a system of two equations in two variables (or *unknowns*, such as x and y). These are the addition method (also known as the *elimination method*) and the substitution method.

These same methods can be used to solve a system of three equations in three variables. Simply perform the above methods on two equations at a time to create equations in only two variables, and then repeat the process until each of the three variables can be isolated.

To solve a linear system in three variables:

1. Select two equations and use the addition method to eliminate a variable. Call the new equation *Result A.*
2. Select another pair of equations and eliminate the *same variable.* Call the new equation *Result B.*
3. Use *Result A* and *Result B* to eliminate a second variable. Solve to find the value of the one remaining variable.
4. Substitute the value of the variable found in step 3 into *Result A* or *Result B.* Solve to find the value of a second variable.
5. Substitute for the two found variables into one of the original equations to find the value of the third variable.

The above steps should be used as a guideline, but there may be quicker solutions in some cases. The objective is to isolate one variable by eliminating the other two. Then, by substitution, find the values of the other two variables.

It is possible for a system of three equations to have **no solutions**. This happens when the three lines in three-dimensional space do not intersect at a single point. When a system has no solutions, following the steps above will lead to an impossibility, such as $0 = 1$.

It is also possible for a system of three equations to have **infinitely many solutions**. This happens when all three equations represent the same line, such as $x + y + z = 1$, $2x + 2y + 2z = 2$, and $3x + 3y + 3z = 3$. Following the steps above for a system with infinitely many solutions may result in an equation that is always true, such $0 = 0$.

One valuable application of this skill is to find the equation of a parabola algebraically, given three points. We do this by substituting each point's coordinates for x and y into the general equation $ax^2 + bx + c = y$ and then solving the resulting system of three linear equations. (It is common to place y on the right side of the equation in this procedure.)

Example: To find the equation of the parabola that passes through the points $(-3,22)$, $(2,-3)$, and $(4,15)$, we can substitute the x and y values of each point into the general equation $ax^2 + bx + c = y$, resulting in these three equations.

$$a(-3)^2 + b(-3) + c = 22$$
$$a(2)^2 + b(2) + c = -3$$
$$a(4)^2 + b(4) + c = 15$$

Simplifying gives us these three equations, which we can number (1) to (3).

(1) $9a - 3b + c = 22$

(2) $4a + 2b + c = -3$

(3) $16a + 4b + c = 15$

Adding equation (1) and -1 times equation (2) gives us *Result A*.

$$9a - 3b + c = 22$$
$$\underline{-4a - 2b - c = 3}$$
$$5a - 5b = 25$$

(A) $a - b = 5$ [divide the whole equation by 5]

Adding equation (1) and -1 times equation (3) gives us *Result B*.

$$9a - 3b + c = 22$$
$$\underline{-16a - 4b - c = -15}$$
$$-7a - 7b = 7$$

(B) $a + b = -1$ [divide the whole equation by -7]

Adding *Result A* and *Result B* gives us

$$a - b = 5$$
$$\underline{a + b = -1}$$
$$2a = 4$$
$$a = 2$$

We can now substitute 2 for a in *Result B* (or A, if you prefer) to find b. Lastly, substitute a and b into equation (1) (or equation 2 or 3) to find c.

$2 + b = -1$ $\quad$ $9(2) - 3(-3) + c = 22$

$b = -3$ $\quad$ $27 + c = 22$

$c = -5$

We found $a = 2$, $b = -3$, and $c = -5$, so the equation of the parabola is

$$y = 2x^2 - 3x - 5$$

CALCULATOR TIP

We can solve a linear system using the calculator's matrix feature:

1. Press [2nd][MATRIX] [MATH] [ALPHA][B] to select the **rref** function (which stands for "reduced row echelon form").
2. The number of rows represents the number of equations, and the number of columns should be one more than the number of variables. So, for a system of 3 equations with 3 variables, create a matrix with 3 rows and 4 columns.
   - On the TI-84, press [ALPHA][F3], select 3 and 4 with the arrow keys, then select OK.
   - On the TI-83, press [2nd][MATRIX][1] and type 3×4 for 3 rows and 4 columns.
3. Each row of the matrix represents an equation. Enter the coefficients of the first variable in column 1, the coefficients of the second variable in column 2, and so on. Enter the constants on the right side of the equations into the last column.
   - On the TI-84, exit the matrix by pressing [▸]. Then, press [)][ENTER].
   - On the TI-83, press [2nd][QUIT][2nd][MATRIX][1][)][ENTER].
4. The resulting matrix will show the values of the variables in the last column.

Example: The screenshots below show how the calculator solves the system used in the previous example,

$$9a - 3b + c = 22$$
$$4a + 2b + c = -3$$
$$16a + 4b + c = 15$$

The solutions matrix shows that $a = 2$, $b = -3$, and $c = -5$.

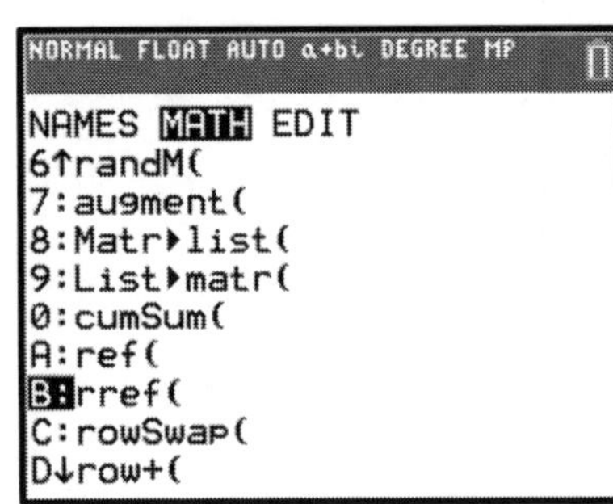

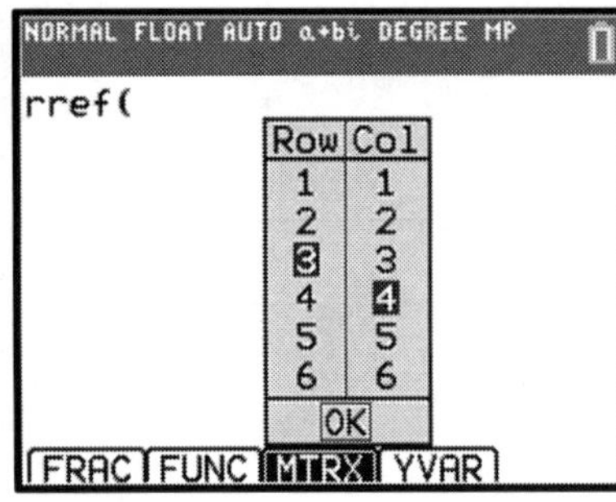

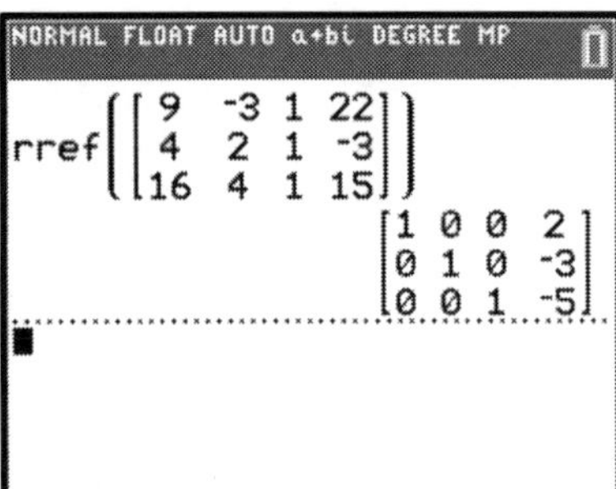

MODEL PROBLEM

Solve the following system of equations:

(1) $x + y + z = 2$

(2) $6x - 4y + 5z = 31$

(3) $5x + 2y + 3z = 14$

Solution:

(A)	$2(5x + 2y + 3z = 14) \rightarrow$	$10x + 4y + 6z = 28$	$2 \times$ equation (3)
		$6x - 4y + 5z = 31$	equation (2)
		$16x \quad + 11z = 59$	Result (A)
(B)	$-2(x + y + z = 2) \rightarrow$	$-2x - 2y - 2z = -4$	$-2 \times$ equation (1)
		$5x + 2y + 3z = 14$	equation (3)
		$3x \quad + z = 10$	Result (B)
(C)		$16x + 11z = 59$	Result (A)
	$-11(3x + z = 10) \rightarrow$	$-33x - 11z = -110$	$-11 \times$ Result (B)
		$-17x \quad = -51$	
		$x = 3$	
(D)	$16(3) + 11z = 59$	substitute for x into Result (A)	
	$48 + 11z = 59$		
	$11z = 11$		
	$z = 1$		
(E)	$(3) + y + (1) = 2$	substitute for x and z into equation (1)	
	$y + 4 = 2$		
	$y = -2$		

Explanation of steps:

(A) Select two equations and use one of the methods to eliminate a variable. *[Multiplying equation (3) by 2 and adding it to equation (2) eliminates y.]*

(B) Select another pair of equations and eliminate the **same variable**. *[Multiplying equation (1) by −2 and adding it to equation (3) also eliminates y.]*

(C) Use Result (A) and Result (B) to eliminate a second variable and solve. This will find the value of one variable. *[Multiplying Result (B) by −11 and adding it to Result (A) eliminates z. We find $x = 3$.]*

(D) Substitute for the variable in Result (A) or Result (B). This will find the value of a second variable. *[Substituting 3 for x in the equation from step (A) gives us $z = 1$.]*

(E) Substitute for the two found variables into one of the original equations to find the value of the third variable. *[We can now substitute x and y into equation (1), chosen because it's the simplest of the three, to find $y = -2$.]*

Practice Problems

1. Solve algebraically: $x = 3$ $5x + 4y = -9$ $-x + 4y - 2z = -25$	2. Solve algebraically: $x + 3y + z = 10$ $x + y + z = 2$ $y - 2z = 2$

3. Solve algebraically: $2x - 4y + 5z = -33$ $4x - y = -5$ $-2x + 2y - 3z = 19$ Check your solutions on the calculator.	4. Solve algebraically: $x - 2y + 3z = 7$ $2x + y + z = 4$ $-3x + 2y - 2z = -10$ Check your solutions on the calculator.

5. Algebraically, find the equation of the parabola through points $(-1,9)$, $(2,3)$, and $(5,15)$.	6. Algebraically, find the equation of the parabola through points $(-1,-2)$, $(1,0)$, and $(2,7)$.

Chapter 2. Irrational Expressions

2.1 Operations with Square Roots [CC]

Key Terms and Concepts

Square roots may be combined by addition or subtraction only if, when expressed in simplest radical form, they are like radicals. **Like radicals** have the **same radicand.** *(Note: Like radicals must also have the same index, but we are only concerned with square roots in this chapter. We will look at radicals with indices other than 2 in Chapter 8.)*

Sometimes unlike radicals may be simplified into like radicals.
Example: $\sqrt{12}$ and $\sqrt{75}$ can be simplified into $2\sqrt{3}$ and $5\sqrt{3}$, which are like radicals.

Combine like radicals just as you would combine like terms: add or subtract the coefficients and keep the radicand unchanged. $m\sqrt{a} + n\sqrt{a} = (m+n)\sqrt{a}$
Example: $2\sqrt{3} + 5\sqrt{3} - \sqrt{3} = (2+5-1)\sqrt{3} = 6\sqrt{3}$

To multiply radicals, separately find the product of their coefficients and the product of their radicands, then simplify if possible. $m\sqrt{a} \cdot n\sqrt{b} = mn\sqrt{ab}$
Example: $(5\sqrt{3})(2\sqrt{7}) = 10\sqrt{21}$

To divide radicals, separately find the quotient of their coefficients and the quotient of their radicands, then simplify if possible. $\frac{m\sqrt{a}}{n\sqrt{b}} = \frac{m}{n} \cdot \sqrt{\frac{a}{b}}$

Example: $\frac{6\sqrt{72}}{3\sqrt{8}} = 2\sqrt{9} = 2 \cdot 3 = 6$

Sometimes, multiple operations involving radicals need to be performed.
Example: Simplify $\sqrt{2}(\sqrt{10} + 4)$ using the distributive property.
$\sqrt{2}(\sqrt{10} + 4) = \sqrt{20} + 4\sqrt{2} = 2\sqrt{5} + 4\sqrt{2}$

MODEL PROBLEM 1: *ADDING OR SUBTRACTING RADICALS*

Express the sum $3\sqrt{8} + 2\sqrt{2}$ in simplest radical form.

Solution:

(A) $3\sqrt{8}$ can be simplified as follows: $3\sqrt{\boxed{2 \cdot 2} \cdot 2} = 3 \cdot 2\sqrt{2} = 6\sqrt{2}$.

(B) So, $3\sqrt{8} + 2\sqrt{2} = 6\sqrt{2} + 2\sqrt{2} = 8\sqrt{2}$.

Explanation of steps:

(A) Express each term in simplest radical form.

(B) Combine like radicals by adding or subtracting their coefficients.

PRACTICE PROBLEMS

1. Add $\sqrt{75} + \sqrt{3}$.	2. Add $\sqrt{27} + \sqrt{12}$.
3. Subtract $\sqrt{48} - \sqrt{12}$.	4. Add $\sqrt{150} + \sqrt{24}$.

5. Add $3\sqrt{2} + \sqrt{8}$.	6. Subtract $5\sqrt{2} - \sqrt{32}$.
7. Subtract $\sqrt{72} - 3\sqrt{2}$.	8. Subtract $2\sqrt{50} - \sqrt{2}$.
9. Add $5\sqrt{7} + 3\sqrt{28}$.	10. Add $6\sqrt{50} + 6\sqrt{2}$.

11. Express $\sqrt{25} - 2\sqrt{3} + \sqrt{27} + 2\sqrt{9}$ in simplest radical form.	12. Express $y\sqrt{3} - \left(\sqrt{32} + y\sqrt{27}\right)$ in simplest radical form.

MODEL PROBLEM 2: ***MULTIPLYING RADICALS***

Express the product $\left(5\sqrt{8}\right)\left(7\sqrt{3}\right)$ in simplest radical form.

Solution:

(A) $\left(5\sqrt{8}\right)\left(7\sqrt{3}\right) = 35\sqrt{24}$

(B) Simplifying, we get $35\sqrt{24} = 35\sqrt{\boxed{2\cdot 2}\cdot 2\cdot 3} = 35 \cdot 2\sqrt{2\cdot 3} = 70\sqrt{6}$

Explanation of steps:

(A) Find the product of the coefficients *[*$5 \times 7 = 35$*]*.
Find the product of the radicands *[*$8 \times 3 = 24$*]*.

(B) Express in simplest radical form.

PRACTICE PROBLEMS

13. Express $\sqrt{6} \cdot \sqrt{15}$ in simplest form.	14. Express $\sqrt{90} \cdot \sqrt{40} - \sqrt{8} \cdot \sqrt{18}$ in simplest form.
15. Express $3\sqrt{20}(2\sqrt{5} - 7)$ in simplest form.	16. Express $3\sqrt{7}(\sqrt{14} + 4\sqrt{56})$ in simplest form.
17. Express the product of $(3 + \sqrt{5})$ and $(3 - \sqrt{5})$ in simplest form.	18. Express $(5 + \sqrt{2})(\sqrt{2} - 4)$ in simplest form.

19. Express $5\sqrt{3x^3} \cdot 2\sqrt{6x^4}$ in simplest form.	20. Express $3\sqrt{2x} \cdot 5\sqrt{x^5y^2}$ in simplest form.
21. Simplify: $(3\sqrt{x} - 3)^2$	22. The legs of a right triangle are represented by $x + \sqrt{2}$ and $x - \sqrt{2}$. Write an expression for the length of the hypotenuse of the right triangle in simplest form.

MODEL PROBLEM 3: ***DIVIDING RADICALS***

Express $\dfrac{9\sqrt{20}}{3\sqrt{5}}$ in simplest radical form.

Solution:

$$\frac{9\sqrt{20}}{3\sqrt{5}} = 3\sqrt{4} = 3 \cdot 2 = 6$$

Explanation of steps:

(A) Find the quotient of the coefficients *[*$9 \div 3 = 3$*]*.
Find the quotient of the radicands *[*$20 \div 5 = 4$*]*.

(B) Express in simplest radical form.

Practice Problems

23. Express $\frac{\sqrt{65}}{\sqrt{5}}$ in simplest form.

24. Express $\frac{\sqrt{84}}{2\sqrt{3}}$ in simplest radical form.

25. Express $\frac{20\sqrt{100}}{4\sqrt{2}}$ in simplest form.

26. Express $\frac{3\sqrt{75}+\sqrt{27}}{3}$ in simplest form.

27. Express in simplest form:

$$\frac{\sqrt{48}-5\sqrt{27}+2\sqrt{75}}{\sqrt{3}}$$

28. Express in simplest form:

$$\frac{\sqrt{27}+\sqrt{75}}{\sqrt{12}}$$

29. Express in simplest form: $\frac{16\sqrt{21}}{2\sqrt{7}} - 5\sqrt{12}$	30. Express in simplest form: $\frac{\sqrt{108x^5y^8}}{\sqrt{6xy^5}}$

2.2 Rationalize Monomial Denominators [CC]

KEY TERMS AND CONCEPTS

When working with algebraic fractions involving square roots, we prefer to change it to an equivalent fraction that does not include any square roots in the denominator. Eliminating radicals from the denominator is called **rationalizing the denominator**. Whenever we simplify fractions, we should rationalize their denominators.

The method for rationalizing a denominator depends on whether the denominator contains a monomial or a binomial.

If a fraction's denominator has a *monomial* containing a square root, we will multiply both the numerator and denominator by that square root. This will work because multiplying a square root by itself will eliminate the radical sign: $\sqrt{x} \cdot \sqrt{x} = x$.

To rationalize a monomial denominator of a fraction:

1. Multiply the fraction by a fraction having the radical part in the numerator and denominator (a form of 1).
2. Simplify.

Example: To simplify $\frac{4}{3\sqrt{6}}$, multiply the fraction by $\frac{\sqrt{6}}{\sqrt{6}}$, which gives us

$$\frac{4}{3\sqrt{6}} \cdot \frac{\sqrt{6}}{\sqrt{6}} = \frac{4\sqrt{6}}{18} = \frac{2\sqrt{6}}{9}.$$

MODEL PROBLEM

Rationalize the denominator of $\frac{10}{\sqrt{5}}$ and write the equivalent expression in simplest form.

Solution:

(A) (B) (C)

$$\frac{10}{\sqrt{5}} \cdot \left(\frac{\sqrt{5}}{\sqrt{5}}\right) = \frac{10\sqrt{5}}{5} = 2\sqrt{5}$$

Explanation of steps:

(A) If the fraction has a single irrational square root term in the denominator *[$\sqrt{5}$]*, create a new fraction with this radical as its numerator and denominator.

(B) Multiply the two fractions. The denominator of the product is now rational.

(C) Simplify.

PRACTICE PROBLEMS

1. Rationalize the denominator and express in simplest form: $\dfrac{3}{\sqrt{7}}$	2. Rationalize the denominator and express in simplest form: $\dfrac{2}{\sqrt{2}}$
3. Rationalize the denominator and express in simplest form: $\dfrac{3\sqrt{5}}{2\sqrt{10}}$	4. Rationalize the denominator and express in simplest form: $\dfrac{3-\sqrt{8}}{\sqrt{3}}$
5. Multiply $\dfrac{2}{\sqrt{3}} \times \dfrac{\sqrt{2}}{5}$ and express the product as a fraction with a rational denominator.	6. Rationalize the denominator and express in simplest form: $\dfrac{\frac{1}{\sqrt{5}}+\frac{1}{\sqrt{5}}}{\sqrt{5}}$

2.3 Rationalize Binomial Denominators

KEY TERMS AND CONCEPTS

If a fraction's denominator has a *binomial* containing a square root, we will need to find the conjugate of the binomial. The **conjugate** of a binomial is the same expression but with the opposite operation (addition or subtraction) between the terms.

Example: The conjugate of $x + \sqrt{2}$ is $x - \sqrt{2}$, and the conjugate of $2\sqrt{3} - 5$ is $2\sqrt{3} + 5$.

If we multiply a binomial containing a square root by its conjugate, the new expression will not contain a square root. Remember, when we multiply the sum and difference of the same two terms, we get the difference of their squares: $(a + b)(a - b) = a^2 - b^2$.

To rationalize a binomial denominator of a fraction:

1. Find the conjugate of the denominator.
2. Multiply the fraction by a fraction having the conjugate in the numerator and denominator (a form of 1).
3. Simplify.

Example: To simplify $\frac{2}{3 - \sqrt{2}}$, we will use the conjugate of the denominator, $3 + \sqrt{2}$.

Multiply the fraction by $\frac{3 + \sqrt{2}}{3 + \sqrt{2}}$. $\frac{2}{3 - \sqrt{2}} \cdot \left(\frac{3 + \sqrt{2}}{3 + \sqrt{2}}\right) = \frac{6 + 2\sqrt{2}}{9 - 2} = \frac{6 + 2\sqrt{2}}{7}$

Model Problem

Rationalize the denominator of $\frac{2-\sqrt{3}}{4+\sqrt{3}}$.

Solution:

$$\frac{2-\sqrt{3}}{4+\sqrt{3}} \cdot \overset{\text{(A)}}{\left(\frac{4-\sqrt{3}}{4-\sqrt{3}}\right)} = \overset{\text{(B)}}{\frac{8-2\sqrt{3}-4\sqrt{3}+3}{16-3}} = \overset{\text{(C)}}{\frac{11-6\sqrt{3}}{13}}$$

Explanation of steps:

(A) Find the conjugate of the denominator $[4-\sqrt{3}]$ and multiply the fraction by a fraction having the conjugate in the numerator and denominator.

(B) Multiply the numerators as a product of binomials. The product of the denominators is the difference of the two squares.

(C) Combine like terms and simplify, if possible.

Practice Problems

1. Simplify: $\frac{\sqrt{5}}{7-\sqrt{5}}$	2. Express the reciprocal of $3-\sqrt{7}$ in simplest radical form with a rational denominator.
3. Simplify: $\frac{5}{4-\sqrt{11}}$	4. Simplify: $\frac{4}{5-\sqrt{13}}$

5. Simplify: $\frac{\sqrt{2}}{\sqrt{14}+4}$	6. Simplify: $\frac{\sqrt{3}}{\sqrt{3}+5}$
7. Simplify: $\frac{2-\sqrt{2}}{2+\sqrt{2}}$	8. Simplify: $\frac{\sqrt{3}+5}{\sqrt{3}-5}$
9. Simplify: $\frac{\sqrt{6}+8}{\sqrt{2}+\sqrt{3}}$	10. Simplify: $\frac{\sqrt{xy}}{\sqrt{x}-\sqrt{y}}$

11. Find the width of a rectangle in simplest radical form if the length of the rectangle is $\sqrt{5}-1$ units and the area is 2 square units.	12. Find the altitude of a triangle in simplest radical form if the base of the triangle measures $6+2\sqrt{2}$ units and the area is $8+12\sqrt{2}$ square units.

Chapter 3. Quadratic Functions

3.1 Factor a Trinomial by Grouping [NG]

Key Terms and Concepts

In Algebra I, we saw the product-sum method for factoring a trinomial for which the lead coefficient is 1. That is, we factored trinomials of the form $x^2 + bx + c$ by finding two integers that multiply to give us c and add to give us b. When the lead coefficient isn't 1 (that is, for $ax^2 + bx + c$ where a is any integer except 0 or 1), that simple method can no longer help us. However, as long as they are not prime, we can still factor these types of trinomials. First, we **expand** them into equivalent polynomials with four terms, and then **factor by grouping**.

Consider the procedure to multiply $(x - 2)(4x + 3)$ by distribution:

$(x - 2)(4x + 3)$

$x(4x + 3) - 2(4x + 3)$ Multiply each term of the first binomial by the second factor.

$4x^2 + 3x - 8x - 6$ The result is a four-term polynomial.

$4x^2 - 5x - 6$ By combining like terms, we end up with a trinomial.

This factoring method will essentially perform these steps in reverse.

To factor a trinomial $ax^2 + bx + c$ where a, b, and c are non-zero integers:

1. Find two integers that multiply to give you ac but add to give you b.
2. Break up the middle term into the sum of two terms with these coefficients.
3. Group the terms into two pairs, where each pair has a common factor.
4. Factor the GCF from each group.
5. Factor out the common binomial (reverse distribute).

MODEL PROBLEM

Factor $4x^2 - 5x - 6$.

Solution:

(A) $4x^2 - 5x - 6$

$ac = 4(-6) = -24$

and $b = -5$

+3 and -8

(B) $4x^2 \boxed{+3x - 8x} - 6$

(C) $\boxed{4x^2 + 3x}\boxed{-8x - 6}$

(D) $x(4x + 3) - 2(4x + 3)$

(E) $(x - 2)(4x + 3)$

Explanation of steps:

(A) Find two integers that multiply to give you *ac* but add to give you *b*. *[$3 \times (-8) = -24$ and $3 + (-8) = -5$.]*

(B) Break up the middle term into the sum of two terms with these coefficients. *[$-5x \rightarrow +3x - 8x$]*

(C) Group the terms into two pairs, where each pair has a common factor.

(D) Factor the GCF from each group. *[x and -2]*

(E) Factor out the common binomial (reverse distribute). *[The factors outside parentheses are rewritten as a binomial factor, $(x - 2)$.]*

PRACTICE PROBLEMS

1. Factor by grouping: $6x^2 + x - 2$	2. Factor by grouping: $12x^2 + 5x - 2$
3. Factor by grouping: $12x^2 - 29x + 15$	4. Factor by grouping: $6x^2 - 11x + 4$

5. Factor by grouping: $15x^2 + 14x - 8$	6. Factor by grouping: $-10x^2 - 29x - 10$
7. Factor $4x^2 + 12x + 9$. What is the square root of this trinomial, written as a binomial?	8. Factor: $-30 - 57x + 6x^2$

9. The product of two factors is $2x^2 + x - 6$. One of the factors is $(x + 2)$. What is the other factor?

3.2 Solve Quadratics with a≠1 [NG]

Key Terms and Concepts

In Algebra I, we learned how to solve quadratic equations of the form $x^2 + bx + c = 0$ (where the leading coefficient is 1). In this chapter, we will look at four methods for solving a quadratic equation of the form $ax^2 + bx + c = 0$, where $a \neq 1$.

Method 1: The Transforming Method

1. Write a new equation of the form $n^2 + bn + ac = 0$, where n is a new variable and a, b, and c are taken from the original equation.
2. Solve the new equation for n.
3. For each solution for n, a solution of the original equation is $x = \frac{n}{a}$.

The Transforming Method is a relatively new method and not as commonly known as the next three. To see why the transforming method works, we can look at the quadratic formula, $x = \frac{-b\pm\sqrt{b^2-4ac}}{2a}$. For the new equation, $n^2 + bn + ac = 0$, we've replaced x with n, a with 1, and c with ac. Substituting these into the quadratic formula results in $n = \frac{-b\pm\sqrt{b^2-4(1)(ac)}}{2(1)} = \frac{-b\pm\sqrt{b^2-4ac}}{2}$. The resulting formula for n is the same as the formula for x except that it is missing the division by a. Therefore, $x = \frac{n}{a} = \frac{-b\pm\sqrt{b^2-4ac}}{2a}$.

Method 2: Factor by Grouping

1. Find two integers that multiply to give you ac but add to give you b.
2. Break up the middle term into the sum of two terms with these coefficients.
3. Group the terms into two pairs, where each pair has a common factor.
4. Factor the GCF from each group.
5. Factor out the common binomial (reverse distribute).
6. Set each factor equal to zero and solve for x.

The first two methods require a trinomial that is factorable, and not all trinomials are. The advantage of the next two methods is that they will also work when factoring isn't possible.

Method 3: Complete the Square

1. Divide both sides (*each term*) by the leading coefficient, resulting in a new equation of the form $x^2 + bx + c = 0$.
2. Add the opposite of c to both sides.
3. Add $\left(\frac{b}{2}\right)^2$ to both sides.
4. Factor the trinomial into a binomial squared.
5. Take the square root of both sides. Use the ± symbol on the right side and simplify radicals.
6. Solve for x, remembering that ± gives two possible solutions.

Method 4: Quadratic Formula

1. Write the quadratic formula: $x = \frac{-b\pm\sqrt{b^2-4ac}}{2a}$.
2. Substitute for a, b, and c, and evaluate.

You can check your solutions using the calculator; however, be aware that the calculator will display irrational numbers as decimal approximations.

CALCULATOR TIP

If you plan on using the quadratic formula more than once, it may be convenient to store the formula on your calculator as a reusable program. *Be aware, however, that if your calculator is placed in Test Mode for an exam, you will no longer have access to the program.*

To create a program to calculate the quadratic formula:

1. To allow for complex roots, press [MODE] and set your calculator to $a + bi$ mode.
2. Press [PRGM] [NEW] [1] for Create New.
3. Type a name, such as QUAD, using the letter keys, and press [ENTER]. Alpha lock is on, so you don't have to press the [ALPHA] key.
4. Press [PRGM] [I/O] [2] for Prompt. Type [ALPHA] [A] [,] [ALPHA] [B] [,] [ALPHA] [C] [ENTER].

continued on the next page ...

5. To create the formula, $m = \frac{-b \pm \sqrt{b^2-4ac}}{2a}$, type [(][(-)][ALPHA][B][+][2nd][√] [ALPHA][B][x^2][-][4][ALPHA][A][ALPHA][C][)][)] [÷] [(][2][ALPHA][A][)][STO▸][ALPHA][M][ENTER].
6. To create the formula, $n = \frac{-b \pm \sqrt{b^2-4ac}}{2a}$, type [(][(-)][ALPHA][B][-][2nd][√] [ALPHA][B][x^2][-][4][ALPHA][A][ALPHA][C][)][)] [÷] [(][2][ALPHA][A][)][STO▸][ALPHA][N][ENTER].
7. Press [PRGM] I/O [3] for Disp. Type the line "X=",M,N using the keys, [ALPHA]["][X,T,Θ,n][2nd][MATH][1][ALPHA]["][,][ALPHA][M][,][ALPHA][N][ENTER].
8. Press [2nd][QUIT]. This will save your program.
9. Press [PRGM], select the QUAD program, and press [ENTER][ENTER].
10. Enter a value for each prompt and press [ENTER]. For example, to find the roots of $2x^2 + 3x + 4 = 0$, enter 2, 3, and 4. The roots are $-\frac{3}{4} \pm \frac{\sqrt{23}}{4}i$ but they are displayed as $-0.75 \pm 1.198957881i$.

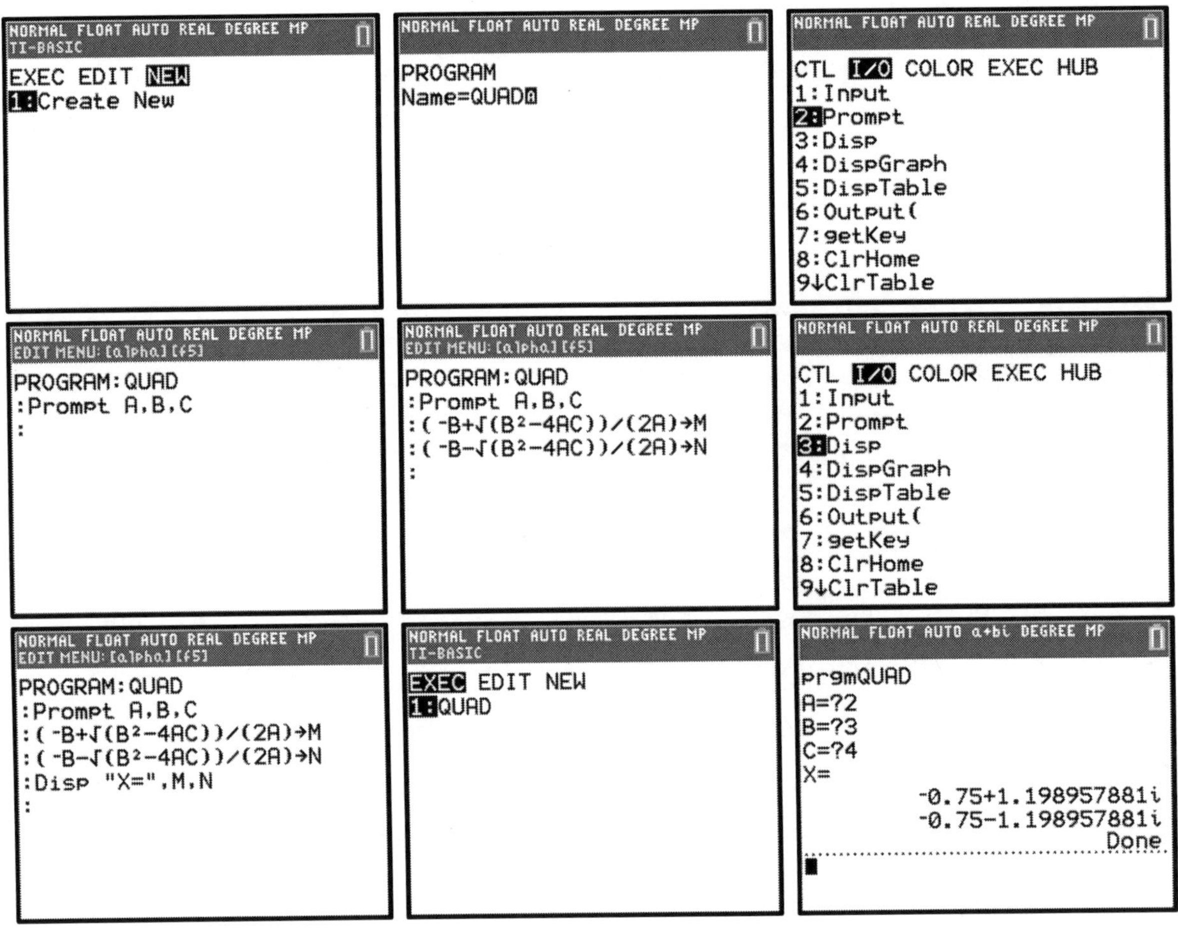

Model Problem 1: *Transforming Method*

Solve $3x^2 + 5x - 2 = 0$ by the transforming method.

Solution:

$3x^2 + 5x - 2 = 0$

(A) $n^2 + 5n - 6 = 0$

(B) $(n + 6)(n - 1) = 0$

$n + 6 = 0 \quad n - 1 = 0$

$n = -6$ or $n = 1$

(C) $x = -\frac{6}{3}$ or $x = \frac{1}{3}$

$\left\{-2, \frac{1}{3}\right\}$

Explanation of steps:

(A) Write a new equation of the form $n^2 + bn + ac = 0$, where n is a new variable and a, b, and c are taken from the original equation. *[$ac = 3(-2) = -6$]*

(B) Solve the new equation for n.

(C) For each solution for n, a solution of the original equation is $x = \frac{n}{a}$. *[$a = 3$]*

Practice Problems

1. Solve by the transforming method: $10x^2 + 9x + 2 = 0$	2. Solve by the transforming method: $2x^2 - 3x - 2 = 0$
3. Solve by the transforming method: $12x^2 + 29x + 15 = 0$	4. Solve by the transforming method: $4x^2 + 109x + 225 = 0$

MODEL PROBLEM 2: *FACTOR BY GROUPING*

Factor by grouping to solve $3x^2 + 5x - 2 = 0$.

Solution:

(A) $3x^2 + 5x - 2 = 0$

$ac = 3(-2) = -6$

and $b = 5$

-1 and +6

(B) $3x^2 \boxed{-x + 6x} - 2 = 0$

(C) $\boxed{3x^2 - x}\boxed{+6x - 2} = 0$

(D) $x(3x - 1) + 2(3x - 1) = 0$

(E) $(x + 2)(3x - 1) = 0$

(F) $x + 2 = 0 \quad 3x - 1 = 0$

$x = -2 \qquad x = \frac{1}{3}$

$\left\{-2, \frac{1}{3}\right\}$

Explanation of steps:

(A) Find two integers that multiply to give you *ac* *[−6]* but add to give you *b [5]*.

(B) Break up the middle term into the sum of two terms with these coefficients. *[$+5x \Rightarrow -x + 6x$]*

(C) Group the terms into two pairs, where each pair has a common factor.

(D) Factor the GCF from each group. *[x and 2]*

(E) Factor out the common binomial (reverse distribute). *[The factors outside parentheses are rewritten as a binomial factor, $(x + 2)$.]*

(F) Set each factor equal to zero and solve for *x*.

PRACTICE PROBLEMS

5. Factor by grouping to solve for *x*: $10x^2 + 9x + 2 = 0$	6. Factor by grouping to solve for *x*: $3x^2 - 14x + 8 = 0$
7. Factor by grouping to solve for *x*: $2x^2 - 3x - 2 = 0$	8. Factor by grouping to solve for *x*: $4x(x - 1) = 15$

MODEL PROBLEM 3: *COMPLETE THE SQUARE*

Solve $4x^2 - 16x - 9 = 0$ by completing the square.

Solution:

$4x^2 - 16x - 9 = 0$

(A) $x^2 - 4x - \frac{9}{4} = 0$

(B) $x^2 - 4x = \frac{9}{4}$

(C) $\left(\frac{b}{2}\right)^2 = \left(-\frac{4}{2}\right)^2 = 4$

$x^2 - 4x + 4 = \frac{9}{4} + \frac{16}{4}$

(D) $(x-2)^2 = \frac{25}{4}$

(E) $x - 2 = \pm\sqrt{\frac{25}{4}} = \pm\frac{5}{2}$

(F) $x = 2 \pm \frac{5}{2} = -\frac{1}{2}$ or $\frac{9}{2}$

$\left\{-\frac{1}{2}, \frac{9}{2}\right\}$

Explanation of steps:

(A) Divide both sides (*each term*) by the leading coefficient. *[Divide both sides – each term – by 4.]*

(B) Add the opposite of c to both sides.

(C) Add $\left(\frac{b}{2}\right)^2$ to both sides. *[Add 4, which is equivalent to $\frac{16}{4}$.]*

(D) Factor the trinomial into a binomial squared.

(E) Take the square root of both sides. Use the $\pm$ symbol on the right side and simplify radicals.

(F) Solve for x, remembering that $\pm$ gives two solutions

PRACTICE PROBLEMS

9. Solve $4x^2 + 8x - 12 = 0$ by completing the square.	10. Solve $3x^2 - 18x - 5 = 16$ by completing the square.
11. Solve $4x^2 + 8x = 45$ by completing the square.	12. Solve $3x^2 - 12x + 2 = 0$ by completing the square.

Model Problem 4: *Quadratic Formula*

Solve $3x^2 + 5x - 2 = 0$ by the quadratic formula.

Solution:

(A) $x = \frac{-b \pm \sqrt{b^2 - 4ac}}{2a}$

(B) $x = \frac{-5 \pm \sqrt{5^2 - 4(3)(-2)}}{2(3)}$

$x = \frac{-5 \pm \sqrt{49}}{6} = \frac{-5 \pm 7}{6}$

$x = \left\{-2, \frac{1}{3}\right\}$

Explanation of steps:

(A) Write the quadratic formula.

(B) Substitute for a, b, and c, and simplify.

Practice Problems

13. Solve $10x^2 + 9x + 2 = 0$ by the quadratic formula.	14. Solve $-3x^2 + 4x - 1 = 0$ by the quadratic formula.
15. Solve $4x^2 + 8x - 9 = 0$ by the quadratic formula.	16. Solve $3x^2 - 12x + 2 = 0$ by the quadratic formula.

3.3 Graphs of Quadratic Functions

Key Terms and Concepts

A **parabola** is a graph of a quadratic function of the form $y = ax^2 + bx + c$ where $a \neq 0$.
Example: For $y = x^2 - 4x + 3$, $a = 1$, $b = -4$, and $c = 3$.

If $a > 0$, then the parabola "**opens up**" like the letter U, but if $a < 0$, then the parabola "**opens down**" like an upside-down U.
Example: For $y = x^2 - 4x + 3$, $a = 1$, so the parabola opens up.

We can **graph the parabola** by:

1. drawing the axis of symmetry as a dashed line for reference only
2. plotting the vertex
3. plotting the y-intercept and its reflection
4. plotting any additional points
5. connecting the points with a solid curve

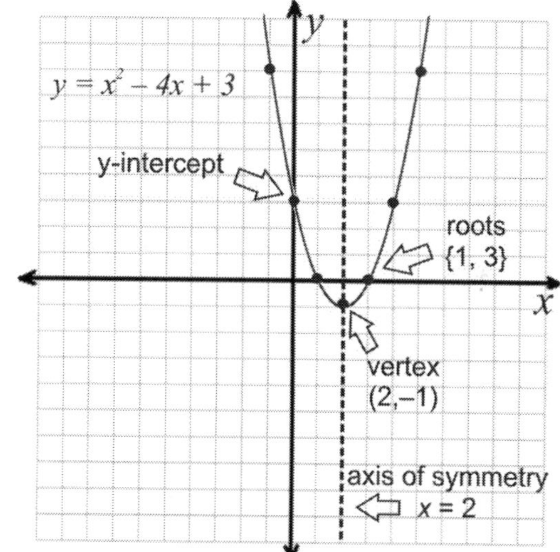

The **axis of symmetry** is a vertical line that divides the parabola into two parts that are mirror images of each other. The **equation for the axis of symmetry** is: $x = \frac{-b}{2a}$.
Example: The axis of symmetry for the parabola whose equation is $y = x^2 - 4x + 3$ is the line $x = \frac{-b}{2a} = \frac{-(-4)}{2(1)} = 2$, or simply $x = 2$.

An alternate method for finding the *equation for the axis of symmetry* is to find the *average of the roots*. This may be easier if the quadratic function is already factored. The **factored form** (or **intercept form**) of a quadratic equation is $y = k(x - m)(x - n)$, where m and n are the real roots.
Example: If we factor $y = x^2 - 4x + 3$, we get $y = (x - 3)(x - 1)$, so the roots are $\{3, 1\}$. Therefore, the equation for the axis of symmetry is $x = \frac{m + n}{2} = \frac{3 + 1}{2} = 2$.

The **vertex** (or turning point) of the parabola is the lowest point (*minimum*) on the curve if the parabola *opens up* ($a > 0$), or the highest point (*maximum*) if the parabola *opens down* ($a < 0$). Since the minimum or maximum point for a parabola is at its vertex, we will often need to find the vertex to determine the largest or smallest value of a real world quadratic function.

The vertex lies on the axis of symmetry. The ***x*-coordinate of the vertex** is determined by the equation for the axis of symmetry. The ***y*-coordinate of the vertex** can be found by substituting for *x* in the quadratic equation.

Example: For $y = x^2 - 4x + 3$, the axis of symmetry is $x = 2$, so substitute 2 for x.
$y = (2)^2 - 4(2) + 3 = -1$, so the vertex is the point $(2, -1)$.

The ***y*-intercept of the parabola** (the *y* coordinate of the point where the parabola crosses the *y*-axis), is *c*. The **reflection of the *y*-intercept** over the axis of symmetry (as long as the axis of symmetry is not $x = 0$) is another point on the parabola.

Example: For $y = x^2 - 4x + 3$, $c = 3$, so the *y*-intercept is 3.
The reflection of (0,3) over the axis of symmetry is (4,3).

We can **plot additional points** on the curve by substituting any integer value of *x* into the equation to find the corresponding value for *y*. We can also plot its reflection.

Example: For $y = x^2 - 4x + 3$, we substitute 5 for x to get $y = (5)^2 - 4(5) + 3 = 8$.
So, the point (5,8) is on the parabola. Its reflection is the point $(-1,8)$.

CALCULATOR TIP

You can also use the graphing calculator to **graph a parabola**.

Example: To graph $y = x^2 - 4x + 3$,

1. Press [Y=], then next to "Y_1 =", type in the quadratic expression $x^2 - 4x + 3$ using [X,T,Θ,n] for the variable, x, and the [x^2] button for the exponent, 2.
2. Press [GRAPH] to view the parabola. Adjust the graph to your liking by using the [ZOOM] and [2nd][FORMAT] features.

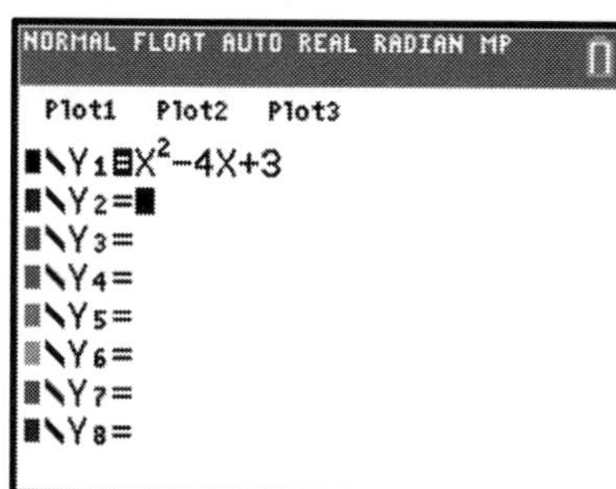

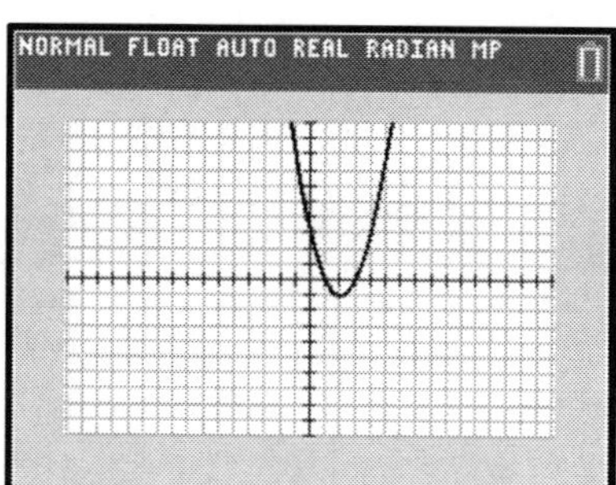

CALCULATOR TIP

The calculator can also be used to find a quadratic equation, given at least three points.

Example: Given the points (1,0), (2,−1), and (3,0) of the parabola shown above,

1. Press [STAT][1] and enter the x-values as L1 and the corresponding y-values as L2.
2. Press [STAT] [CALC] [5] for QuadReg (the quadratic regression).
3. On the TI-84 models, press [2nd][LIST][1] to select L1 for XList and [2nd][LIST][2] to select L2 for YList. To store the resulting equation in Y1, press [ALPHA][F4] to select Y1 for Store RegEQ. *[The TI-83 skips this screen; press [ENTER] when QuadReg appears.]*
4. The calculator will show the values of a, b, and c for the equation $y = ax^2 + bx + c$.

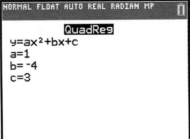

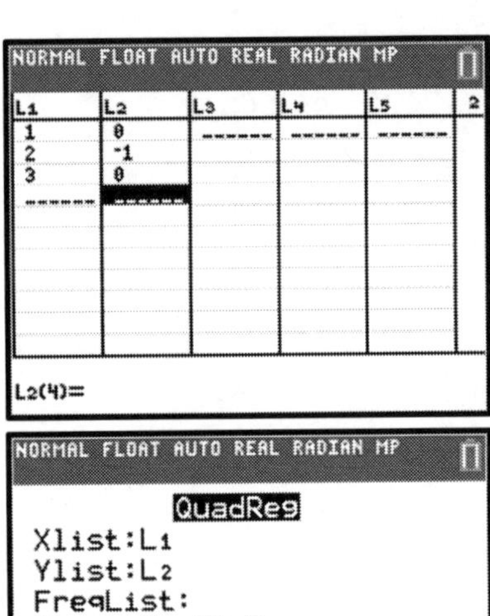

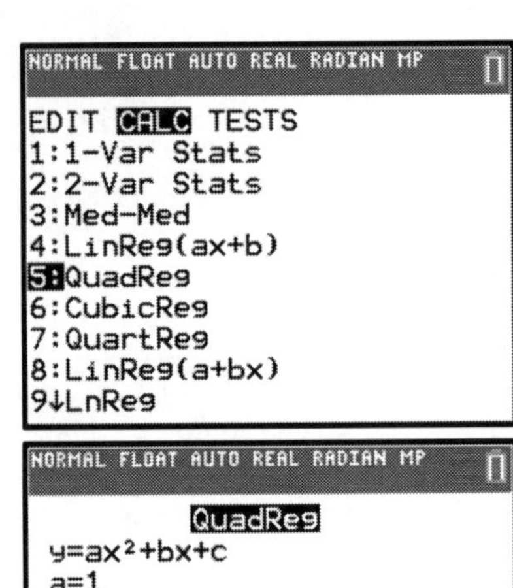

NORMAL FLOAT AUTO REAL RADIAN MP
QuadReg
Xlist:L1
Ylist:L2
FreqList:
Store RegEQ:Y1
Calculate

NORMAL FLOAT AUTO REAL RADIAN MP
QuadReg
y=ax²+bx+c
a=1
b=-4
c=3

Model Problem

Find the axis of symmetry and vertex for the parabola with equation $y = x^2 + 12x + 32$.

Solution:

(A) $a = 1$ and $b = 12$

(B) $x = \frac{-b}{2a} = \frac{-(12)}{2(1)} = -6$

(C) $y = x^2 + 12x + 32$

$y = (-6)^2 + 12(-6) + 32$

$y = -4$

(D) Axis of symmetry: $x = -6$.

Vertex: $(-6, -4)$.

Explanation of steps:

(A) The values for a and b come from the coefficients of the x^2 and x terms of the equation.

(B) Substitute for a and b in the axis of symmetry equation $x = \frac{-b}{2a}$ and evaluate.

(C) Substitute the value of x found in the previous step into the original equation to find the value of y. You now have the coordinates of the vertex.

(D) State your answers.

Practice Problems

1. Find the axis of symmetry and vertex of the parabola whose equation is $y = -2x^2 - 8x + 3$	2. Find the axis of symmetry and vertex of the parabola whose equation is $y = -x^2 - 2x + 1$
3. Find the maximum value of the function $f(x) = -3x^2 + 6x - 2$.	4. Find the minimum value of the function $g(x) = 5x^2 - 20x + 14$.

5. Graph the parabola whose equation is $y = 2x^2 - 8x + 4$.

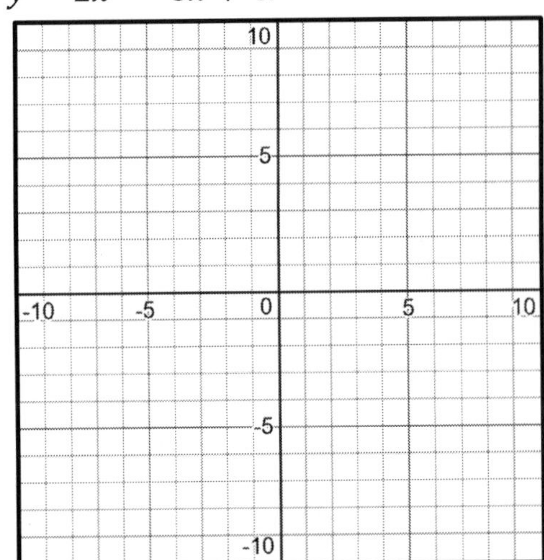

6. Graph the parabola whose equation is $y = -x^2 + 6x - 5$.

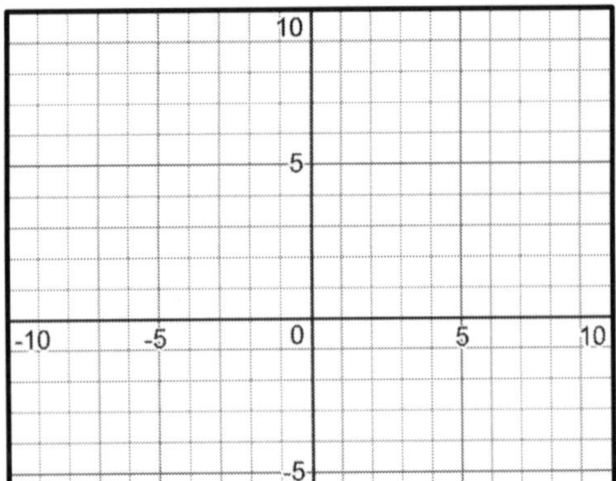

7. Graph the parabola whose equation is $y = -x^2 + 4x - 8$.

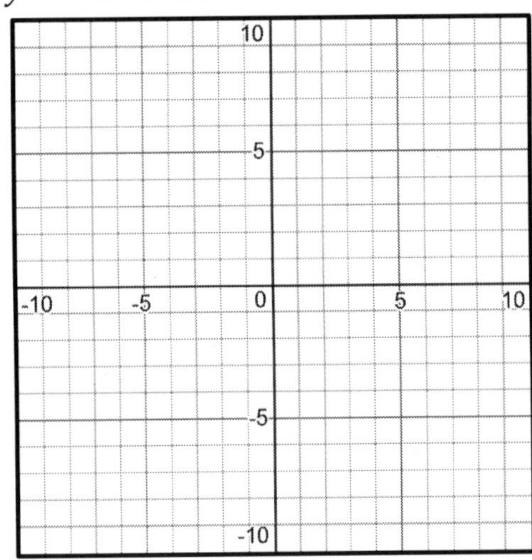

8. Graph the parabola whose equation is $y = x^2 - 6x + 10$.

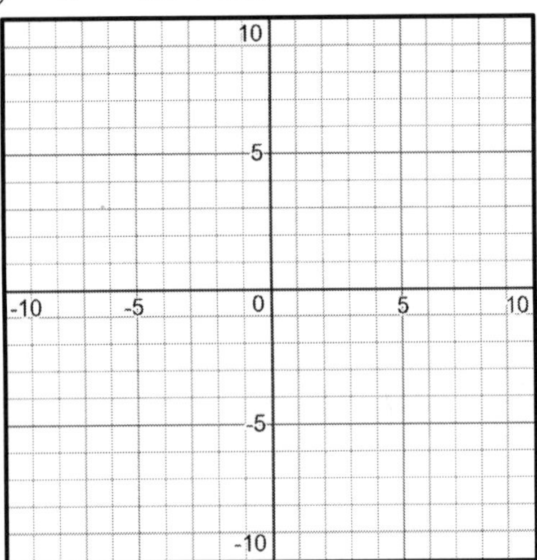

3.4 Vertex Form and Transformations

KEY TERMS AND CONCEPTS

Quadratic equations may be written in **vertex form**: $y = a(x-h)^2 + k$. In vertex form, the values of h and k represent the coordinates of the vertex (h, k) of the parabola. Just like in standard form, if a is positive then the parabola opens up and if a is negative then the parabola opens down.

To convert from standard form, $y = ax^2 + bx + c$, to vertex form, $y = a(x-h)^2 + k$:

1. The value of a is the same as the leading coefficient of the equation in standard form.
2. Since the axis of symmetry is $x = -\dfrac{b}{2a}$ and the vertex lies on the axis of symmetry, we know $h = -\dfrac{b}{2a}$.
3. Once we find h, we can find $k = f(h)$. The vertex is (h, k).
4. Write the equation in vertex form.

Example: To convert $y = 2x^2 - 4x + 5$ into vertex form,

$a = 2$ (the leading coefficient)

$h = -\dfrac{b}{2a} = 1$

$k = f(h) = f(1) = 2(1)^2 - 4(1) + 5 = 3$

Vertex is (1,3).

This gives is $y = 2(x-1)^2 + 3$.

An alternate method for converting a quadratic equation from standard form into vertex form is by completing the square and then isolating y.

Examples:

(a)

$y = x^2 - 2x - 3$

$y + 3 = x^2 - 2x$

$y + 3 + 1 = x^2 - 2x + 1$

$y + 4 = (x-1)^2$

$y = (x-1)^2 - 4$

Vertex is at $(1, -4)$

(b)

$y = 2x^2 - 4x + 5$

$\frac{1}{2}y = x^2 - 2x + \frac{5}{2}$

$\frac{1}{2}y - \frac{5}{2} = x^2 - 2x$

$\frac{1}{2}y - \frac{5}{2} + 1 = x^2 - 2x + 1$

$\frac{1}{2}y - \frac{3}{2} = (x-1)^2$

$\frac{1}{2}y = (x-1)^2 + \frac{3}{2}$

$y = 2(x-1)^2 + 3$

Vertex is at (1,3)

To convert from vertex form to standard form, simply square the binomial and simplify.

Examples:

(a)

$y = (x - 1)^2 - 4$

$y = (x - 1)(x - 1) - 4$

$y = x^2 - 2x + 1 - 4$

$y - x^2 - 2x - 3$

(b)

$y = 2(x - 1)^2 + 3$

$y = 2(x - 1)(x - 1) + 3$

$y = 2(x^2 - 2x + 1) + 3$

$y = 2x^2 - 4x + 2 + 3$

$y = 2x^2 - 4x + 5$

One reason that the vertex form is appealing is that it allows us to easily see any parabola as a transformation of its parent function by inspecting its equation.

A **parent function** is the most basic function in a family of functions. For the family of quadratic functions, the parent function is $f(x) = x^2$, which has its vertex at the origin and opens up. All other quadratic functions are **transformations** of the parent function. Transformations include translations, reflections, stretches, or any combination of these.

For a quadratic function in vertex form $f(x) = a(x - h)^2 + k$, first look at the value of a. If a is negative, then the parabola is reflected over the x-axis. Also, if $|a| \neq 1$, then the parabola is vertically stretched by a factor of $|a|$. If $|a| > 1$ then the parabola is stretched and becomes more narrow. If $0 < |a| < 1$, then the parabola is shrunk and becomes wider. Remember that in a vertical stretched by a factor of r, each point is mapped $(x, y) \rightarrow (x, ry)$.

Then, look at the values of h and k, which represent the coordinates of the new vertex, (h, k). These values tell us how many units to shift (or translate) the graph of $f(x) = x^2$. The value of h tells us how far to shift to the right (or left, if h is negative) and the value of k tells us how far to shift up (or down, if k is negative).

Let's consider the function $g(x) = -4(x-1)^2 + 3$ as a transformation of $f(x) = x^2$.

1. Since a is negative, reflect the function over the x-axis, as shown in Step 1 below.
2. Since $|a| = 4$, vertically stretch the function by a factor of 4, as shown in Step 2. For example, the points $(1,-1) \rightarrow (1,-4)$ and $(2,-4) \rightarrow (2,-16)$.
3. Since $h = 1$ and $k = 3$, shift to the right 1 and up 3 so the vertex is $(1,3)$, as in Step 3.

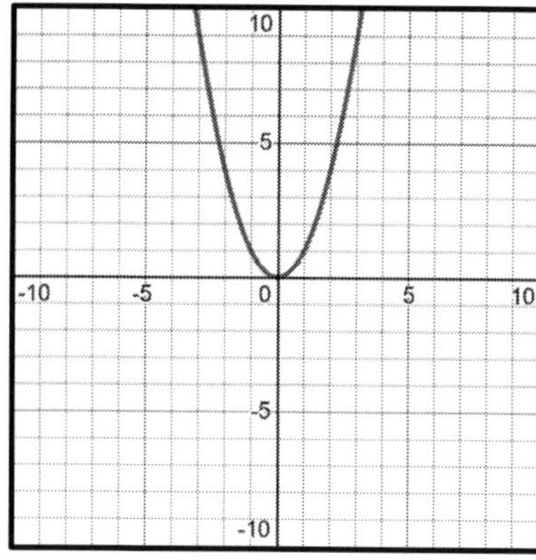

parent function $f(x) = x^2$

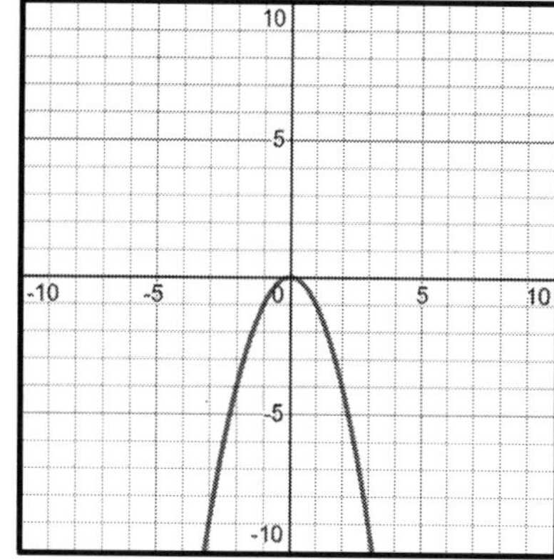

Step 1: reflect over x-axis

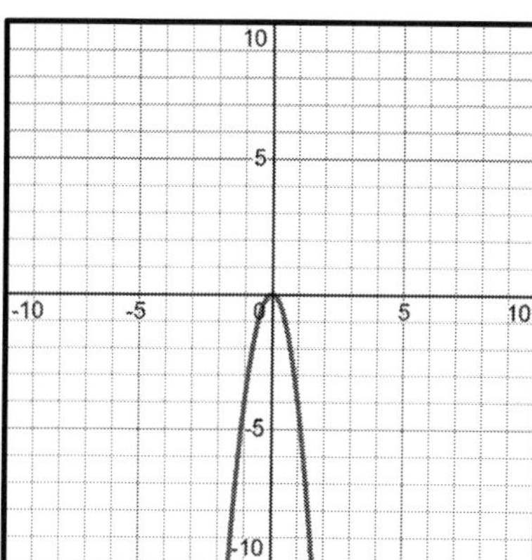

Step 2: stretch by factor of 4

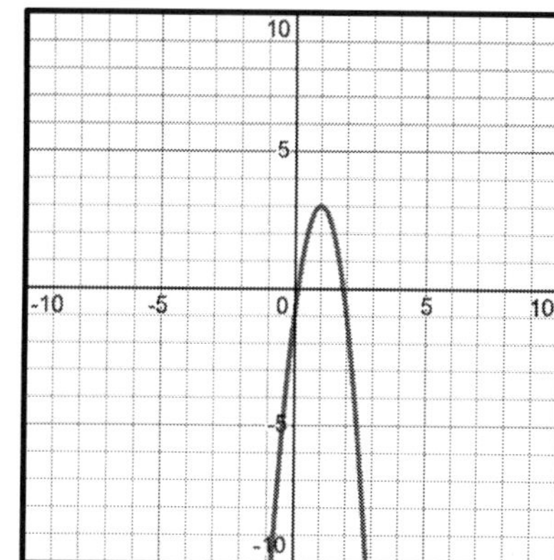

Step 3: shift right 1 and up 3

Model Problem

Graph the function $f(x) = \frac{1}{2}(x-4)^2 - 2$.

Solution:

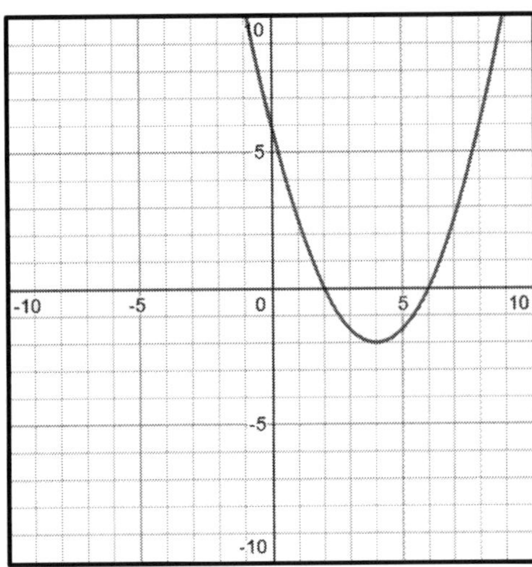

Explanation of steps:

(A) To graph a parabola written in vertex form $f(x) = a(x-h)^2 + k$, we can use our knowledge of transformations. Start with the parent function, $y = x^2$ *[shown as a dashed curve to the right].*

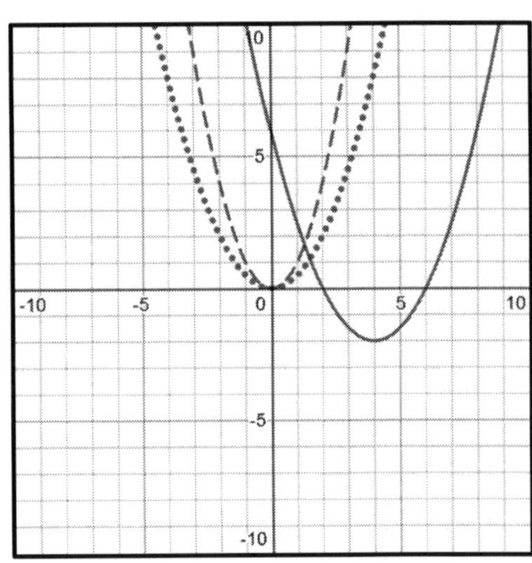

(B) The value of a determines if there is a reflection (if $a < 0$) and/or a vertical stretch (if $|a| \neq 1$). *[The dotted curve shows a vertical stretch of* $\frac{1}{2}$*; for example,* $(2,4) \rightarrow (2,2)$ *and* $(3,9) \rightarrow (3,4.5)$*.]*

(C) The values of h and k determines the new vertex after a translation. *[*$h = 4$ *and* $k = -2$*, so the parabola is shifted 4 units to the right and 2 units down. Graph a few simple point translations; for example,* $(0,0) \rightarrow (4,-2)$*,* $(2,2) \rightarrow (6,0)$*, and* $(4,8) \rightarrow (8,6)$*.]*

PRACTICE PROBLEMS

1. Which of the following represents the graph of $f(x) = 2(x+3)^2 - 4$?

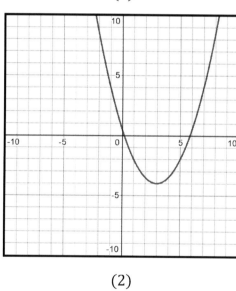

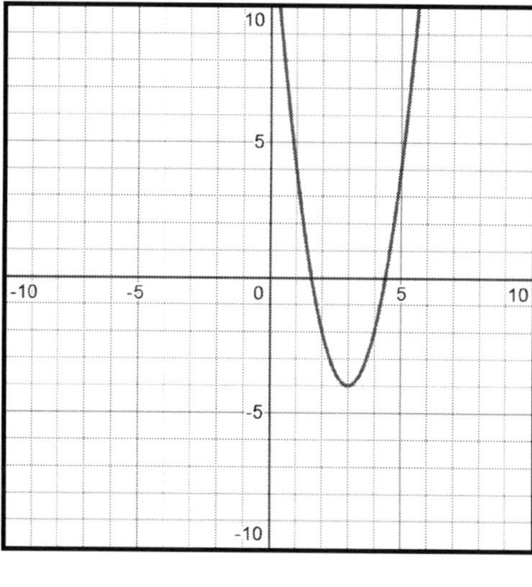

(1)

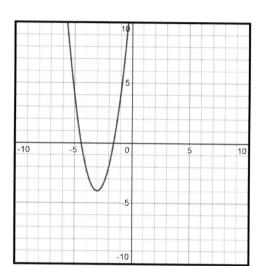

(3)

(2)

(4)

2. Translate $y = x^2 + 6x + 10$ into vertex form and state the coordinates of the vertex.	3. Translate $f(x) = x^2 + 10x + 21$ into vertex form by completing the square. State the coordinates of the vertex.
4. Translate the equation $y = (x + 5)^2 + 3$ from vertex form to standard form.	5. Translate the equation $y = -2(x - 4)^2 - 5$ from vertex form to standard form.

3.5 Focus and Directrix [CC]

Key Terms and Concepts

A parabola is the set of all points in a plane equidistant from a given line, called the **directrix**, and a given point not on the line, called the **focus**.

Example: In the graph below, the point $(-2,4)$, which is on the parabola, is both 5 units away from the focus $(2,7)$ and 5 units away from the directrix, $y = -1$. The points on the parabola represent all the points that are equidistant to the focus and directrix.

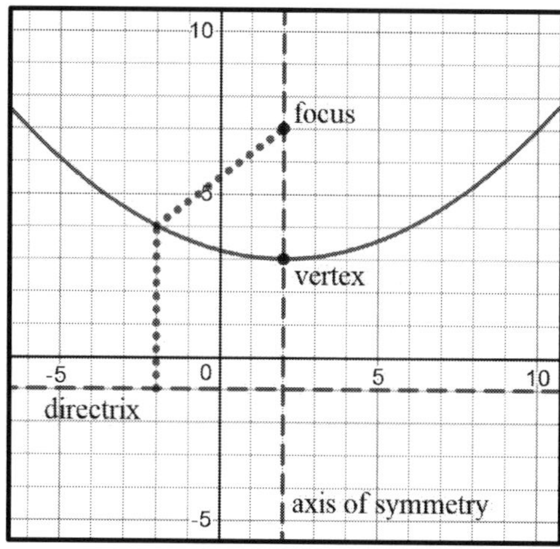

The directrix is always perpendicular to the axis of symmetry. If the parabola represents a quadratic function, then it opens up or down, and the directrix is horizontal.

If the vertex is (h, k), then the focus is $(h, k + p)$ for some p and the equation of the directrix is $y = k - p$.

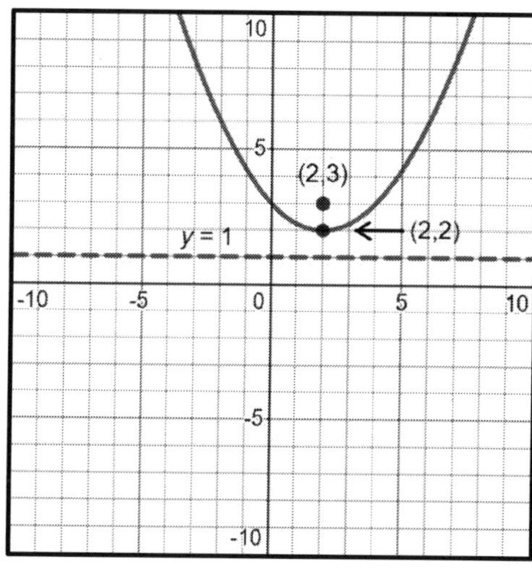

If we are given the coordinates of the focus and the equation of the directrix, we can find the coordinates of the vertex (h, k):

1) h is equal to the x-coordinate of the focus, because the focus and vertex lie on the same vertical axis of symmetry, and

2) k is equal to the average of the y-values of the focus and directrix, because the vertex is located midway between the focus and directrix on the axis of symmetry.

If $p > 0$, the focus is above the vertex, which is above the directrix, and the parabola opens up $(a > 0)$. If $p < 0$, the focus is below the vertex, which is below the directrix, and the parabola opens down $(a < 0)$. The focus is always "inside" the parabola and the directrix is always "outside" the parabola.

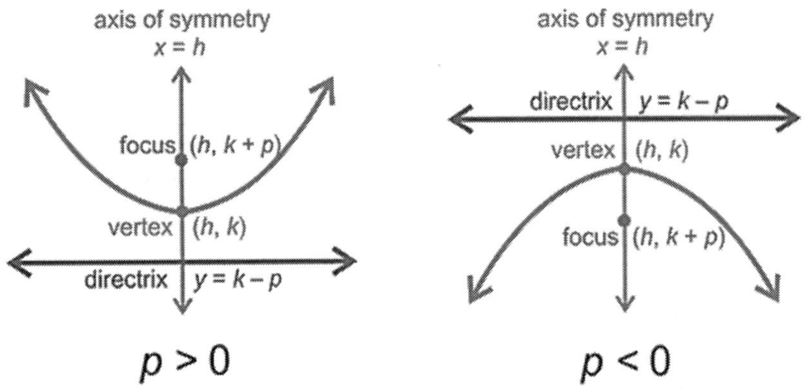

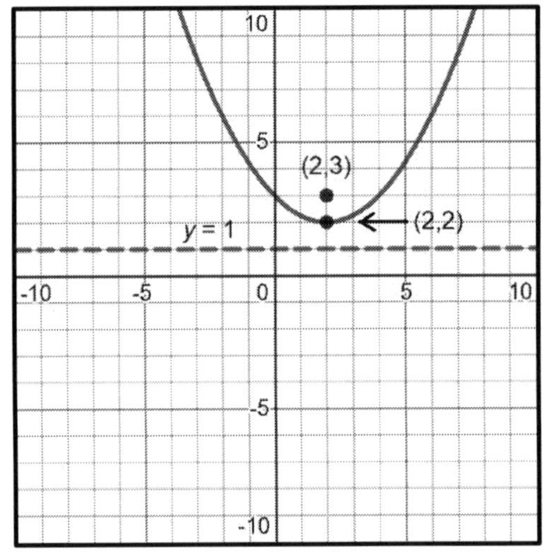

If we know the focus and vertex, we can find the value or p by subtracting their y-coordinates:

$$p = y_{focus} - y_{vertex}$$

Example: In the graph to the right, the focus is (2,3) and the vertex is (2,2), so we can calculate p as $p = 3 - 2 = 1$.

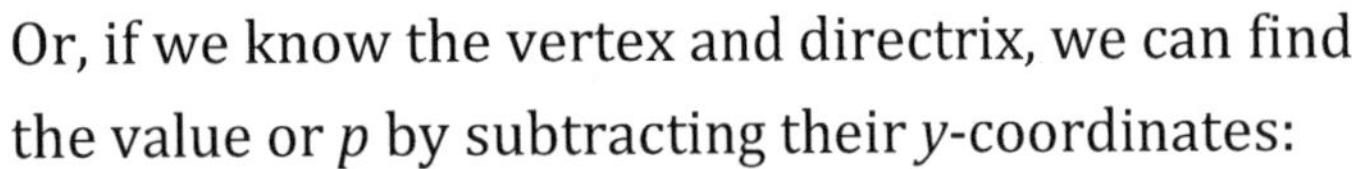

Or, if we know the vertex and directrix, we can find the value or p by subtracting their y-coordinates:

$$p = y_{vertex} - y_{directrix}$$

Examples: In the graph to the right, the vertex is (2,2) and the directrix is $y = 1$, we can calculate p as $p = 2 - 1 = 1$.

There is a relationship between a (the leading coefficient of $y = ax^2 + bx + c$) and p (the distance from the focus to the vertex, sometimes called the *focal length*). That relationship can be written as:

$$a = \frac{1}{4p} \quad \text{or} \quad p = \frac{1}{4a}$$

Recall that quadratic equations may be written in **vertex form**: $y = a(x - h)^2 + k$.

Since $a = \frac{1}{4p}$, the vertex form may also be written as $y = \frac{1}{4p}(x - h)^2 + k$.

If the vertex of the parabola is at the origin, this simplifies to $y = \frac{1}{4p}x^2$. In this case, the focus is located at $(0, p)$ and the equation of the directrix is $y = -p$.

Quadratic equations may also be written in conics form. The **conics form** of a parabolic equation is $(x-h)^2 = 4p(y-k)$. Note that in conics form, the opposite signs of both coordinates of the vertex are used.

We can convert from vertex form into conics form by isolating the binomial squared.

Example: Given $y = 2(x-1)^2 + 3$, we can isolate $(x-1)^2$ by the following steps.

$y - 3 = 2(x-1)^2$	Subtract 3 from both sides.
$\frac{1}{2}(y-3) = (x-1)^2$	Divide both sides by 2.
$(x-1)^2 = \frac{1}{2}(y-3)$	Flip the equation (this is optional).

We can write a quadratic equation in both vertex and conics forms if given any two of the following: the coordinates of the focus, the coordinates of the vertex, and/or the equation of the directrix.

Example: Given the parabola with a vertex of $(5, -2)$ and a directrix of $y = 3$. From the coordinates of the vertex, let $h = 5$ and $k = -2$. Since the directrix is above the vertex, the parabola opens downward, so p is negative. We can get the value of p by subtracting the y value of the vertex minus the y value of the directrix: $p = (-2) - (3) = -5$. Therefore, $\frac{1}{4p} = -\frac{1}{20}$ and $4p = -20$.

1) In *vertex form*, the equation can be written as $y = -\frac{1}{20}(x-5)^2 - 2$.
2) In *conics form*, the equation can be written as $(x-5)^2 = -20(y+2)$.

The distance between the focus and vertex is $|p|$, and the distance between the vertex and directrix is $|p|$. Therefore, the distance between the focus and directrix is $|2p|$. If we draw a horizontal line through the focus, called a **latus rectum**, the line will intersect the parabola at two points, as shown below. Since the directrix and latus rectum are parallel, and the distance between the focus and directrix is $|2p|$, it stands to reason that each of these two points is also $|2p|$ units from the directrix. Therefore, each point is also $|2p|$ units from the focus.

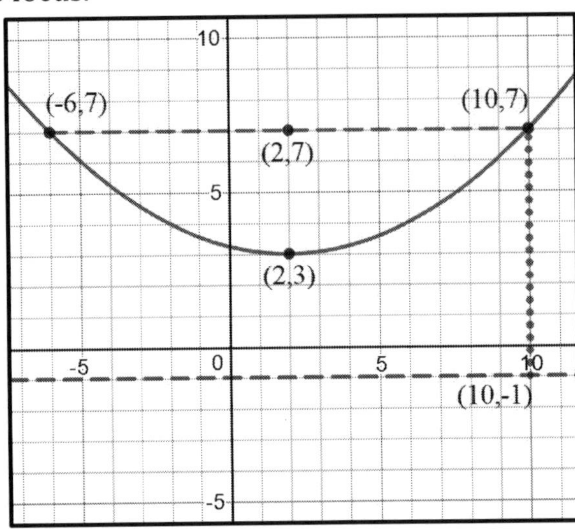

MODEL PROBLEM 1: ***VERTEX FORM***

Write an equation, in vertex form, of the parabola with a focus of (4,1) and a directrix of $y = -3$.

Solution:

(A) $y = a(x-h)^2 + k$

(B) $h = 4, k = \frac{1-3}{2} = -1$

Vertex is $(4,-1)$

(C) $p = 1 - (-1) = 2$

(D) $y = \frac{1}{4(2)}(x-4)^2 - 1$

$y = \frac{1}{8}(x-4)^2 - 1$

Explanation of steps:

(A) Write the general equation of a parabola in vertex form.

(B) To find the vertex (h, k), $h = x_{focus}$ and $k = \frac{y_{focus} + y_{directrix}}{2}$.

(C) To find p, use $p = y_{focus} - k$.

(D) Substitute for h, k, and $a = \frac{1}{4p}$ in the general equation.

PRACTICE PROBLEMS

1. State the vertex and directrix of the parabola having the equation $y = \frac{1}{16}(x-2)^2 - 4$	2. A parabola has a focus of $(0,3)$ and a directrix of $y = 5$. Write an equation for the parabola in vertex form.
3. A parabola has a focus of $(4,2)$ and a directrix of $y = -4$. Write an equation for the parabola in vertex form.	4. State the vertex, focus, and directrix of the parabola whose equation is $x^2 + 6x + 4y + 5 = 0$.

MODEL PROBLEM 2: *CONICS FORM*

Write an equation, in conics form, of the parabola with a vertex of $(-3,-1)$ and a directrix of $y = 1$.

Solution:

(A) $(x-h)^2 = 4p(y-k)$

(B) $h = -3$ and $k = -1$

(C) $p = (-1) - (1) = -2$

(D) $(x+3)^2 = -8(y+1)$

Explanation of steps:

(A) Write the general equation of a parabola in conics form.

(B) Vertex (h,k) gives us the values of h and k.

(C) To find p, use $p = y_{vertex} - y_{directrix}$.
[The directrix is above the vertex, so $p < 0$.]

(D) Substitute for h, k, and p in the general equation, simplifying the factor involving p.

PRACTICE PROBLEMS

5. Write the equation for the directrix for the parabola $(x+1)^2 = 8(y-2)$.	6. State the vertex and directrix of the parabola having the equation $(x+3)^2 = -20(y-1)$
7. Write an equation, in conics form, of the parabola with a vertex of $(2,1)$ and a directrix of $y = -3$.	8. Find the coordinates of the focus and the equation of the directrix for the parabola $(x-2)^2 = 2(y-1)$.

Chapter 4. Imaginary Numbers

4.1 Set of Complex Numbers

Key Terms and Concepts

Given an equation such as $x^2 = -9$, there are no real numbers in the domain of x that are a solution to the equation. To solve an equation such as this, we need to expand our domain to include not only real numbers, but also **imaginary numbers**.

The **imaginary unit *i*** is defined as the square root of -1; that is, $i = \sqrt{-1}$. The set of imaginary numbers is built by using this unit as a factor.

Examples: The value $\sqrt{-9}$ can be rewritten as $\sqrt{9}\sqrt{-1}$, and therefore as $3i$.

The value $\sqrt{-18}$ can be rewritten as $\sqrt{9}\sqrt{-1}\sqrt{2}$, and therefore as $3i\sqrt{2}$.

Rational factors are usually written before the *i* and radicals are written after the *i*.

If we allow for imaginary solutions, we can now solve the equation $x^2 = -9$:

$$x^2 = -9$$
$$x = \pm\sqrt{-9}$$
$$x = \pm\sqrt{9}\sqrt{-1}$$
$$x = \pm 3i$$

In general, if n is a positive number, then $\sqrt{-n} = i\sqrt{n}$.

$i^2 = -1$

Note that, since $i = \sqrt{-1}$, it follows that $i^2 = \left(\sqrt{-1}\right)^2 = -1$. Therefore, we can simplify any expression involving powers of i by replacing i^2 with -1.

Examples: The product $(6i)(2i) = 12i^2 = -12$.

The expression $(2i)^4$ can be rewritten as $2^4 i^4 = 16i^2 i^2 = 16(-1)(-1) = 16$.

The product $\left(\sqrt{-6}\right)\left(\sqrt{-2}\right) = \left(i\sqrt{6}\right)\left(i\sqrt{2}\right) = i^2\sqrt{12} = (-1)\left(2\sqrt{3}\right) = -2\sqrt{3}$.

It should be noted here that the rule for multiplying radicals, $\sqrt{a} \cdot \sqrt{b} = \sqrt{ab}$, only applies when $\sqrt{a}$ or $\sqrt{b}$ is a real number.

Example: $\left(\sqrt{-2}\right)\left(\sqrt{-8}\right)$ does *not* equal $\sqrt{(-2)(-8)}$ or $\sqrt{16}$ or 4.

In fact, $\left(\sqrt{-2}\right)\left(\sqrt{-8}\right) = \left(i\sqrt{2}\right)\left(i\sqrt{8}\right) = i^2\sqrt{16} = -4$.

Powers of *i*

If i is raised to an integer power, the possible results are 1, i, –1, or –i. In fact, the results will cycle among these four values as the power increases, with each power of i that is a multiple of 4 resulting in 1, as shown below.

$i^0 = 1$	$i^4 = 1$	$i^8 = 1$	*continue*: $1, i, -1, -i, \ldots$
$i^1 = i$	$i^5 = i$	$i^9 = i$	
$i^2 = -1$	$i^6 = -1$	$i^{10} = -1$	
$i^3 = -i$	$i^7 = -i$	$i^{11} = -i$	

A **complex number** is a number in the form $a + bi$, where a and b are real numbers and i is the imaginary unit. The a is called the real part and the bi is called the imaginary part.

As long as a complex number includes an imaginary part (that is, $b \neq 0$), then it is also called an **imaginary number**.

Examples: $-7 + 2i$ and $3 - i\sqrt{5}$ are imaginary numbers.

If there is no real part ($a = 0$), then the number is called a **pure imaginary number**.

Examples: $2i$ and $-i\sqrt{5}$ are pure imaginary numbers because they consist of only an imaginary part.

The set of complex numbers includes the real numbers and imaginary numbers as subsets.

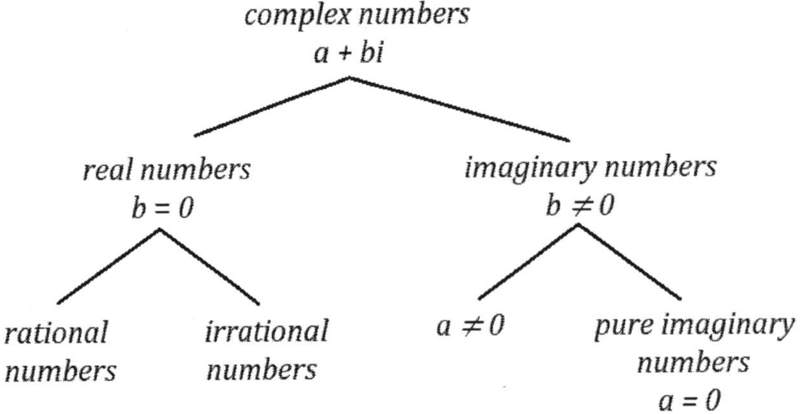

Using the set of complex numbers as the domain, any polynomial equation with a degree of at least 1 can now be solved. In fact, a polynomial function of degree n has exactly n roots in the set of complex numbers, including repeated roots. We will learn about solving polynomial equations in Section 7.1.

Example: The polynomial function $f(x) = x^5 - x^4 + x^3 - x^2 - 12x + 12$ has exactly five roots, since it has a degree of 5. Its roots are 1, $\sqrt{3}$, $-\sqrt{3}$, $2i$, and $-2i$. The function has three real roots and two imaginary roots.

CALCULATOR TIP

Imaginary numbers can be entered into the calculator. The imaginary i is shown above the decimal point button.

Example: To enter $3 + 2i$ on the calculator, press [3][+][2][2nd][i].

Imaginary numbers cannot be graphed on the coordinate plane used for real numbers (known as the Cartesian plane). However, we can graph them on the complex plane. The **complex plane** is like the Cartesian plane except that the horizontal axis represents the real part and the vertical axis represents the imaginary part.

Example: The graph below shows $2 + 3i$ plotted on the complex plane.

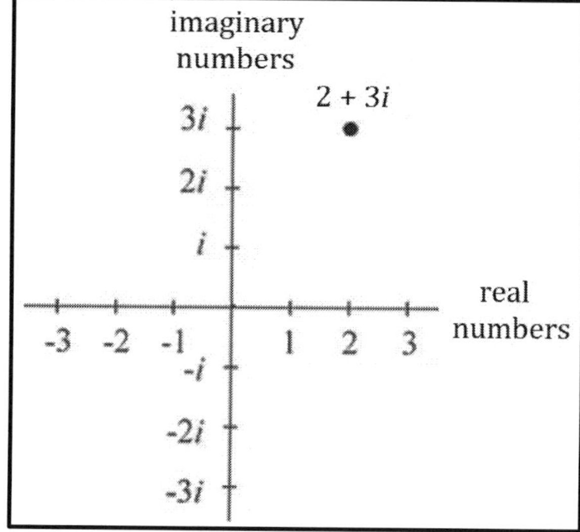

Model Problem

Write $3\sqrt{-8x^2}$ in simplest form.

Solution:

$3\sqrt{-8x^2} = 3\sqrt{4}\sqrt{x^2}\sqrt{-1}\sqrt{2} = 6xi\sqrt{2}$

Explanation of steps:

Simplify the radical by pulling out the square roots of perfect squares *[$\sqrt{4}$ and $\sqrt{x^2}$]* as well as $\sqrt{-1}$, which is written as *i*. Any remaining factors will stay under the radical sign.

Practice Problems

1. Write $\sqrt{-25}$ using the imaginary unit, *i*.	2. Write $\sqrt{-300}$ in simplest form.
3. Write $2 + \sqrt{-12}$ in simplest form.	4. Write $-8 + \frac{3}{4}\sqrt{-80}$ in simplest form.

5. Write $\sqrt{-180x^{16}}$ in simplest form.	6. Write $(3i)^3$ in simplest form.
7. Write $(2i)^4$ in simplest form.	8. Write $(-3)^3 i^{10}$ in simplest form.

4.2 Operations with Complex Numbers

Key Terms and Concepts

We can **add or subtract complex numbers** by separately adding or subtracting their real parts and their imaginary parts.

Example: $(3 + 4i) + (2 - 7i) = 5 - 3i$

We can **multiply complex numbers** as we would multiply binomials, by distribution. We can then substitute –1 for i^2, combine the real terms, and combine the imaginary terms.

Example: $(3 + 4i)(2 - 7i) = 6 - 21i + 8i - 28(-1) = 34 - 13i$

CALCULATOR TIP

Operations with complex numbers can be performed on the calculator.

Example: To calculate $(3 + 4i)(2 - 7i)$, enter [(] [3] [+] [4] [2nd] [i] [)] [(] [2] [-] [7] [2nd] [i] [)].

Of course, **squaring a complex number** is the same as multiplying it by itself.

Example: $(3 + 4i)^2 = (3 + 4i)(3 + 4i) = 9 + 24i + 16(-1) = -7 + 24i$

When we allow for complex numbers as factors, it's possible to factor some expressions that would be prime when working in the domain of real numbers only.

Example: We can factor $x^2 + 1$ as $(x + i)(x - i)$. We can see that this works by multiplying the factors: $(x + i)(x - i) = x^2 - ix + ix - i^2 = x^2 + 1$. Essentially, $x^2 + 1$ is a difference of two squares, $x^2 - i^2$ since $i^2 = -1$.

When we factor over the set of complex numbers, we can factor the **sum of two squares**: $a^2 + b^2 = (a + bi)(a - bi)$.

Example: $x^2 + 16$ is not factorable in the domain of real numbers, but it can be factored into $(x + 4i)(x - 4i)$ if we allow for imaginary factors.

Model Problem

Express $(3i)(3-2i)+(4-3i)$ in simplest form.

Solution:

(A) $(9i-6i^2)+(4-3i)$

(B) $(9i+6)+(4-3i)$

(C) $10+6i$

Explanation of steps:

(A) Multiply by distribution.

(B) Wherever an expression contains i^2, we can replace it with -1. *[*$-6i^2=-6(-1)=6$.*]*

(C) Add or subtract the real parts *[*$6+4$*]* and the imaginary parts *[*$9i-3i$*]*.

Practice Problems

1. What is the sum of $5-4i$ and $3+2i$?	2. What is the product of $3+2i$ and $2-i$?
3. Express $(3-7i)^2$ in $a+bi$ form.	4. Express the product of $2\sqrt{2}+5i$ and $5\sqrt{2}-2i$ in simplest form.

5. Express $(-1+i)^3$ in simplest form.	6. If $A = 3i$, $B = 2i$, and $C = 2 + i$, express AB^2C in simplest form.
7. Express $(x+i)^2 - (x-i)^2$ in simplest form.	8. Express $2xi(i - 4i^2)$ in simplest form.
9. Factor $3x^2 + 48$ over the set of complex numbers.	10. Factor $-9x^2 - 81$ over the set of complex numbers.

4.3 Imaginary Roots

Key Terms and Concepts

We have learned several methods for solving quadratic equations that have real roots. These same methods can also be used to solve quadratic equations with imaginary roots.

We can now find the imaginary solutions of $x^2 = a$, where a is a negative number.

Example: $x^2 = -81$. By taking the square root of both sides, $x = \pm\sqrt{-81}$.
Since $\sqrt{-81} = \sqrt{81}\sqrt{-1} = 9i$, the solutions are $x = \pm 9i$.

We may find imaginary solutions of other quadratic equations by completing the square.

Example: To solve $x^2 - 4x + 13 = 0$,

$x^2 - 4x = -13$ *[move the constant term to the other side]*

$x^2 - 4x + 4 = -13 + 4$ *[add* $\left(\frac{b}{2}\right)^2 = 4$ *to both sides]*

$(x-2)^2 = -9$ *[express the left side as a binomial squared]*

$x - 2 = \pm\sqrt{-9}$ *[take the square root of both sides]*

$x - 2 = \pm 3i$ *[express the right side as imaginary numbers]*

$x = 2 \pm 3i$ *[isolate x and express the roots in complex form]*

If the leading coefficient $a \neq 1$, it is usually easiest to use the "transforming method" for solving quadratics.

Example: To solve $2x^2 + 4x + 3 = 0$ by the transforming method,

$n^2 + 4n + 6 = 0$ *[write a new equation of the form* $n^2 + bn + ac = 0$*]*

$n^2 + 4n = -6$ *[solve the new equation for n by completing the square]*

$n^2 + 4n + 4 = -6 + 4$

$(n+2)^2 = -2$

$n + 2 = \pm\sqrt{-2}$

$n = -2 \pm i\sqrt{2}$

$x = -1 \pm \frac{i\sqrt{2}}{2}$ *[to get the solutions for x, divide solutions for n by a]*

We can also choose to use the quadratic formula to solve any quadratic equation.

Example: To solve the same equation $2x^2 + 4x + 3 = 0$ using the quadratic formula,

$$x = \frac{-b \pm \sqrt{b^2 - 4ac}}{2a} = \frac{-(4) \pm \sqrt{4^2 - 4(2)(3)}}{2(2)} = \frac{-4 \pm \sqrt{-8}}{4} = -1 \pm \frac{i\sqrt{2}}{2}$$

In the quadratic formula, the **discriminant** is the part of the formula that is under the radical, namely $b^2 - 4ac$, shown in the box below.

$$x = \frac{-b \pm \sqrt{\boxed{b^2 - 4ac}}}{2a}$$

When we use the quadratic formula to solve quadratic equations over the domain of complex numbers, we will end up with imaginary roots when the discriminant is negative.

Example: For the equation above, the discriminant was –36. Therefore, we know that the roots are imaginary.

In general, for a quadratic function, if the discriminant is

a) **negative**, then there are two imaginary roots (conjugates) and the parabola has no x-intercepts because there are no real roots.

b) **zero**, then there is one real root (called a double root) and the parabola has one x-intercept at its vertex, where $x = -\frac{b}{2a}$.

c) **positive**, then there are two distinct real roots and the parabola has two x-intercepts. *[If the discriminant is a perfect square, then the roots are rational.]*

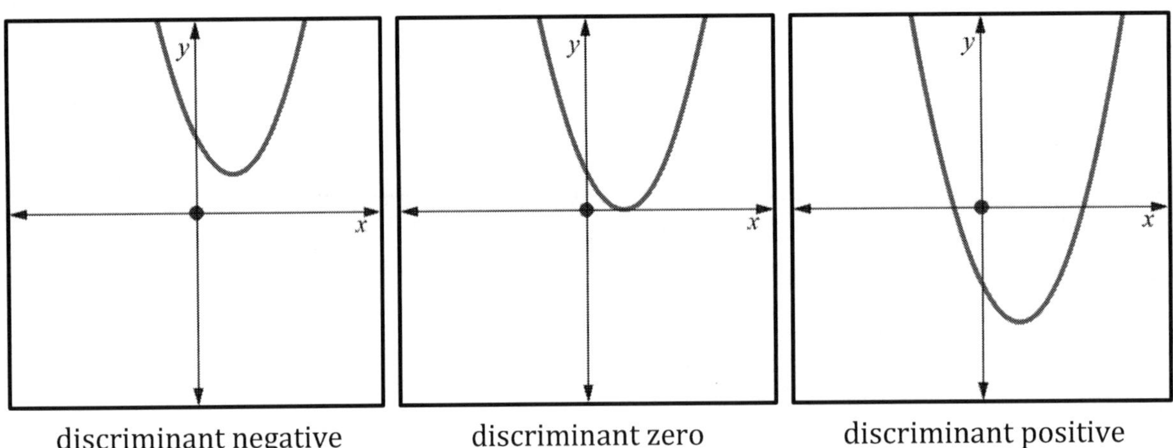

discriminant negative discriminant zero discriminant positive

If $a + bi$ is a complex number, then $a - bi$ is called its **conjugate** (or *complex conjugate*).

Examples: The following are pairs of conjugates:

$6 + 3i$ and $6 - 3i$ $\quad$ $2 + i\sqrt{5}$ and $2 - i\sqrt{5}$ $\quad$ $7i$ and $-7i$

Complex Conjugate Root Theorem:

If a polynomial function with real coefficients has an imaginary root, then its conjugate is also a root.

Example: Notice that in each of the equations solved on the previous page, the roots were a pair of conjugates. For the equation $x^2 = -81$, the roots were $\pm 9i$, and for the equation $x^2 - 4x + 13 = 0$, the roots were $2 \pm 3i$.

MODEL PROBLEM 1: *COMPLETE THE SQUARE*

Solve $2x^2 + 4x + 3 = 0$ by completing the square.

Solution:

$2x^2 + 4x + 3 = 0$

(G) $x^2 + 2x + \frac{3}{2} = 0$

(H) $x^2 + 2x = -\frac{3}{2}$

(I) $x^2 + 2x + 1 = -\frac{3}{2} + 1$

(J) $(x + 1)^2 = -\frac{1}{2}$

(K) $x + 1 = \pm\sqrt{-\frac{1}{2}}$

(L) $x + 1 = \pm i\sqrt{\frac{1}{2}}$

(M) $x + 1 = \pm\frac{i\sqrt{2}}{2}$

(N) $x = -1 \pm \frac{i\sqrt{2}}{2}$

Explanation of steps:

(G) Divide the equation by the leading coefficient. *[Divide every term by 2.]*

(H) Add the opposite of c to both sides.

(I) Add $\left(\frac{b}{2}\right)^2$ to both sides. *[*$\left(\frac{2}{2}\right)^2 = 1$*.]*

(J) Factor the trinomial into a binomial squared.

(K) Take the square root of both sides. Use the $\pm$ symbol on the right side and simplify radicals.

(L) For the square root of a negative number, rewrite it as an imaginary number.

(M) For square roots of fractions, rationalize the denominator. *[*$\sqrt{\frac{1}{2}} = \frac{\sqrt{1}}{\sqrt{2}} = \frac{1}{\sqrt{2}}$. $\frac{1}{\sqrt{2}}\left(\frac{\sqrt{2}}{\sqrt{2}}\right) = \frac{\sqrt{2}}{2}$*]*

(N) Solve for x, remembering that $\pm$ gives two solutions

Practice Problems

1. What are the roots of $x^2 = -25$?	2. What are the roots of $x^2 + 16 = 0$?
3. What are the roots of the equation $\frac{2}{3}x^2 + 18 = 0$?	4. What are the roots of the equation $(x+2)^2 = -9$?
5. Solve $x^2 - 12x + 40 = 0$ by completing the square.	6. Solve $x^2 + 8x + 25 = 0$ by completing the square.

7. Solve $x^2 - 4x + 9 = 0$ by completing the square.	8. Solve $3x^2 + 6x + 4 = 0$ by the transforming method.

MODEL PROBLEM 2: *QUADRATIC FORMULA*

Solve $3x^2 - x + 2 = 0$ by the quadratic formula.

Solution:

(A) $\dfrac{-(-1) \pm\sqrt{(-1)^2 - 4(3)(2)}}{2(3)}$

(B) $= \dfrac{1 \pm\sqrt{-23}}{6} = \dfrac{1 \pm i\sqrt{23}}{6}$

Explanation of steps:

(A) Substitute the values of a, b, and c into the quadratic formula, $x = \dfrac{-b \pm\sqrt{b^2 - 4ac}}{2a}$.

[$a = 3$, $b = -1$, and $c = 2$.]

(B) Simplify.

PRACTICE PROBLEMS

9. Solve $x^2 - 3x + 7 = 0$ by the quadratic formula.	10. Solve $4x^2 - 12x + 25 = 0$ by the quadratic formula.
11. Solve $3x^2 + 5 = 4x$.	12. Solve $-3x^2 + 2x = 2$.

MODEL PROBLEM 3: *CONJUGATES*

One of the roots of the quadratic function $g(x)$ is $5+i$. Write the definition of $g(x)$ as a polynomial with integer coefficients.

Solution:

(A) The 2 roots are $5+i$ and $5-i$.
(B) $(x-(5+i))(x-(5-i))=0$
(C) $((x-5)-i)((x-5)+i)=0$
(D) $(x-5)^2-i^2=0$
(E) $(x^2-10x+25)-(-1)=0$
(F) $x^2-10x+26=0$
(G) $g(x)=x^2-10x+26$

Explanation of steps:

(A) A quadratic function has a degree of 2, so it has two roots. If an imaginary number *[*$5+i$*]* is one of its roots, then its conjugate *[*$5-i$*]* is the other root.
(B) Given roots a and b, we can write $(x-a)(x-b)=0$.
(C) Apply the distributive and associative properties to regroup the terms.
(D) When we multiply the difference and sum of the same two operands *[*$(x-5)$ *and* i*]*, we get the difference of two squares.
(E) Square the binomial, and rewrite i^2 as -1.
(F) Combine like terms *[*$25+1$*]*.
(G) Write the function definition.

PRACTICE PROBLEMS

13. One root of a polynomial function is $-2+3i$. State another root of the function.	14. Express the product of $a+bi$ and its conjugate in simplest form.

15. One of the roots of the quadratic function $f(x)$ is $1 - 2i\sqrt{2}$. Write the definition of $f(x)$ as a polynomial with integer coefficients.

Chapter 5. Circles

5.1 Equations of Circles

Key Terms and Concepts

Circles do not represent quadratic functions (their graphs can't pass the vertical line test), but they are similar in that their equations contain an x^2 term. They are different, however, in that they also contain a y^2 term. There are other types of graphs whose equations have both an x^2 and a y^2 term, such as an ellipse or a hyperbola, but these will be examined in a later course.

The **center-radius form** for the equation of a circle is:

$(x-h)^2+(y-k)^2=r^2$, where (h,k) is the center and r is the radius of the circle

Example: $(x-6)^2+(y+5)^2=4$ is the equation of a circle with a center at $(6,-5)$ and a radius of 2.

You can consider any circle as a **translation** of $x^2+y^2=r^2$ for a given r. The circle with the equation $(x-h)^2+(y-k)^2=r^2$ is a translation of $x^2+y^2=r^2$ by h units to the right and k units up. Changing the radius of a circle, r, will **dilate** the circle.

If the equation of a circle is not written in center-radius form, you may be able to convert the equation into this form by completing the square twice; once for the x terms and once for the y terms.

Start by writing the equation in standard form as $x^2+mx+y^2+ny+c=0$. In a circle, the coefficients of the x^2 and y^2 will be the same; if they are not 1, you can divide the entire equation by the coefficient so that they are both 1.

To convert an equation into center-radius form by completing the square twice:

1. Write the equation in the form, $x^2 + mx + y^2 + ny + c = 0$.
   [If the coefficients of the x^2 and y^2 terms are not 1, divide the entire equation by the coefficient so that they are both 1.]
2. Add the opposite of c to both sides.
3. Add $\left(\frac{m}{2}\right)^2$ and $\left(\frac{n}{2}\right)^2$ to both sides.
4. Factor each trinomial into a binomial squared.

MODEL PROBLEM

What is the center and radius of the circle whose equation is $x^2 - 2x + y^2 + 6y - 6 = 0$?

Solution:

(A) $x^2 - 2x + y^2 + 6y - 6 = 0$

(B) $x^2 - 2x + y^2 + 6y = 6$

(C) $\left(\frac{-2}{2}\right)^2 = 1 \quad \left(\frac{6}{2}\right)^2 = 9$

$(x^2 - 2x + 1) + (y^2 + 6y + 9) = 6 + 1 + 9$

(D) $(x - 1)^2 + (y + 3)^2 = 16$

(E) Center is $(1, -3)$ and radius is 4.

Explanation of steps:

(A) Write the equation in the form, $x^2 + mx + y^2 + ny + c = 0$.
[The equation was already in this form.]

(B) Add the opposite of c to both sides. *[Add 6 to both sides.]*

(C) Add $\left(\frac{m}{2}\right)^2$ and $\left(\frac{n}{2}\right)^2$ to both sides. Write the first value *[1]* next to the x terms to create a trinomial, and the second value *[9]* next to the y terms to create a trinomial.

(D) Factor each trinomial into a binomial squared. *[On the right side, $6 + 1 + 9 = 16$.]*

(E) Now that the equation is in the form $(x - h)^2 + (y - k)^2 = r^2$, state the center (h,k) and the radius r. *[$r^2 = 16$, so $r = \sqrt{16} = 4$.]*

Practice Problems

1. State the center and radius of the circle whose equation is $(x-5)^2+(y+2)^2=36$.	2. State the center and radius of the circle whose equation is $3(x+1)^2+3y^2=27$.
3. Write an equation, in center-radius form, of the circle with a center at $(-8,6)$ and a radius of 5.	4. Circle B is the image of circle A after a translation 2 units left and 3 units down. If the equation of circle B is $x^2+(y+5)^2=1$, what are the coordinates of the center of circle A?

5. Find the center and radius of the circle whose equation is $x^2 - 6x + y^2 + 14y + 42 = 0.$

6. Find the center and radius of the circle whose equation is $2x^2 + 4x + 2y^2 - 20y - 46 = 0$

7. Graph the circle with the equation $(x + 3)^2 + (y - 4)^2 = 4.$

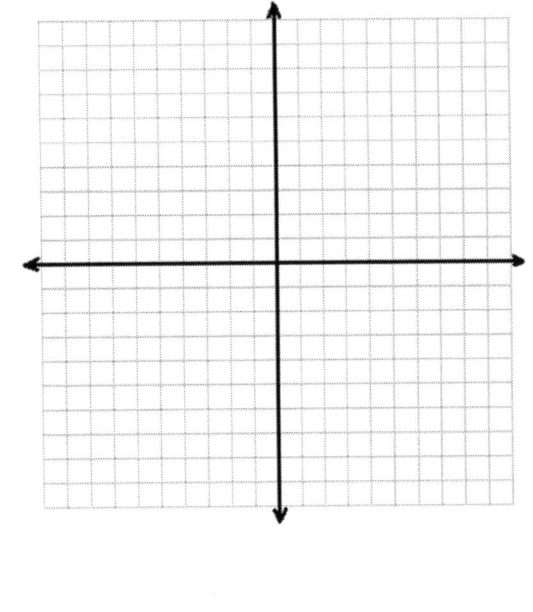

8. Graph the circle with the equation $x^2 + y^2 - 6y - 7 = 0.$

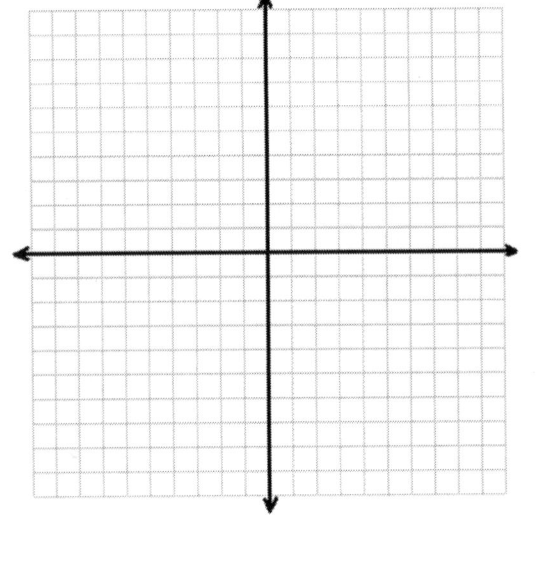

5.2 Circle-Linear Systems

Key Terms and Concepts

To solve a system of equations in which one of the equations represents a circle and the other is a linear function, we can substitute the expression for *y* into the circle equation.

Example: To solve the system $x^2 + y^2 = 90$ and $y = 3x$, substitute $3x$ for y in the first equation to get $x^2 + (3x)^2 = 90$, or $x^2 + 9x^2 = 90$. Solving for *x* gives us $x = \pm 3$, giving us solutions of $(3,9)$ and $(-3,-9)$.

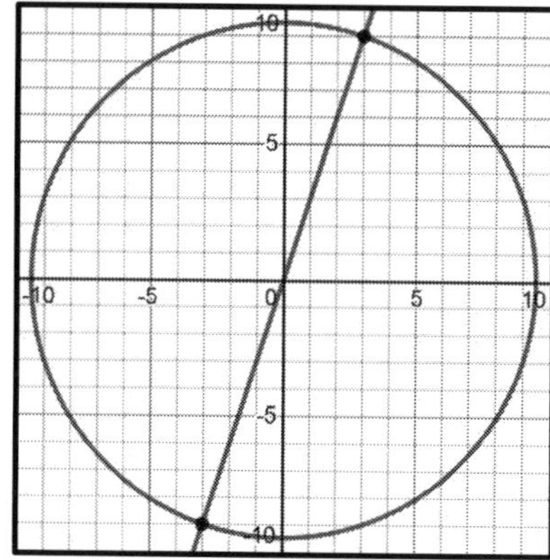

The number of solutions to the system depends on the number of points of intersection.

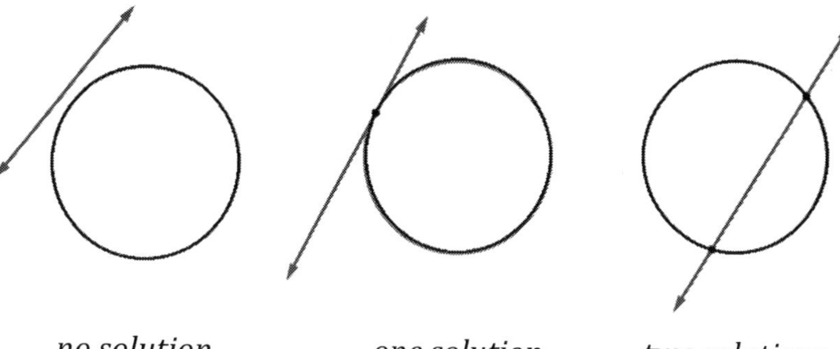

To solve a circle-linear system:

1. Substitute the linear expression for *y* in the circle equation.
2. Solve for *x*.
3. For each value of *x*, find the corresponding *y* value by substituting for *x* in the linear equation.
4. State the solutions as a set of points.

You can check your solutions by graphing both equations on your calculator.

Model Problem

Algebraically, find the solutions to the system $(x+1)^2+(y-2)^2=4$ and $y=x+1$.

Solution:

(A) $(x+1)^2+\left(\boxed{x+1}-2\right)^2=4$

(B) $(x+1)^2+(x-1)^2=4$

$x^2+2x+1+x^2-2x+1=4$

$2x^2-2=0$

$x^2-1=0$

$(x+1)(x-1)=0$

$x=\{-1,1\}$

(C) For $x=-1, y=-1+1=0$

For $x=1, y=1+1=2$

(D) $(-1,0)$ and $(1,2)$

Explanation of steps:

(A) Substitute the linear expression *[x – 1]* for *y* in the circle equation.

(B) Solve for *x*.

(C) For each value of *x*, find the corresponding *y* value by substituting for *x* in the linear equation.

(D) State the solutions as a set of points.

Practice Problems

1. Solve algebraically: $x^2+y^2=36$; $y=-x$	2. Solve algebraically: $(x-1)^2+y^2=9$; $y=3x$

3. Solve algebraically:

$$x^2 + y^2 = 16$$
$$y = -x + 2$$

4. Solve algebraically:

$$(x + 2)^2 + (y - 1)^2 = 25$$
$$y = 2x - 5$$

5. Solve algebraically:

$$y = \frac{1}{2}x$$
$$(x - 2)^2 + (y - 4)^2 = 9$$

6. Given the equations of a line and a circle below, algebraically determine the number of points of intersection.

$$-3y - 6 = 6x$$
$$(x - 4)^2 + y^2 = 20$$

7. Given the equations of a line and a circle below, algebraically determine the number of points of intersection.

   $y = x - 6$

   $x^2 + y^2 = 16$

Chapter 6. Polynomials

6.1 Operations with Functions

Key Terms and Concepts

Whenever there are multiple functions defined in a problem, it is possible to combine them using any of the four basic operations. In Algebra I, we learned how to add, subtract and multiply polynomials. In the next sections, we'll also learn how to divide polynomials.

Example: Suppose $f(x) = x^2 + 2x - 35$ and $g(x) = x - 5$. Then,
$f(x) + g(x) = (x^2 + 2x - 35) + (x - 5) = x^2 + 3x - 40$,
$f(x) - g(x) = (x^2 + 2x - 35) - (x - 5) = x^2 + x - 30$,
$f(x) \bullet g(x) = (x^2 + 2x - 35)(x - 5) = x^3 - 3x^2 - 45x + 175$, and
$\frac{f(x)}{g(x)} = \frac{x^2 + 2x - 35}{x - 5} = x + 7$, for $x \neq 5$.

Another common notation is to place the operation symbol between the function letters:

$(f + g)(x)$	means	$f(x) + g(x)$
$(f - g)(x)$	means	$f(x) - g(x)$
$(f \bullet g)(x)$	means	$f(x) \bullet g(x)$
$\left(\frac{f}{g}\right)(x)$	means	$\frac{f(x)}{g(x)}$ for $g(x) \neq 0$

Of course, it's possible to perform multiple operations on functions in a single expression.

Example: Suppose $f(x) = x^2 + 2x - 35$ and $g(x) = x - 5$. Then,
$f(x) \bullet g(x) - 2f(x)$ means $(x^2 + 2x - 35)(x - 5) - 2(x^2 + 2x - 35)$,
which simplifies to $x^3 - 5x^2 - 49x + 245$.

Model Problem

The revenue, $R(x)$, from selling x units of a product is represented by the equation $R(x) = 35x$, while the total cost, $C(x)$, of making x units of the product is represented by the equation $C(x) = 20x + 500$. The total profit, $P(x)$, is represented by the equation $P(x) = R(x) - C(x)$. For the values of $R(x)$ and $C(x)$ given above, what is $P(x)$?

Solution:	**Explanation of steps:**
$P(x) = 15x - 500$	Subtract the functions' expressions. *[$P(x) = R(x) - C(x) = (35x) - (20x + 500) = 15x - 500.$]*

Practice Problems

1. $f(x) = 9x - 5$ and $g(x) = 4x + 1$.	2. $f(x) = 3x$ and $g(x) = 3x^2 + 12x$,
a) Write an expression for $s(x) = f(x) + g(x)$ in simplest form.	a) Write an expression for $d(x) = g(x) - f(x)$ in simplest form.
b) Write an expression for $p(x) = f(x) \bullet g(x)$ in simplest form.	b) Write an expression for $q(x) = \frac{g(x)}{f(x)}$ where $f(x) \neq 0$, in simplest form.

3. $f(x) = 2x + \sqrt{3}$ and $g(x) = 5\sqrt{3}$.
   Write in simplest form:

   a) $(f + g)(x)$

   b) $(f \cdot g)(x)$

   c) $\left(\frac{f}{g}\right)(x)$

4. $f(x) = x^2 - 1$ and $g(x) = 2x + 4$.

   a) Write an expression for $h(x) = 2f(x) - 4g(x) + 8$ in simplest form.

   b) For what values of x is $h(x) = 0$?

5. A company calculates its profit by finding the difference between revenue and cost. The cost function of producing x headphones is $C(x) = 4x + 170$. If each unit is sold for \$10, the revenue function for selling x headphones is $R(x) = 10x$.

   a) Write an expression for the profit, $P(x) = R(x) - C(x)$.

   b) How many headphones must be sold to make a profit?

6.2 Long Division

Key Terms and Concepts

When dividing whole numbers as children, we learned a method called long division.

```
                 173  ◄—— Quotient
Divisor ——► 4)692  ◄—— Dividend
             -4
             29
            -28
             12
            -12
              0  ◄—— Remainder
```

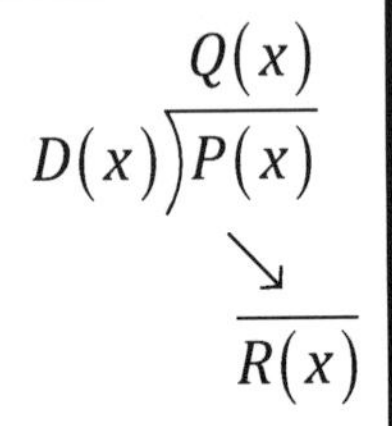

To divide polynomials, we'll use a similar procedure which is also called **long division**. In this chapter, we'll refer to $P(x)$ as the dividend, $D(x)$ as the divisor, $Q(x)$ as the quotient, and $R(x)$ as the remainder. These are all polynomials.

To perform long division of polynomials:

1. Divide the first term of the dividend by the first term of the divisor and write this quotient above the long division bar.
2. Multiply this term by the divisor and write this product below the dividend. Be sure to write each term below its corresponding term of the same degree.
3. Subtract this from the terms of the dividend, up to the last corresponding term and bring down the next term.
4. Repeat steps 1-3 for the remaining terms.
5. Write the answer. If the remainder is *not* zero, the answer may be written in $quotient + \frac{remainder}{divisor}$ form, but *not* as a polynomial expression.

When dividing polynomials, if the remainder is zero, then they divide exactly. This means the quotient $Q(x)$ is a polynomial expression and $P(x) = D(x) \cdot Q(x)$.

Example: To divide $x^2 + 2x - 15$ by $x - 3$ using long division,

$$\begin{array}{r} x+5 \\ x-3\overline{)x^2+2x-15} \\ (-)\ x^2-3x \\ \hline 5x-15 \\ (-)\ 5x-15 \\ \hline 0 \end{array}$$

1. Divide the first terms of the polynomials and write this quotient above the long division bar. Since $x^2 \div x = x$, write x above its corresponding term.
2. Multiply this term by the divisor, $x(x - 3)$, and write the product, $x^2 - 3x$, below the dividend.
3. Subtract this from the corresponding terms of the dividend, $(x^2 + 2x) - (x^2 - 3x) = 5x$, and bring down the next term, –15.
4. Repeat steps 1-3 for the remaining terms: $5x \div x = 5$, so write +5 above the bar. Then write the product, $5(x - 3) = 5x - 15$, below and subtract.
5. The quotient is $x + 5$ and the remainder (below the bottom line) is 0.

In the example above, the dividend $P(x) = x^2 + 2x - 15$, the divisor $D(x) = x - 3$, the quotient $Q(x) = x + 5$, and the remainder $R(x) = 0$. Since the remainder is zero, $P(x) = D(x) \cdot Q(x)$, or $x^2 + 2x - 15 = (x - 3)(x + 5)$, which we can confirm by multiplying the binomial factors.

When the remainder of a polynomial division is zero, then both the divisor and the quotient are **factors** of the dividend.

Example: In the example above, $x - 3$ and $x + 5$ are both factors of $x^2 + 2x - 15$.

Sometimes the polynomials do not divide exactly and we get a remainder that is not zero. In these cases, we need the more general statement, $P(x) = D(x) \cdot Q(x) + R(x)$.

Example: When dividing $x^2 - 3x + 5$ by $x - 2$ as shown below, the quotient is $x - 1$ and the remainder is 3.

$$\begin{array}{r} x-1 \\ x-2\overline{)\,x^2-3x+5} \\ \underline{(-)\,x^2-2x} \\ -x+5 \\ \underline{(-)\;-x+2} \\ 3 \end{array}$$

This means $P(x) = D(x) \cdot Q(x) + R(x)$,
or $x^2 - 3x + 5 = (x - 2)(x - 1) + 3$.
We can confirm this by multiplying and simplifying:

$(x - 2)(x - 1) + 3 =$
$x^2 - x - 2x + 2 + 3 =$
$x^2 - 3x + 5$

When the remainder is not zero, we can write the answer as $quotient + \frac{remainder}{divisor}$.

Example: For the previous example, we can write the answer as $x - 1 + \frac{3}{x-2}$.

The degree of the remainder, $R(x)$, will be less than the degree of the divisor, $D(x)$.

To be safe, when working through a long division of polynomials, always write terms directly above or below the like terms of the dividend, even if this leaves a space. Also, if the dividend is "missing" a term, write the missing term with a coefficient of zero. This will help with lining up the terms.

Examples: (a) Divide: $\frac{x^3 - 2x^2 + 5x - 2}{x^2 + 1}$ (b) Divide: $\frac{2x^3 - 2x + 12}{x + 2}$

(a)

$$\begin{array}{r} x-2 \\ x^2+1\overline{)\,x^3-2x^2+5x-2} \\ \underline{(-)\,x^3+x} \\ -2x^2+4x-2 \\ \underline{(-)-2x^2-2} \\ 4x \end{array}$$

Leave a space to line up like terms

(b)

$$\begin{array}{r} 2x^2-4x+6 \\ x+2\overline{)\,2x^3+0x^2-2x+12} \\ \underline{(-)\,2x^3+4x^2} \\ -4x^2-2x \\ \underline{(-)-4x^2-8x} \\ 6x+12 \\ \underline{(-)6x+12} \\ 0 \end{array}$$

Placeholder for "missing" term in dividend

MODEL PROBLEM

Divide $x^3 + x^2 - 5x + 2$ by $x^2 + 2$ using long division.

Solution:

$$
\begin{array}{lrl}
 & & \text{(A)} \\
 & & \quad\quad\quad\;\; x+1 \\
 & x^2+2 & \overline{)\,x^3+x^2-5x+2} \\
\text{(B)} & & \underline{(-)\,x^3 \qquad\;\; +2x} \\
\text{(C)} & & \qquad\quad\;\; x^2-7x+2 \\
\text{(D)} & & \qquad\;\; \underline{(-)\,x^2 \qquad\;\; +2} \\
 & & \qquad\qquad\quad -7x
\end{array}
$$

(E) The quotient is $x + 1$ with a remainder of $-7x$, which may be written as $x + 1 - \frac{7x}{x^2+2}$.

Explanation of steps:

(A) Divide the first terms of the polynomials and write the quotient above the long division bar.
[Since $x^3 \div x^2 = x$, write x above its corresponding term of the same degree.]

(B) Multiply this term by the divisor *[$x(x^2 + 2)$]* and write this product *[$x^3 + 2x$]* below the dividend. Be sure to write each term below its corresponding term of the same degree.

(C) Subtract this from the terms of the dividend, up to the last corresponding term *[$(x^3 + x^2 - 5x) - (x^3 + 2x) = x^2 - 7x$]* and bring down the next term *[+2]*.

(D) Repeat steps A-C for the remaining terms.
[$x^2 \div x^2 = 1$, so write +1 above the bar. Then write the terms of the product, $1(x^2 + 2) = x^2 + 2$ below their corresponding terms, and subtract.]

(E) Write the answer in $quotient + \frac{remainder}{divisor}$ form. *[The quotient is $x + 1$ and the remainder is $-7x$, so the answer may be written as $x + 1 - \frac{7x}{x^2+2}$.]*

Practice Problems

1. Divide $5x^2 + 3x - 2$ by $x + 1$.	2. Divide $3x^2 - 5x + 1$ by $3x + 1$.
3. Divide $x^3 + 3x^2 - 4x - 12$ by $x^2 + x - 6$.	4. Divide: $\dfrac{2x^3 + 3x^2 - 8x - 12}{x^2 - 4}$

5. Divide $2x^3 + 7x^2 + 2x + 9$ by $2x + 3$.

6. Divide: $\frac{x^3 - 2x^2 - 4}{x - 3}$

7. Divide: $\frac{x^6 + 2x^5 + 5x^3 + 4x^2 + 6}{x^3 + 2}$

8. Divide: $\frac{2x^3 - 5x^2 + x - 10}{x^2 - 4x + 1}$

6.3 Synthetic Division

Key Terms and Concepts

When the divisor of a polynomial division is of the form $x - a$, where a is a constant, then the long division process can be shortened by a method called **synthetic division**.

Example: To divide $x^2 + 2x - 15$ by $x - 3$ using synthetic division, follow the steps given below this example.

$$\begin{array}{r|rrr} \boxed{3} & 1 & 2 & -15 \\ & & 3 & 15 \\ \hline & 1 & 5 & | \; 0 \end{array}$$

So, the quotient is $x + 5$.

To perform synthetic division of $P(x)$ by $x - a$:

1. In a box at the top left, write the value of a.
2. On the top row, write the coefficients of the terms of $P(x)$ in standard order. If the term for a power of x is missing, write the coefficient of zero.
3. Skip a row and draw a horizontal line. Copy the first coefficient below the line.
4. Multiply this number by a (the number written in the box) and write this product above the line in the next column. Add the two values in this column and write the sum below the line.
5. Repeat step 4, multiplying each new sum by a, and then adding.
6. The numbers written below the line, except the last number, represent the coefficients of the polynomial quotient. The last number represents the remainder. Draw a vertical line to the left of the remainder to separate it from the quotient.
7. Write the answer in $quotient + \frac{remainder}{divisor}$ form. The degree of the quotient will always be one less than the degree of the dividend.

MODEL PROBLEM

Divide the polynomial $2x^4 - 3x^3 - 5x^2 + 3x + 9$ by $x - 2$.

Solution:

(A) (B) →

(C)

2	2	−3	−5	3	9
		4	2	−6	−6
	2	1	−3	−3	| 3

(D) (E) →

(F) Quotient of $2x^3 + x^2 - 3x - 3$ and a remainder of 3.

Explanation of steps:

(A) In a box at the top left, write the value of a.
[Since we are dividing by $x - 2$, the value of a is 2. Note that if we were dividing by $x + 2$ instead, then the value of a would have been −2.]

(B) On the top row, write the coefficients of the terms of $P(x)$ in standard order. If the term for a power of x is missing, write the coefficient of zero.
[The coefficients of the divisor are 2, −3, −5, 3, and 9.]

(C) Skip a row and draw a horizontal line. Copy the first coefficient *[2]* below the line.

(D) Multiply this number by a (the number written in the box) and write this product on the line, below the next coefficient. *[$2 \times 2 = 4$.]* Add the two values in this column and write the sum below the line. *[$-3 + 4 = 1$.]*

(E) Repeat step 4, multiplying each new sum by a, and then adding.
[$1 \times 2 = 2$, then $-5 + 2 = -3$, and repeat: $-3 \times 2 = -6$, $3 + (-6) = -3$, $-3 \times 2 = -6$, and $9 + (-6) = 3$.]

(F) The numbers written below the line, except the last number, represent the coefficients of the polynomial quotient. *[The quotient row is shifted left; since the divisor begins with an x^4 term, the quotient begins with an x^3 term.]* The last number represents the remainder *[3]*. Draw a vertical line to the left of the remainder to separate it from the quotient.

PRACTICE PROBLEMS

1. Divide $3x^3 - 4x^2 - 7x + 6$ by $x - 2$ using synthetic division.	2. Divide $2x^3 - 5x^2 - 11x - 4$ by $x - 4$ using synthetic division.
3. Divide $3x^3 + x^2 - 6x + 2$ by $x - 1$ using synthetic division.	4. Divide $3x^4 + 7x^3 - x^2 - 5x + 5$ by $x + 2$ using synthetic division.
5. Divide $x^4 - 4x^2 - 4x + 8$ by $x - 2$ using synthetic division.	6. Divide $\frac{x^4 - 11x^2 - x + 7}{x + 3}$ using synthetic division.

6.4 Remainder Theorem

Key Terms and Concepts

The **Remainder Theorem** states that, when a polynomial $P(x)$ is divided by $(x - a)$, the remainder is always $P(a)$.

For example, when we divide the polynomial $P(x) = x^2 + 3x - 4$ by the binomial $(x - 2)$, we can find the remainder simply by evaluating $P(a)$, where $a = 2$ in this case:
$P(2) = 2^2 + 3(2) - 4 = 6$, so the remainder is 6.

We can see that this is true by synthetic division. Note that the remainder below is 6.

2	1	3	−4
		2	10
	1	5	| 6

Why does this work? Let's refer to the parts of polynomial division using function notation:

$$\begin{array}{r} Q(x) \\ D(x)\overline{)P(x)} \\ \searrow \\ \hline R(x) \end{array}$$

Quotient is $Q(x)$
Divisor is $D(x)$
Dividend is $P(x)$
Remainder is $R(x)$

Recall the general equation, $P(x) = D(x) \cdot Q(x) + R(x)$. When the divisor is of the form $(x - a)$, this becomes $P(x) = (x - a) \cdot Q(x) + R(x)$. Now, what happens when we evaluate $P(a)$? We get $P(a) = (a - a) \cdot Q(a) + R(a) = 0 \cdot Q(a) + R(a) = R(a)$. So, $P(a)$ is equal to the remainder, $R(a)$.

Therefore, by the Remainder Theorem, when the divisor is of the form $(x - a)$, we can derive this: $P(x) = (x - a) \cdot Q(x) + P(a)$.

The Factor Theorem, which is closely related to the Remainder Theorem, allows us to check whether a binomial is a factor of a polynomial.

The **Factor Theorem** states that if $P(a) = 0$, then $(x - a)$ is a factor of $P(x)$. On the other hand, if $P(a) \neq 0$, then $(x - a)$ is *not* a factor of $P(x)$.

Example: Is $(x - 3)$ a factor of $P(x) = x^2 + 2x - 15$? For $a = 3$, we can check $P(a)$:
$P(3) = 3^2 + 2(3) - 15 = 0$, so it is a factor.

This follows from the Remainder Theorem, since $P(a) = 0$ indicates that the remainder of $\frac{P(x)}{(x-a)}$ is zero, so $(x - a)$ must be a factor of $P(x)$.

Model Problem

Determine whether $x = 10$ is a root of the function $f(x) = -x^3 + 17x^2 - 110x + 400$.

Solution:

$f(10) = -(10^3) + 17(10^2) - 110(10) + 400 = 0$, so $x = 10$ is a real root.

Explanation of steps:

If $x = a$ is a root, then $(x - a)$ is a factor. We can test whether $(x - a)$ is a factor by checking whether $f(a) = 0$. *[Since $f(10) = 0$, $(x - 10)$ is a factor, so 10 is a root.]*

Practice Problems

1. Given $f(x) = x^3 + 2x^2 + x + 6$ and $g(x) = x + 2$, what is the remainder of $\frac{f(x)}{g(x)}$?	2. Which of the following is *not* a factor of $P(x) = x^4 - 2x^3 - 7x^2 + 8x + 12$? (1) $(x + 1)$ (3) $(x - 1)$ (2) $(x + 2)$ (4) $(x - 2)$

3. The 5 roots of the polynomial function $P(x) = x^5 + x^4 - 11x^3 - 9x^2 + 18x$ are integers between -3 and 3, inclusive. Use your calculator to find the roots without graphing.

4. When $a(x)$ is divided by $x - 2$, the quotient is $3x + 13 + \frac{6}{x-2}$. State $a(x)$ in standard form.

5. If $x + 4$ is a factor of $x^3 + 3x^2 + kx - 24$,

   a) what is the value of k?

   b) what are all the zeros of the polynomial?

6.5 Factor Polynomials

Key Terms and Concepts

In Section 3.1, we saw how to factor a quadratic trinomial by grouping. We can also factor cubic (or higher) expressions by grouping.

Example: To factor $x^3 - 4x^2 - 3x + 12$,

$(x^3 - 4x^2) + (-3x + 12)$	Group the terms into pairs.
$x^2(x - 4) - 3(x - 4)$	Factor out the GCF of each pair.
$(x - 4)(x^2 - 3)$	The terms have $x - 4$ in common, so factor it out.

Note that we can't always factor by grouping.

Example: If we try to factor $x^3 + x^2 - x - 10$ by grouping, we'd get $x^2(x + 1) - 1(x + 10)$. Since $(x + 1)$ and $(x + 10)$ are not a common factor, factoring by grouping won't work here.

Always factor out any GCF first.

Example: $x^4 + 2x^3 - x^2 - 2x$ has a GCF of x, so we'll factor that out first.

$x(x^3 + 2x^2 - x - 2)$

We can now factor the polynomial factor by grouping.

$x[x^2(x + 2) - (x + 2)]$

$x(x^2 - 1)(x + 2)$

$x(x + 1)(x - 1)(x + 2)$

Sometimes, if there are an odd number of terms, it may be helpful to "split" one of the terms into two terms. For polynomials of the form $ax^4 + bx^2 + c$, try to split the middle term using coefficients that add to b and multiply to ac.

Example: For $2x^4 + 3x^2 - 2$, we need two integers that add to 3 and multiply to –4. The integers –1 and 4 work, so split the middle term into two terms with these coefficients.

$2x^4 - x^2 + 4x^2 - 2$

$x^2(2x^2 - 1) + 2(2x^2 - 1)$

$(x^2 + 2)(2x^2 - 1)$

This method may also work for a polynomial with two variables if it is of the form $ax^4 + bx^2y^2 + cy^4$.

Example: To factor $x^4 - 6x^2y^2 + 5y^4$, we can split the middle term $-6x^2y^2$ into $-x^2y^2 - 5x^2y^2$. This gives us $x^4 - x^2y^2 - 5x^2y^2 + 5y^4$.
Now we can factor by grouping.
$x^2(x^2 - y^2) - 5y^2(x^2 - y^2)$
$(x^2 - 5y^2)(x^2 - y^2)$
Finally, we can factor the difference of squares to get
$(x^2 - 5y^2)(x + y)(x - y)$

The steps to factor by grouping are:

1. Write the polynomial in standard form.
2. If a GCF exists (for all the terms), factor it out.
3. For a trinomial, split the middle term (coefficients that add to b and multiply to ac).
4. Factor out the GCF of each pair.
5. If there is a common factor, rewrite in factored form.
6. Be sure to factor completely.

[Note: For this section, we will factor polynomials over the set of rational numbers only. This means that $(x^2 - 2)$ should not be factored into $(x + \sqrt{2})(x - \sqrt{2})$, nor should $(x^2 + 4)$ be factored into $(x + 2i)(x - 2i)$. The expression $(x^2 - 2)(x^2 + 4)$ would be considered completely factored.]

For two special cases, the **sum of two cubes** or the **difference of two cubes**, we can apply the factoring rules below:

$$a^3 + b^3 = (a + b)(a^2 - ab + b^2)$$
$$a^3 - b^3 = (a - b)(a^2 + ab + b^2)$$

Examples: (a) To factor $x^3 + 8$, let a be the cube root of the first term, x, and let b be the cube root of the second term, 2. Substituting x for a and 2 for b in the expression $(a + b)(a^2 - ab + b^2)$ gives us $(x + 2)(x^2 - 2x + 4)$.

(b) To factor $8y^3 - 27$, let a be the cube root of the first term, $2y$, and let b be the cube root of the second term, 3. (Always ignore the minus sign before the second term.) Substituting $2y$ for a and 3 for b in the expression $(a - b)(a^2 + ab + b^2)$ gives us $(2y - 3)(4y^2 + 6y + 9)$.

To help you remember the correct operations to use in the rules for the sum or difference of two cubes, some students like to use the acronym SOAP, which stands for Same, Opposite, and Always Plus. The first operation is always the same (plus for the addition of cubes and minus for the subtraction of cubes), the second operation is the opposite, and the third operation is always plus.

$$a^3 + b^3 = (a + b)(a^2 - ab + b^2)$$
$$a^3 - b^3 = (a - b)(a^2 + ab + b^2)$$

Same Opposite Always Plus

Note that when you factor a sum or difference of cubes, the resulting trinomial factor will always be prime, so you won't need to check whether it can be factored any further.

To recognize terms that are perfect cubes, it would be helpful to become familiar with the cubes of the first ten whole numbers: 1, 8, 27, 64, 125, 216, 343, 512, 729, 1000.

What should you do if an expression is both a difference of squares and a difference of cubes, such as $x^6 - y^6$? In this case, factor the difference of squares first:

$$x^6 - y^6 = (x^3 + y^3)(x^3 - y^3) =$$
$$(x + y)(x^2 - xy + y^2)(x - y)(x^2 + xy + y^2)$$

To factor more complicated expressions, look for ways to substitute a new variable for parts of the expression to find simple patterns such as a factorable trinomial or a difference of two squares. By convention, the variable u is often used for this purpose.

Example: In the expression $(x - 2)^2 + 4(x - 2) + 3$, we see $x - 2$ appearing twice. By letting $u = x - 2$ and substituting, we get $u^2 + 4u + 3$, which we recognize as a factorable trinomial. It can be factored as $(u + 3)(u + 1)$. Replacing $x - 2$ for u, we get $(x - 2 + 3)(x - 2 + 1)$, which we can simplify to $(x + 1)(x - 1)$.

CALCULATOR TIP

We can use the calculator to check whether two expressions appear to be equivalent by using an arbitrary value of the variable and testing the equality of the two expressions. Although this method doesn't confirm that the expressions are equivalent for *all* values, it can help us recognize a likely error if they are not equal for the arbitrary value.

1. Store an arbitrary, preferably non-integer, value into x. For example, we can store 12.3 into x by pressing [1][2][.][3][STO▶][X,T,Θ,n].
2. Now, test whether the two expressions are equal using [2nd][TEST][1] for the equal sign. A result of 1 means they are equal, or a result of 0 means they are not.

Example: The screenshots below show how to test whether $2x^4 + 3x^2 - 2$ and its factored form, $(x^2 + 2)(2x^2 - 1)$, are equal for the arbitrary value of x.

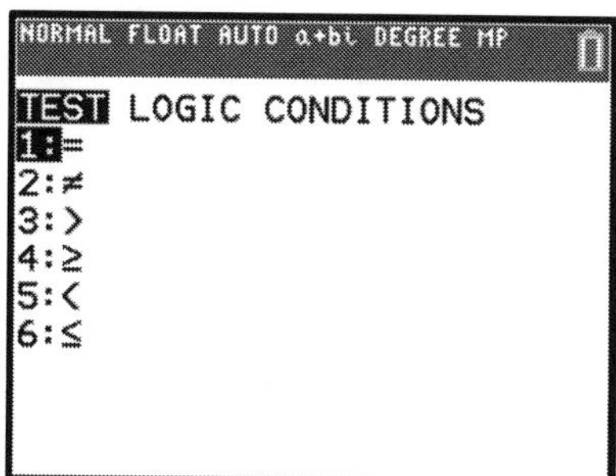

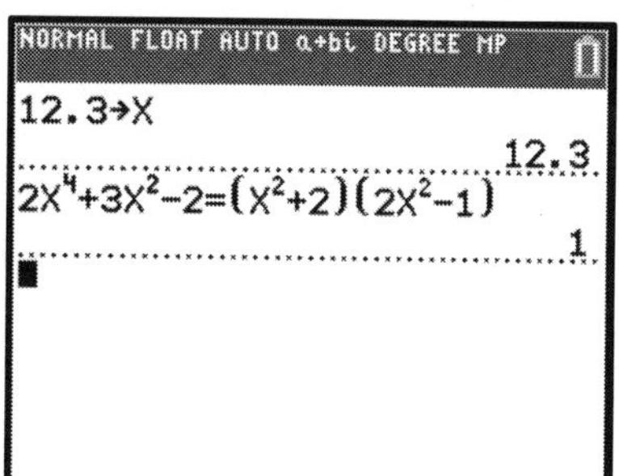

MODEL PROBLEM 1: *FACTOR BY GROUPING*

Factor $x^3 + 5x^2 - 4x - 20$ completely.

Solution:

(A) $x^2(x + 5) - 4(x + 5)$
$(x^2 - 4)(x + 5)$

(B) $(x + 2)(x - 2)(x + 5)$

Explanation of steps:

(A) Factor by grouping. *[For $x^3 + 5x^2$, the GCF is x^2. For $-4x - 20$, the GCF is -4. We get the common factor of $(x + 5)$.]*

(B) Continue to factor further if possible. *[$x^2 - 4$ is a difference of two squares.]*

Practice Problems

1. Factor completely: $x^3 + 3x^2 + 2x + 6$	2. Factor completely: $4x^3 + 10x^2 - 10x - 25$
3. Factor completely: $x^3 + 3x^2 - 4x - 12$	4. Factor completely: $x^3 - 2x^2 - 9x + 18$
5. Factor completely: $x^3 + 2x^2 - x - 2$	6. Factor completely: $3x^3 - 5x^2 - 48x + 80$
7. Factor completely: $a^2 + ab + ac + bc$	8. Factor completely: $a^3 - ab + a^2 - b$

MODEL PROBLEM 2: *SPLIT THE MIDDLE TERM*

Factor by grouping: $3x^4 + 8x^2y^2 + 5y^4$

Solution:

$3x^4 + 8x^2y^2 + 5y^4$

(A) $3x^4 + 3x^2y^2 + 5x^2y^2 + 5y^4$

(B) $3x^2(x^2 + y^2) + 5y^2(x^2 + y^2)$

(C) $(3x^2 + 5y^2)(x^2 + y^2)$

Explanation of steps:

(A) For polynomials of the form $ax^4 + bx^2y^2 + cy^4$, split the middle term using coefficients that add to b and multiply to ac. *[The two integers that add to 8 and multiply to 15 are 3 and 5.]*

(B) Factor out the GCF from each pair of terms. *[$3x^2$ for the first pair and $5x^2$ for the second pair.]*

(C) If the pairs have a common factor *[they both have a factor of $(x^2 + y^2)$]*, then we were able to factor by grouping. Write the other binomial factor by adding the terms outside the parentheses *[$(3x^2 + 5y^2)$]*.

PRACTICE PROBLEMS

9. Factor completely: $x^4 - 10x^2 + 9$	10. Factor completely: $8y^4 - 38y^2 + 35$

11. Factor completely: $a^4 - 5a^2b^2 + 4b^4$	12. Factor completely: $2x^4 - x^2y^2 - y^4$
13. Factor completely: $2x^5 - x^4 + 10x^3 - 5x^2 + 8x - 4$	

Model Problem 3: *Factor the sum or difference of two cubes*

Factor completely: $8x^3 + 27$

Solution:

(A) $a^3 + b^3 = (a + b)(a^2 - ab + b^2)$

(B) $(2x + 3)(4x^2 - 6x + 9)$

Explanation of steps:

(A) Write the rule for factoring the sum of two cubes.

(B) Substitute for a and b. *[$a = 2x$ and $b = 3$.]*

PRACTICE PROBLEMS

14. Factor $x^3 + 125$	15. Factor $b^3 + 64$
16. Factor $y^3 - 216$	17. Factor $27x^3 - 8$
18. Factor $16x^3 + 54$	19. Factor $x^3y^6 - 64$
20. Factor $512x^3 - 343y^3$	21. Factor $2x^3 + 128y^3$

MODEL PROBLEM 4: *INTRODUCE A NEW VARIABLE*

Rewrite $(2x^2)^2 + 2(2x^2) - 15$ as a product of binomial factors.

Solution:

(A) Let $u = 2x^2$.
$u^2 + 2u - 15$

(B) $(u + 5)(u - 3)$

(C) $(2x^2 + 5)(2x^2 - 3)$

Explanation of steps:

(A) By recognizing a factorable pattern *[a trinomial]*, introduce a new variable u to substitute for the repeated expression *[*$2x^2$*]*.

(B) Factor.

(C) Replace the original expression for u. Continue to factor further if possible.

PRACTICE PROBLEMS

22. Factor completely:

$$2(2x^2)^2 - 3(2x^2) - 2$$

23. Factor completely:

$$x^4 - (x-6)^2$$

24. Factor completely:

$$(x^5 + 2x)^2 + 2x^5 + 4x + 1$$

6.6 Polynomial Identities [CC]

Key Terms and Concepts

A **polynomial identity** is an equation that demonstrates two different ways to express the same thing. You can prove that an identity is true by rewriting the polynomial on one side of the equation, using the rules of algebra, to show that it can be expressed in the same way as the other side. Each step should be justified along the way.

Example: The identity $(x^2 + y^2)^2 = (x^2 - y^2)^2 + (2xy)^2$ is a known rule that is used to find Pythagorean triples. It can be proven to be an identity by these steps.

$(x^2 - y^2)^2 + (2xy)^2$	select the right side
$= x^4 - 2x^2y^2 + y^4 + (2xy)^2$	expand the square of the binomial
$= x^4 - 2x^2y^2 + y^4 + 4x^2y^2$	apply the power rule
$= x^4 + 2x^2y^2 + x^4$	combine like terms
$= (x^2 + y^2)^2$	rewrite as the square of a binomial

We now have the same expression as the left side, so the identity is proven.

Sometimes it is easier to work from just one side of the equation, but there are also times when it is easier to work with both sides simultaneously until they are both written the same way.

Model Problem

Prove the identity $x^6 + y^6 = (x^2 + y^2)(x^4 - x^2y^2 + y^4)$.

Solution:

$(x^2 + y^2)(x^4 - x^2y^2 + y^4)$	
$= x^6 - x^4y^2 + x^2y^4 + x^4y^2 - x^2y^4 + y^6$	multiply the polynomial factors
$= x^6 + y^6$	combine like terms

Explanation of steps:

Use the rules of algebra to rewrite either side until it is expressed in the same way as the other side. It usually makes sense to select the more complex side and to try to simplify.

PRACTICE PROBLEMS

1. Prove the identity, $(x+a)(x+b) = x^2 + (a+b)x + ab$. Explain each step.

2. Prove the identity, $(x^2+1)^2 = (x^2-1)^2 + (2x)^2$. Explain each step.

3. Prove the identity, $(a+b)^3 = a^3 + 3a^2b + 3ab^2 + b^3$. Explain each step.

4. Prove the identity, $2(a^2 + b^2) = (a + b)^2 + (a - b)^2$. Explain each step.

5. Prove the identity, $a^4 - b^4 = (a + b)(a - b)[(a + b)^2 - 2ab]$. Explain each step.

6. Prove the identity, $x^4 + y^4 = \left(x^2 + y^2 + \sqrt{2}xy\right)\left(x^2 + y^2 - \sqrt{2}xy\right)$. Explain each step.

Chapter 7. Polynomial Functions

7.1 Find Roots by Factoring

Key Terms and Concepts

A **polynomial function** is a function of a single variable that involves only real coefficients and whole number exponents of that variable. The degree of the polynomial function is the largest exponent of the variable. Polynomial functions are named by their degrees:

Degree	Function
0	constant
1	linear
2	quadratic
3	cubic
4	quartic
5	quintic
n	nth degree

For a given polynomial function $P(x)$ and real number a, the following statements are all equivalent:

- a is a zero of the function $P(x)$
- a is a solution or root of the equation $P(x) = 0$
- $x - a$ is a factor of $P(x)$
- a is an x-intercept of the graph of $P(x)$
- $(a, 0)$ is a point on the graph of $P(x)$

The **Fundamental Theorem of Algebra** states that every nth degree polynomial function has exactly n roots. This assumes that the roots may be real or imaginary, and that some roots may be repeated.

A root's **multiplicity** is the number of times the root appears.

Examples: (a) $f(x) = x^2 - 6x + 9$ can be factored as $(x-3)^2 = (x-3)(x-3)$, so the root 3 has a multiplicity of 2. $f(x)$ is quadratic, so it has 2 roots.

(b) $g(x) = x^4 + x^3$ can be factored as $x^3(x+1) = x \cdot x \cdot x \cdot (x+1)$, so the root 0 has a multiplicity of 3. The other root is –1, with a multiplicity of 1. $g(x)$ is quartic, so it has 4 roots.

If we can convert an equation into factored form, we will be able to determine its roots. Any polynomial expression can be converted into a product of constants, linear factors and/or irreducible quadratic factors.

For example, if the factored form of a cubic function $f(x)$ is $k(x-a)(x-b)(x-c) = 0$, then we know that a, b, and c are roots. Likewise, if the factored form of $f(x)$ is $k(x-a)(x^2-b) = 0$, then a, $\sqrt{b}$, and $-\sqrt{b}$ are roots.

Example: $x^3 + 5x^2 - 4x - 20 = 0$

$x^2(x+5) - 4(x+5) = 0$ Factor by grouping

$(x^2-4)(x+5) = 0$

$(x+2)(x-2)(x+5) = 0$ Factor difference of squares

$x = \{-5, -2, 2\}$

Of course, the factors aren't always as simple. However, as long as we can set each factor equal to zero and solve, we can find the roots.

Example: $6x^3 + 4x^2 - 3x - 2 = 0$

$2x^2(3x+2) - (3x+2) = 0$ Factor by grouping

$(3x+2)(2x^2-1) = 0$

$3x + 2 = 0 \quad 2x^2 - 1 = 0$ Set each factor equal to 0

$x = -\frac{2}{3} \quad x = \pm\sqrt{\frac{1}{2}} = \pm\frac{\sqrt{2}}{2}$

Roots are $\left\{-\frac{2}{3}, -\frac{\sqrt{2}}{2}, \frac{\sqrt{2}}{2}\right\}$

If we know the roots of a polynomial function, we can find an equation for one possible function with these roots by multiplying the factors. If a is a root, we can find the corresponding linear factor by setting $x = a$ and then getting zero to the right side of the equation.

Example: If a cubic function has roots of –3, 5, and $\frac{2}{3}$, we can find the factors by setting

(a) $x = -3$, leading to $x + 3 = 0$, giving us a factor of $(x + 3)$,

(b) $x = 5$, leading to $x - 5 = 0$, giving us a factor of $(x - 5)$, and

(c) $x = \frac{2}{3}$, leading to $3x - 2 = 0$, giving us a factor of $(3x - 2)$.

Multiplying these factors gives us, $(x + 3)(x - 5)(3x - 2) =$ $(x^2 - 2x - 15)(3x - 2) = 3x^3 - 8x^2 - 41x + 30$. So, the equation of one possible cubic function with these roots would be $y = 3x^3 - 8x^2 - 41x + 30$.

Of course, this method gives us only one possible function with these roots. Multiplying the polynomial expression by a constant other than 1 would give us a different function with the same roots: that is, the image of this function after a vertical stretch and/or a reflection over the x-axis.

MODEL PROBLEM 1: *REAL ROOTS ONLY*

Find the zeros of $f(x) = x^4 - 5x^3 - 4x^2 + 20x$.

Solution:

(A) $x^4 - 5x^3 - 4x^2 + 20x = 0$

(B) $x^3(x - 5) - 4x(x - 5) = 0$

$(x^3 - 4x)(x - 5) = 0$

(C) $x(x^2 - 4)(x - 5) = 0$

$x(x + 2)(x - 2)(x - 5) = 0$

(D) $\{0, -2, 2, 5\}$

Explanation of steps:

(A) Set the expression equal to zero.

(B) Factor by grouping.

(C) Factor further, if possible *[for the first factor, factor out the GCF, x, then factor the difference of squares].*

(D) State the zeros (roots).

Practice Problems

1. Solve for x: $x^3 + x^2 - 2x = 0$	2. Solve for x: $2x^3 - x^2 - 8x + 4 = 0$
3. Solve for x: $8x^3 + 4x^2 - 18x - 9 = 0$	4. Solve for x: $x^3 + 5x^2 = 4x + 20$

Model Problem 2: *Real and Imaginary Roots*

In factored form, $f(x) = 2x(x^2 + 27)(x - 1)$. What are the roots of $f(n)$?

Solution:

$2x = 0$ yields $x = 0$

$x^2 + 27 = 0$ yields $x = \pm\sqrt{-27} = \pm 3i\sqrt{3}$

$x - 1 = 0$ yields $x = 1$

Roots are $\{0, 1, \pm 3i\sqrt{3}\}$

Explanation of steps:

Set each factor equal to 0 and solve each equation for x.

Practice Problems

5. Solve for x: $x^4 - 3x^2 - 4 = 0$	6. Solve for x: $3x^5 - 48x = 0$
7. Solve for x: $x^4 + 4x^3 + 4x^2 + 16x = 0$	8. Solve for x: $9x^3 - 90x^2 + 64x - 640 = 0$

MODEL PROBLEM 3: *FIND EQUATION FROM GIVEN ROOTS*

The quartic function $f(x)$ has a leading coefficient of 2 and its roots are $\pm 3i$, -1, and $\frac{1}{2}$. Write an equation of the function.

Solution:

(A) $(x-3i)(x+3i)(x+1)(2x-1) =$

(B) $(x^2+9)(x+1)(2x-1) =$

$(x^3+x^2+9x+9)(2x-1) =$

$2x^4+x^3+17x^2+9x-9.$

(C) So, $f(x) = 2x^4+x^3+17x^2+9x-9.$

Explanation of steps:

(A) Use the roots to determine the factors by setting x equal to each root and then transforming each equation to get zero to one side.

[From the roots, we have equations of $x = 3i$, $x = -3i$, $x = -1$, *and* $x = \frac{1}{2}$.
Getting zero to one side of each equation gives us $x - 3i = 0$, $x + 3i = 0$, $x + 1 = 0$, *and* $2x - 1 = 0$. *Therefore, the factors are* $(x-3i)$, $(x+3i)$, $(x+1)$, *and* $(2x-1)$.*]*

(B) Multiply the factors.

(C) If the leading coefficient is correct, then write the equation of the function.

[The leading coefficient is 2, so this is the correct function. If the leading coefficients didn't match, we would have to multiply the polynomial expression by a constant to get them to match.]

Practice Problems

9. The quartic function $f(x)$ has a leading coefficient of 9 and its roots are $-1, 2,$ $-\frac{1}{3}$ and $\frac{2}{3}$. Write an equation of the function.	10. The cubic function $g(x)$ has a leading coefficient of 4 and its roots are $-1 \pm 2i$ and -2. Write an equation of the function.

7.2 Root Theorems

Key Terms and Concepts

For functions that are not easily factorable, the Rational Root Theorem can help us find rational factors, if any exist.

The **Rational Root Theorem** states that If the polynomial function $f(x)$ has integer coefficients, then every rational root of $f(x)$ will have the reduced form $\frac{p}{q}$, where p is a factor of the constant term, and q is a factor of the leading coefficient.

Example: For $f(x) = x^3 + 3x^2 - 13x - 15$, the possible rational roots are $\pm\frac{1}{1}$, $\pm\frac{3}{1}$, $\pm\frac{5}{1}$, or $\pm\frac{15}{1}$. The numerators are the factors of the constant term and the denominators are the factors of the leading coefficient. So, the numerators may be 1, 3, 5, or 15 (the factors of 15) and the denominator must be 1 (since the leading coefficient here is 1). This gives us eight possibilities: $1, -1, 3, -3, 5, -5, 15$, and -15.

Once we know the possible rational roots, we can use the Factor Theorem to test each one until we find a factor. Remember that $(x - a)$ is a factor if $f(a) = 0$. Once a factor is found, we can stop checking and use synthetic division to find the other factor.

Example: For $f(x) = x^3 + 3x^2 - 13x - 15$ above, we can start by testing if 1 is a root.
$f(1) = (1)^3 + 3(1)^2 - 13(1) - 15 = -24$, so $(x - 1)$ is *not* a factor.
We then test for -1.
$f(-1) = (-1)^3 + 3(-1)^2 - 13(-1) - 15 = 0$, so $(x + 1)$ is a factor.
Divide $f(x)$ by $(x + 1)$ using synthetic division:

$$\begin{array}{r|rrrr} \boxed{-1} & 1 & 3 & -13 & -15 \\ & & -1 & -2 & 15 \\ \hline & 1 & 2 & -15 & \mid\; 0 \end{array}$$

The quotient, $x^2 + 2x - 15$, is the other factor, which we can factor further.
$(x + 1)(x^2 + 2x - 15)$
$(x + 1)(x + 5)(x - 3)$

CALCULATOR TIP

Reminder: We can use the calculator to test the possible factors.

1. Press [Y=] and enter the equation as Y1, then press [2nd] [QUIT].
2. On the TI-84 models, press [ALPHA][F4][1] to select Y1.

   [On the TI-83, press [VARS] and select [Y-VARS] [1][1].]
3. Type the value in parentheses and press [ENTER]. If the result is 0, it is a factor.

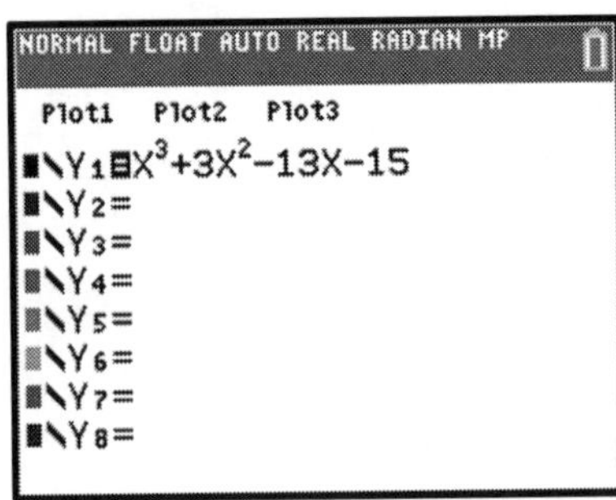

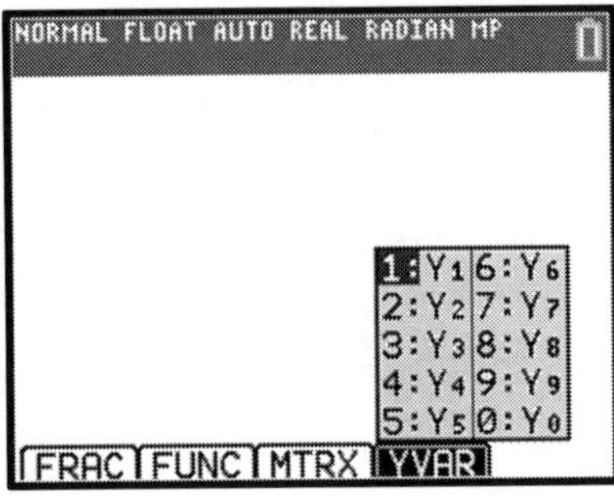

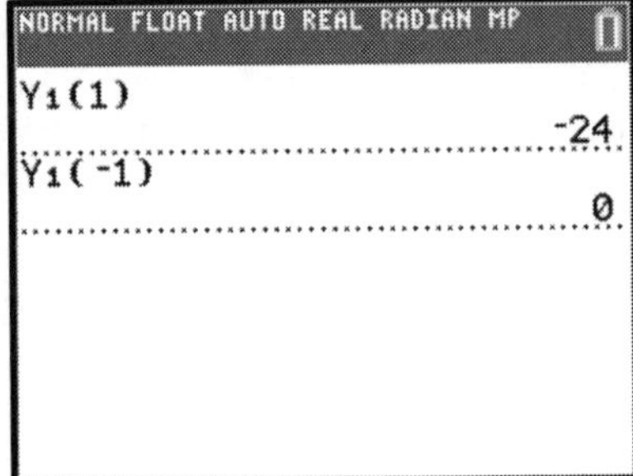

We know from the **Complex Conjugate Root Theorem** that imaginary roots come in pairs. If a polynomial function with real coefficients has an imaginary root $(a + bi)$, then its conjugate $(a - bi)$ is also a root of the equation.

It is also true that irrational roots come in pairs. The **Irrational Conjugates Theorem** states that if $a + \sqrt{b}$ is a root of a polynomial function with rational coefficients (where a is rational and $\sqrt{b}$ is irrational), then its conjugate $a - \sqrt{b}$ is also a root of the function.

We can use Descartes' Rule of Signs to determine the possible combinations of types of roots that a polynomial function may have: positive reals, negative reals, or imaginary.

Descartes' Rule of Signs states:

(a) The number of **positive** real roots equals the number of sign changes in consecutive coefficients of $\boldsymbol{f(x)}$, or an even number less than this (to account for the possibility of imaginary roots).

(b) The number of **negative** real roots equals the number of sign changes in consecutive coefficients of $\boldsymbol{f(-x)}$, or an even number less than this (to account for possibility of imaginary roots).

In both cases, use the multiplicity of the roots when counting.

Example: Given the cubic function $P(x) = -3x^3 + 2x^2 + 5x - 2$, there are three roots, matching the degree of the function.

For $P(x)$, the signs of the coefficients are $(-, +, +, -)$, so there are two sign changes, from negative to positive and then from positive to negative. So, there are either 2 or 0 positive real roots.

For $P(-x) = -3(-x)^3 + 2(-x)^2 + 5(-x) - 2$, the expression for $P(-x)$ is $3x^3 + 2x^2 - 5x - 2$. The signs of the coefficients are $(+, +, -, -)$, so there is only one sign change, from positive to negative, meaning that there is exactly 1 negative real root.

Therefore, the Rule of Signs tells us the possible combinations of types of roots: either there are 2 positive real, 1 negative real, and 0 imaginary roots; or there are 0 positive real, 1 negative real, and 2 imaginary roots, as shown.

Positive Real Roots	Negative Real Roots	Imaginary Roots	Total Roots
2	1	0	3
0	1	2	3

As it turns out, for this function, all three roots are real.

MODEL PROBLEM 1: *FINDING RATIONAL ROOTS*

Find the roots of $f(x) = 2x^3 - x^2 - 13x - 6$.

Solution:

(A) Possible rational roots: $1, -1, 2, -2, 3, -3, 6, -6, \frac{1}{2}, -\frac{1}{2}, \frac{3}{2}$, and $-\frac{3}{2}$

(B)
$$\begin{array}{r|rrrr} \boxed{-2} & 2 & -1 & -13 & -6 \\ & & -4 & 10 & 6 \\ \hline & 2 & -5 & -3 & | \; 0 \end{array}$$

(C) $(x + 2)(2x^2 - 5x - 3)$

(D) $(x + 2)(2x^2 - 6x + x - 3)$
$(x + 2)[2x(x - 3) + (x - 3)]$
$(x + 2)(x - 3)(2x + 1)$

(E) The roots are $-2, 3$, and $-\frac{1}{2}$.

Explanation of steps:

(A) List the possible rational roots, using the factors of the constant term [6] for the numerators and factors of the leading coefficient [2] for the denominators.
[The possible roots are $\pm\frac{1}{1}, \pm\frac{2}{1}, \pm\frac{3}{1}, \pm\frac{6}{1}, \pm\frac{1}{2}, \pm\frac{2}{2}, \pm\frac{3}{2}$, *and* $\pm\frac{6}{2}$, *which gives us, after eliminating duplicates,* $1, -1, 2, -2, 3, -3, 6, -6, \frac{1}{2}, -\frac{1}{2}, \frac{3}{2}$, *and* $-\frac{3}{2}$*.]*

(B) Test each possible root by synthetic division until a factor is found.
[−2 is the first in the list to yield a remainder of 0, signifying that $(x + 2)$ *is a factor.]*

(C) Once a factor is found, the other factor is the quotient produced by synthetic division *[*$2x^2 - 5x - 3$*].*

(D) Continue to factor until all factors are constant, linear or irreducible quadratics.
[The second factor can be factored further by grouping.]

(E) States the roots.
[We can find the roots by setting each factor equal to zero and solving for x.]

PRACTICE PROBLEMS

1. Find the rational roots of $f(x) = x^3 - 8x^2 + 11x + 20$.	2. Find the rational roots of $g(x) = 3x^3 - 10x^2 + x + 6$.
3. Find the rational roots of $h(x) = 2x^3 - x^2 - 22x - 24$.	4. Find the rational roots of $f(x) = 2x^4 + x^3 - 19x^2 - 9x + 9$.

MODEL PROBLEM 2: *USING DESCARTES' RULE OF SIGNS*

Determine the possible number of positive and negative real roots of the function $f(x) = x^5 - x^4 + 3x^3 + 2x^2 - x + 9$.

Solution:

(A) $f(x)$ is a fifth-degree (quintic) polynomial, so there are 5 roots.

(B) $f(x)$ has 4 sign changes, so there are at most 4 positive real roots.

(C) $f(-x)$ has 1 sign change, so there is exactly 1 negative real root.

(D)

Positive Real Roots	Negative Real Roots	Imaginary Roots	Total Roots
4	1	0	5
2	1	2	5
0	1	4	5

Explanation of steps:

(A) The degree of the polynomial function tells you how many total roots it has *[5]*.

(B) Count the changes in signs of $f(x)$. This tells you the maximum number of positive real roots it may have, although it may have an even number less than this. *[There are 4 sign changes as shown below, so $f(x)$ has at most 4 positive real roots. That is, it may have 4, 2, or 0 positive real roots.]*

$$\boxed{+}\,x^5 \quad \boxed{-}\,x^4 \quad \boxed{+}\,3x^3 \quad \boxed{+}\,2x^2 \quad \boxed{-}\,x \quad \boxed{+}\,9$$

1 2 3 4

(C) Count the changes in signs of $f(-x)$. This tells you the maximum number of negative real roots it may have, although it may have an even number less than this. *[Write an expression for $f(-x)$:*

$f(-x) = (-x)^5 - (-x)^4 + 3(-x)^3 + 2(-x)^2 - (-x) + 9$, *so*

$f(-x) = -x^5 - x^4 - 3x^3 + 2x^2 + x + 9$.

For $f(-x)$, there is only 1 sign change as shown below, so there is exactly 1 negative real root.]

$$\boxed{-}\,x^5 \quad \boxed{-}\,x^4 \quad \boxed{-}\,3x^3 \quad \boxed{+}\,2x^2 \quad \boxed{+}\,x \quad \boxed{+}\,9$$

1

(D) Use a table to list all possible combinations of the types of roots. *[There are either 4, 2, or 0 positive real roots, so write these in column 1. There must be 1 negative real root, so write this down column 2. The total roots must be 5, so write that down column 4. The number of imaginary roots in column 3 can be determined by subtracting the number of real roots in each row from the total.]*

PRACTICE PROBLEMS

5. Use Descartes' Rule of Signs to list all possible combinations of types of roots for the function $P(x) = 2x^3 + 3x^2 - 10x + 1$.

Positive Real Roots	Negative Real Roots	Imaginary Roots	Total Roots

6. Use Descartes' Rule of Signs to list all possible combinations of types of roots for the function $h(x) = -x^5 + x^4 + x^2 - x + 1$.

Positive Real Roots	Negative Real Roots	Imaginary Roots	Total Roots

7.3 Properties of Polynomial Graphs

Key Terms and Concepts

The **degree** of a polynomial function tells us the maximum number of times that the function may intersect the x-axis.

Examples: A linear function (degree 1) can have at most one x-intercept, a quadratic function (degree 2) can have at most two x-intercepts, and a cubic function (degree 3) can have at most three x-intercepts, etc.

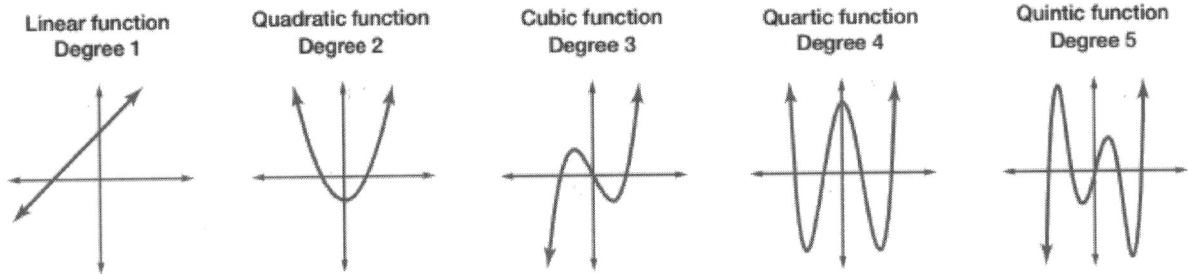

Multiplying the highest degree terms of all the factors will give us the leading term of the polynomial.

Example: Given $h(x) = -2(x^2 - 3)(x + 1)(5x + 3)(2x - 1)^2$, multiplying the highest degree terms of the factors gives us $(-2)(x^2)(x)(5x)(2x)^2 = -40x^6$, so the degree of $h(x)$ is 6 and the leading coefficient is -40.

The **x-intercepts** are the x-coordinates of the point(s), if there are any, where the graph intersects the x-axis. The x-intercepts represent the real roots (or zeros) of the function, so we can find the x-intercepts by solving for x. If a graph of a polynomial function has no x-intercepts, then the function has no real roots – it has only imaginary roots.

Example: Given $f(x) = (x + 4)(x + 1)(x - 1)(x - 3)$, the roots are –4, –1, 1, and 3. Therefore, the x-intercepts of the graph of $f(x)$ are $(-4,0)$, $(-1,0)$, $(1,0)$, and $(3,0)$. *[In standard form, this function is $f(x) = x^4 + x^3 - 13x^2 - x + 12$.]*

CALCULATOR TIP

To find the x-intercepts of a graph using the calculator:

1. Press [Y=] and enter the equation.
2. Press [2nd][CALC][2] to select zero.
3. For the "Left Bound?" prompt, move the cursor to a point on the curve to the left of the x-intercept and press [ENTER].
4. For the "Right Bound?" prompt, move the cursor to a point on the curve to the right of the x-intercept and press [ENTER].
5. For the "Guess?" prompt, use the arrow keys to move the cursor near the x-intercept and press [ENTER].
6. The coordinates of the x-intercept will be shown (y will be zero).
7. If there are multiple x-intercepts, repeat steps 2-6 for each additional point.

Example: To find the real roots of $y = x^3 - 4x^2 + 2x + 2$, follow these steps as shown in the screenshots below to find the first root as approximately –0.481. Repeat steps 2-6 to find the other two x-intercepts at approximately 1.311 and 3.17.

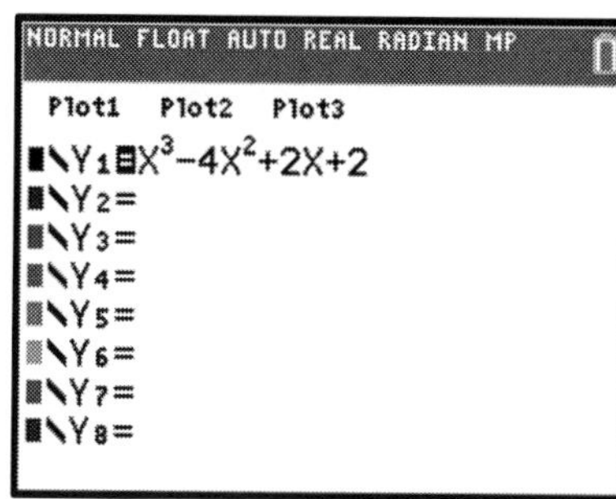

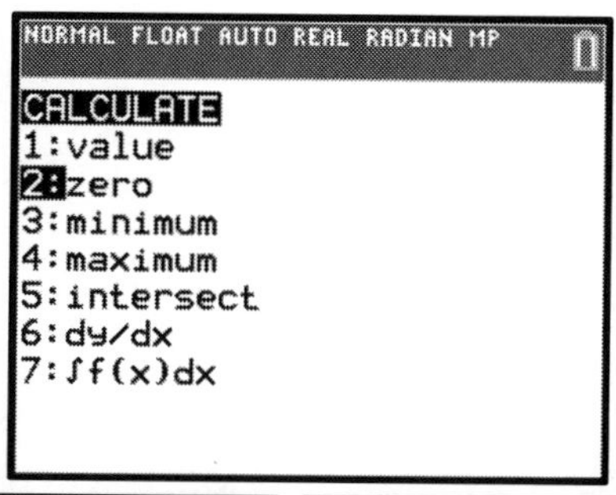

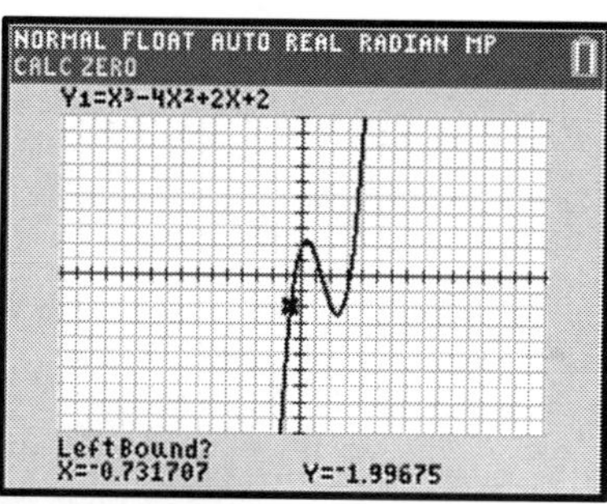

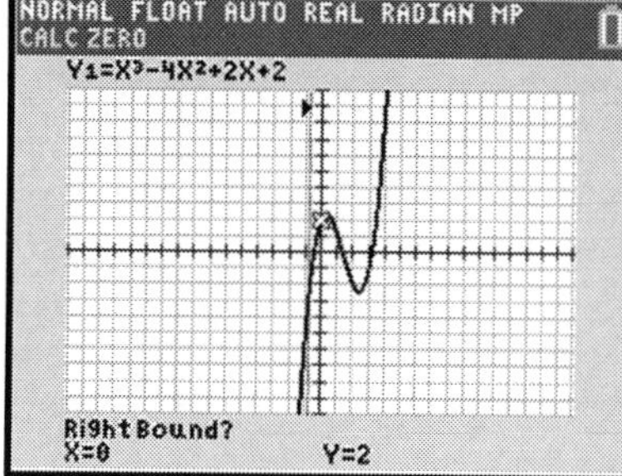

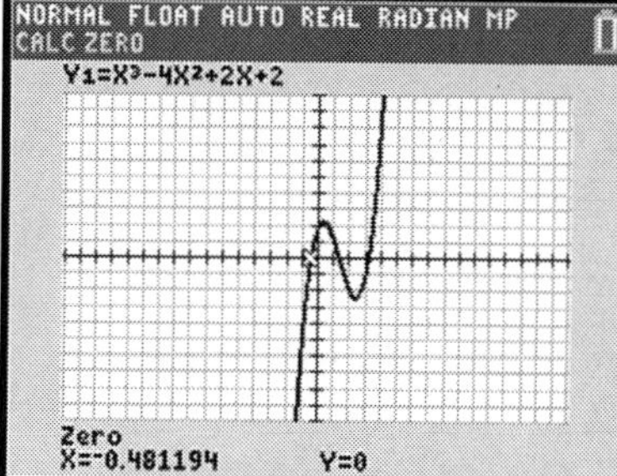

When a real root appears twice (known as **double roots**), then instead of crossing the x-axis at that x-intercept, the graph will "bounce off" the x-axis.

Example: The function $g(x) = \frac{1}{4}(x+1)(x-2)(x-2)$ has roots of −1, 2, and 2. Notice that the root 2 appears twice. The graph of this function will have only two x-intercepts at −1 and 2, but the graph will "bounce off" the x-axis at (2,0).

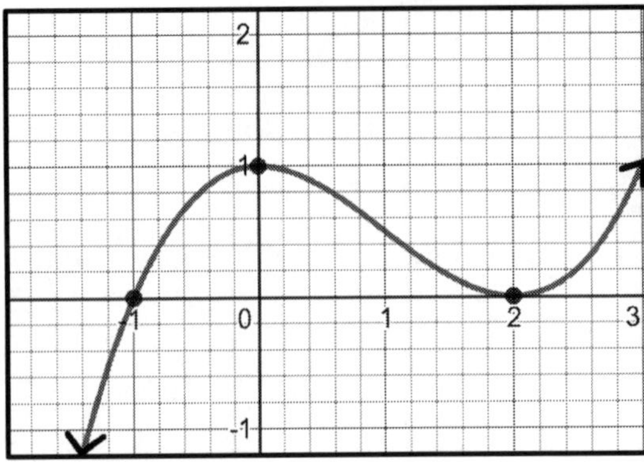

In fact, the graph will *bounce off* the axis when the root appears an *even* number of times and *pass through* when it appears an *odd* number of times.

As the multiplicity of a root increases, the graph appears flatter at that root on the x-axis.

Example: Below are the graphs of functions that have 2 as a root, but with different multiplicities. That is, they have $(x-2)$ as a factor to different powers.

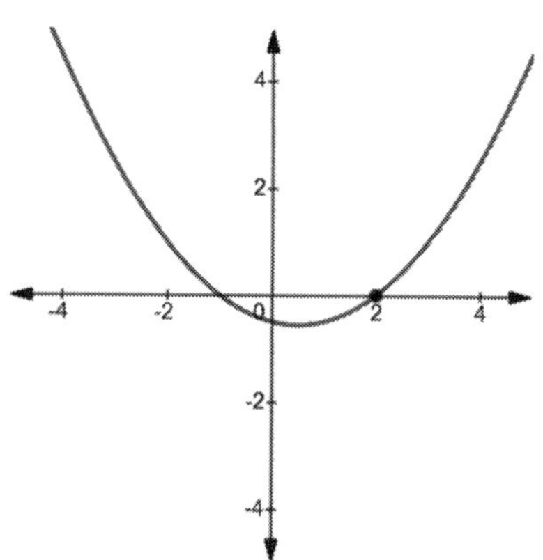

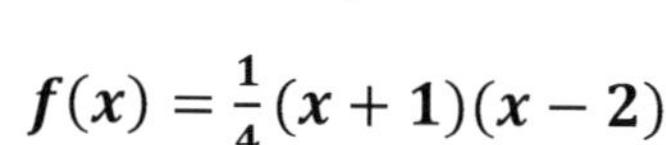
$\boldsymbol{f(x)=\frac{1}{4}(x+1)(x-2)}$

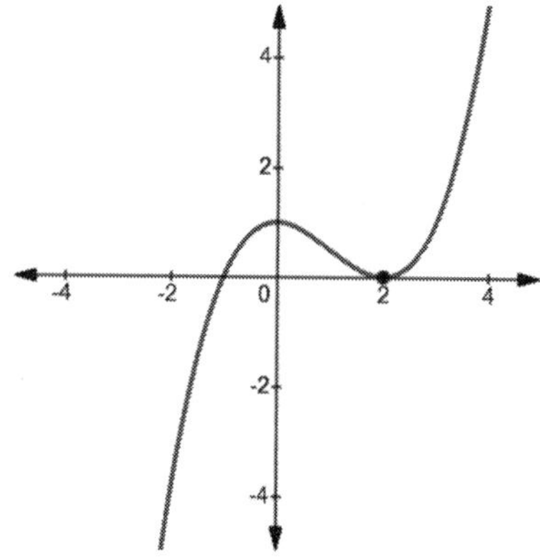

$\boldsymbol{f(x)=\frac{1}{4}(x+1)(x-2)^2}$

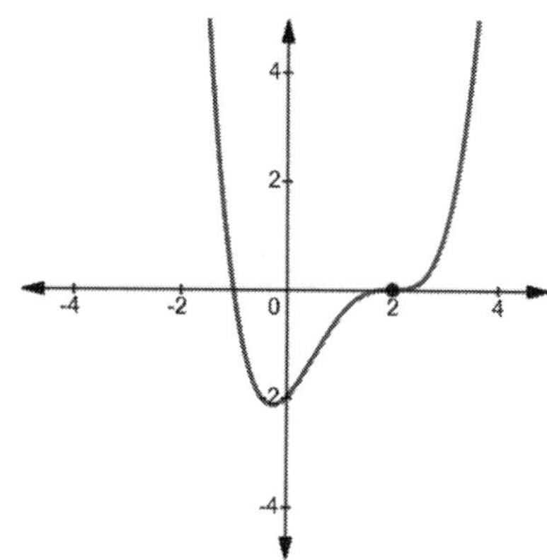

$\boldsymbol{f(x)=\frac{1}{4}(x+1)(x-2)^3}$

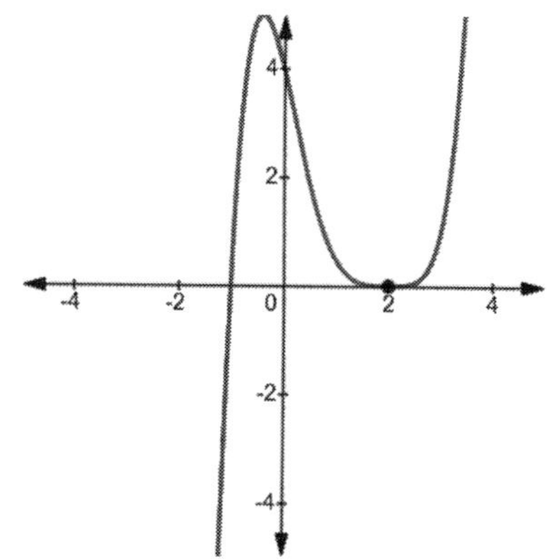

$\boldsymbol{f(x)=\frac{1}{4}(x+1)(x-2)^4}$

The ***y*-intercept** is the y-coordinate of the point, if there is one, where the graph intersects the y-axis. We can find the ***y*-intercept** of a function by evaluating $f(0)$. For a polynomial function, this will be equal to the constant term.

Example: For the function, $f(x)=x^4+x^3-13x^2-x+12$, the y-intercept is $f(0)=12$.

CALCULATOR TIP

To find the y-intercept of a graph using the calculator:

1. Press [Y=] and enter the equation.
2. Press [GRAPH] to view the graph.
3. Press [TRACE][0][ENTER] to evaluate the function for $x = 0$. The y value is shown.

Example: For the equation $y = x^3 - 4x^2 + 2x + 2$, the y-intercept is 2, as shown below.

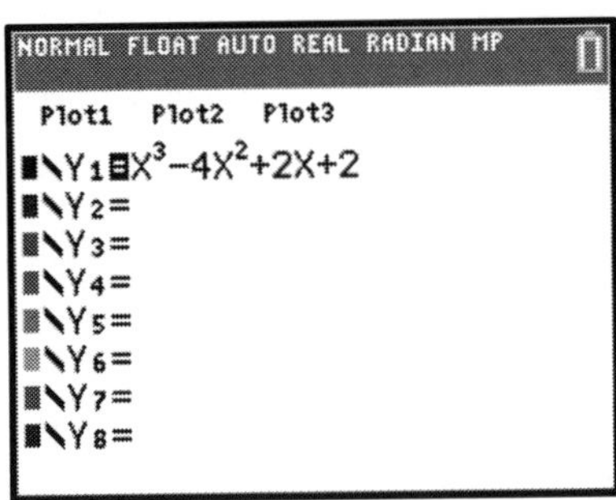

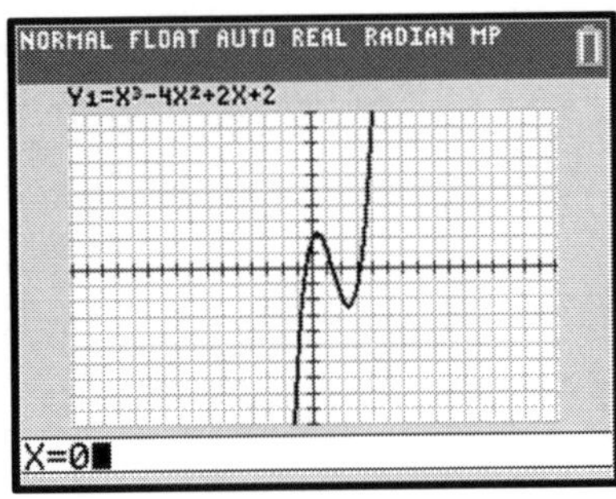

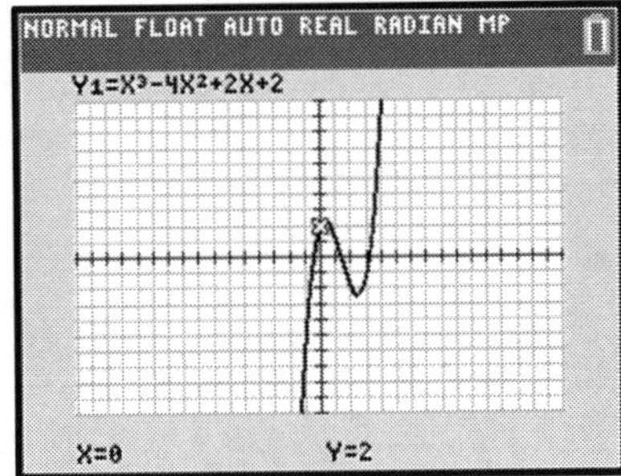

A function may have intervals in which its graph is **increasing** or **decreasing**. As we move from left to right (i.e., as x increases), if the value of the function increases over the entire interval then it is increasing and if the value of the function decreases over the entire interval then it is decreasing.

Example: In the graph of $g(x)$ below, we see that $g(x)$ is increasing over the intervals $x < 0$ and $x > 2$ and it is decreasing over the interval $0 < x < 2$.

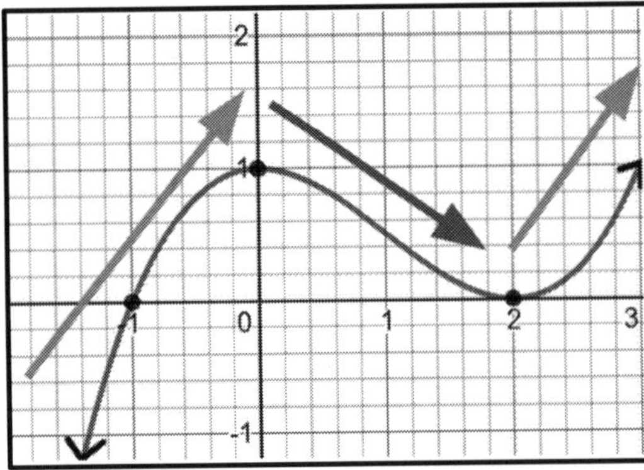

The **minimum** and **maximum** of a function $f(x)$ are the lowest and highest possible values of $f(x)$, respectively.

A function may also have points that appear at the bottom or top of a portion of a curve but may not be the lowest or highest point of a function over its complete domain. Each of these is called a **relative minimum** (*local minimum*) or a **relative maximum** (*local maximum*). Collectively, these are called **turning points.**

Example: This graph has a relative minimum at $(2,0)$ and a relative maximum at $(0,1)$.

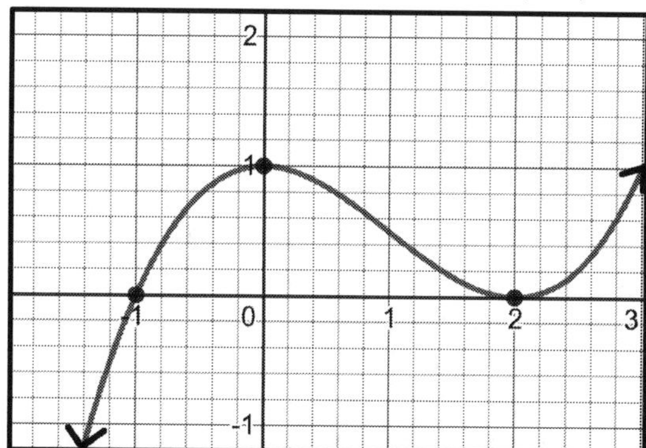

Although $(2,0)$ is not the lowest point on the curve (points where $x < -1$ will be lower) it is lower than the points directly beside it on both sides, so it appears at the bottom of a "valley".

Likewise, $(0,1)$ is not the highest point on the curve (points where $x > 3$ will be higher), but it is higher than the points directly beside it on both sides, so it appears at the top of a "hill".

The graph of a polynomial of degree n will have at most $n - 1$ turning points. If all of the roots are real and distinct (have a multiplicity of 1), then the graph will have exactly $n - 1$ turning points.

For some functions, finding the turning points algebraically may require the use of derivatives, which is taught in calculus. However, we can use the calculator's functions to find them, as well.

CALCULATOR TIP

To find the turning points of a graph using the calculator:

1. Press [Y=] and enter the equation.
2. Press [2nd][CALC][3] or [2nd][CALC][4] for minimum or maximum.
3. For the "Left Bound?" prompt, move the cursor to a point on the curve to the left of the minimum or maximum and press [ENTER].
4. For the "Right Bound?" prompt, move the cursor to a point on the curve to the right of the minimum or maximum and press [ENTER].
5. For the "Guess?" prompt, use the arrow keys to move the cursor near the minimum or maximum point and press [ENTER].
6. The coordinates of the minimum or maximum will be shown.

Example: For $y = x^3 - 4x^2 + 2x + 2$, the relative maximum at approximately $(0.279, 2.268)$ may be found using the calculator, as shown below.

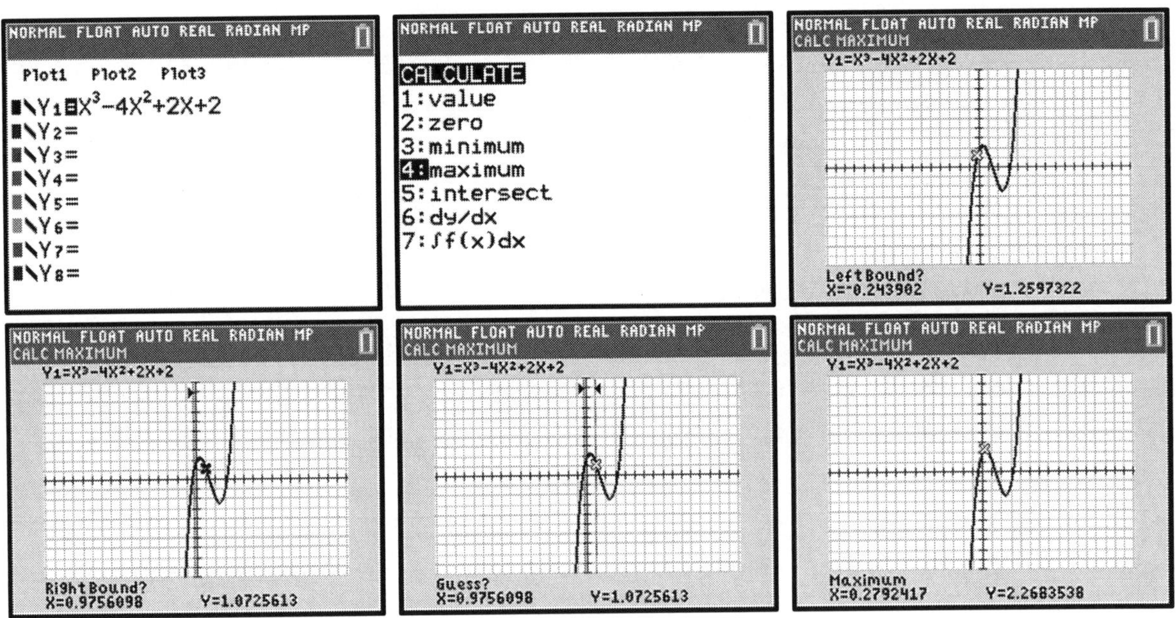

CALCULATOR TIP

Alternatively, you can find the minimum or maximum value of a function between two values of x by using the calculator's fMin or fMax function.

1. Press [Y=] and enter the equation, and [2nd][QUIT].
2. Press [MATH][6] or [MATH][7] for fMin or fMax.
3. Enter [ALPHA][F4][1] *[or* [VARS] [Y-VARS] [1][1] *on the TI-83]* to select Y1. Then press [,][X,T,Θ,n][,], type a lower bound for x, [,], an upper bound for x, and [)][ENTER].
4. The x-value is shown. To find the y-value, press [ALPHA][F4][1][(][2nd][ANS][)][ENTER] *[or* [VARS] [Y-VARS] [1][1][(][2nd][ANS][)][ENTER] *on the TI-83]*.

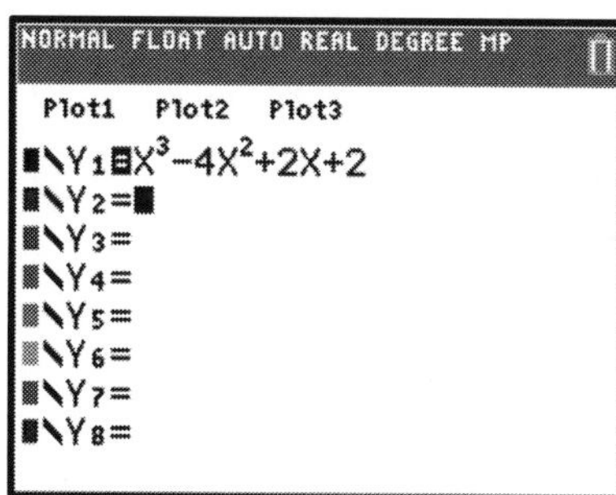

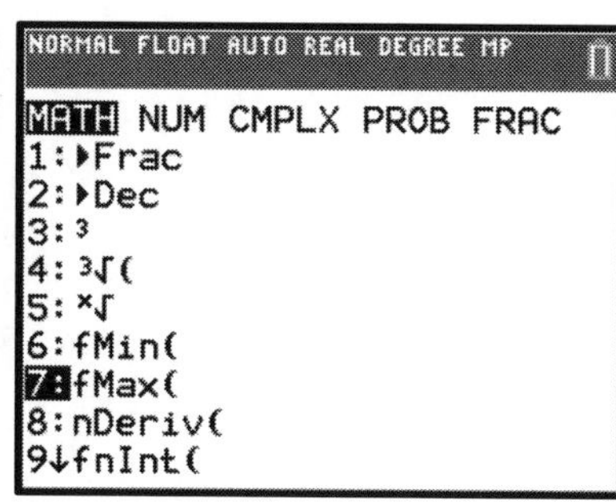

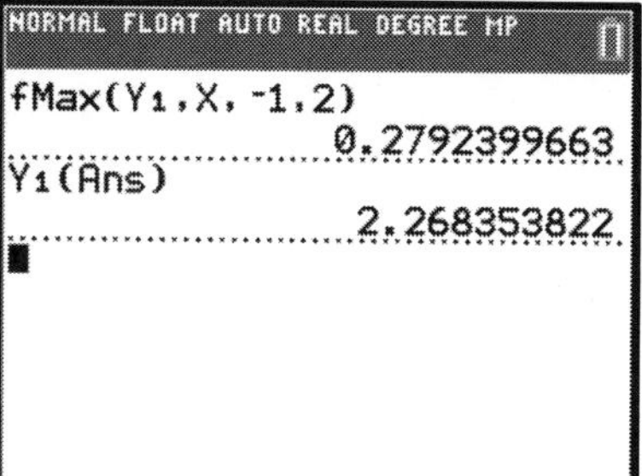

The **end behavior** of the graph of $f(x)$ describes what happens to the values of $f(x)$ as x decreases towards $-\infty$ and as x increases towards ∞.

The end behavior of a polynomial function depends on its **degree** (odd or even) and its **leading coefficient** (positive or negative).

	Odd Degree	Even Degree
Positive Leading Coefficient	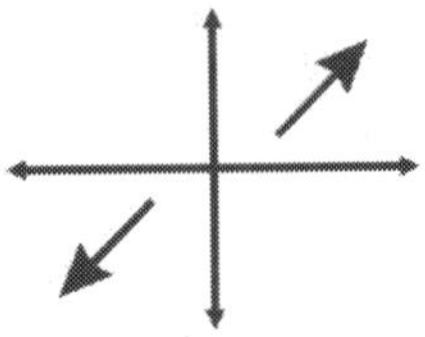 As $x \to -\infty$, $f(x) \to -\infty$. As $x \to \infty$, $f(x) \to \infty$. Example: 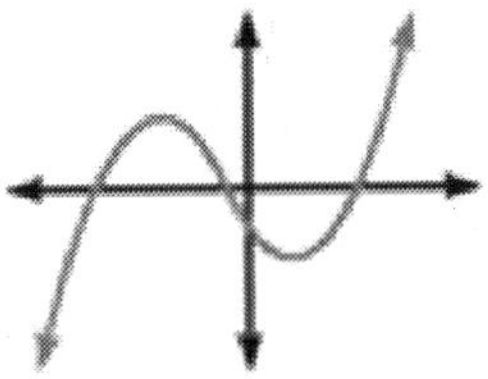	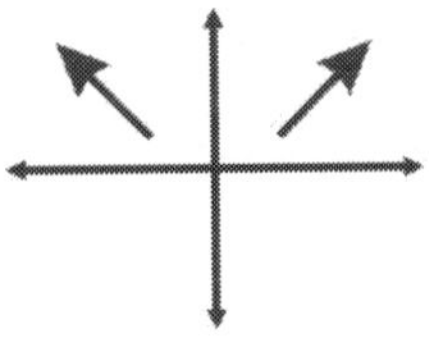 As $x \to -\infty$, $f(x) \to \infty$. As $x \to \infty$, $f(x) \to \infty$. Example: 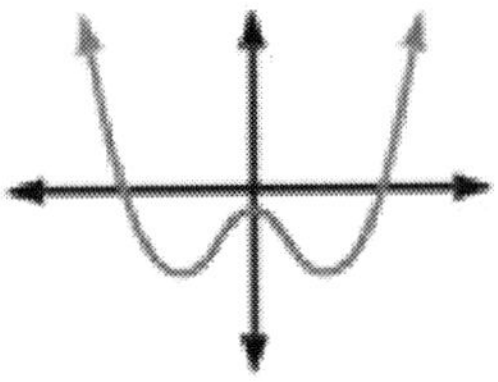
Negative Leading Coefficient	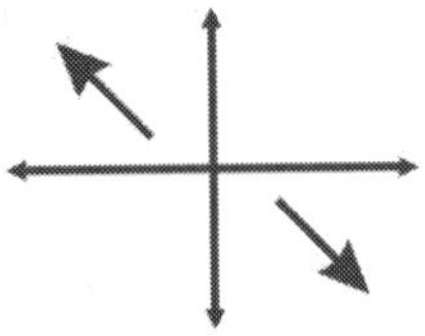 As $x \to -\infty$, $f(x) \to \infty$. As $x \to \infty$, $f(x) \to -\infty$. Example: 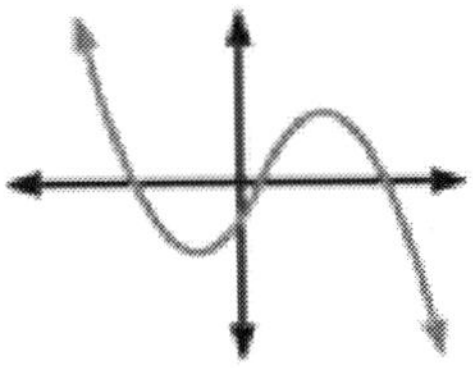	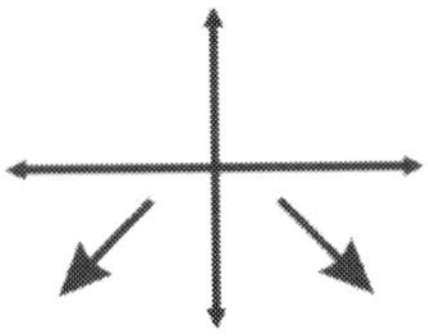 As $x \to -\infty$, $f(x) \to -\infty$. As $x \to \infty$, $f(x) \to -\infty$. Example: 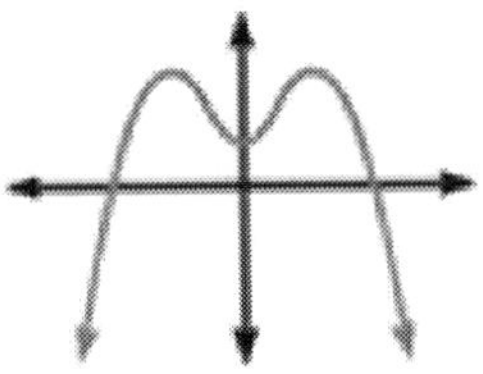

To memorize the end behaviors of polynomial functions, you just have to remember how the graphs of linear and quadratic functions look.

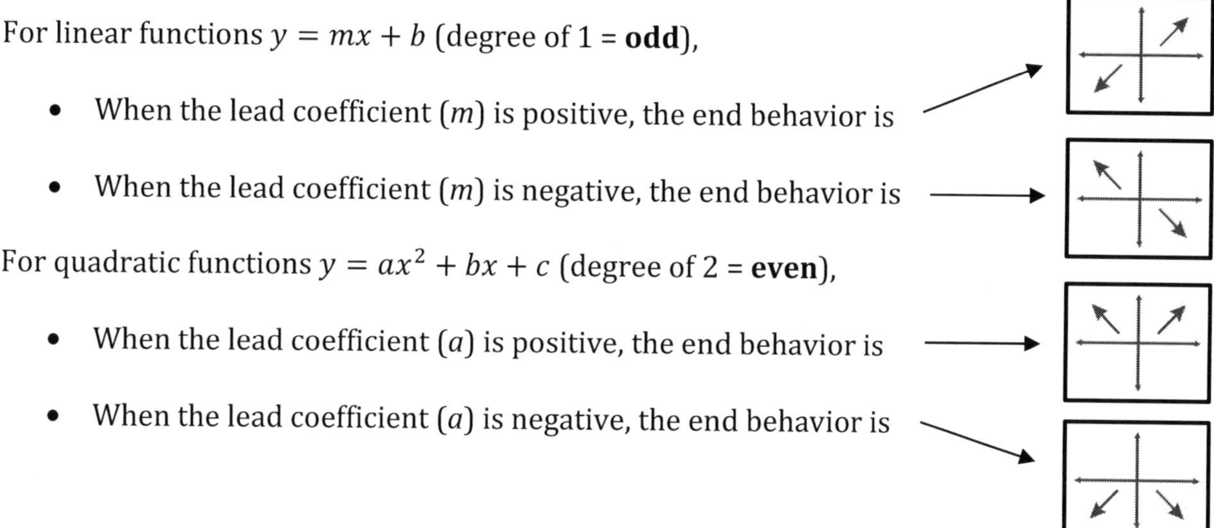

For linear functions $y = mx + b$ (degree of 1 = **odd**),

- When the lead coefficient (m) is positive, the end behavior is
- When the lead coefficient (m) is negative, the end behavior is

For quadratic functions $y = ax^2 + bx + c$ (degree of 2 = **even**),

- When the lead coefficient (a) is positive, the end behavior is
- When the lead coefficient (a) is negative, the end behavior is

All odd degree polynomial graphs will have the same end behaviors as linear functions (lines) and all even degree polynomial graphs will have the same end behaviors as quadratic functions (parabolas).

MODEL PROBLEM

For the function $f(x) = (2x + 1)(3x - 2)(x + 1)(x + 1)$,

a) state the degree and leading coefficient
b) state the end behavior
c) state the y-intercept
d) state the x-intercepts

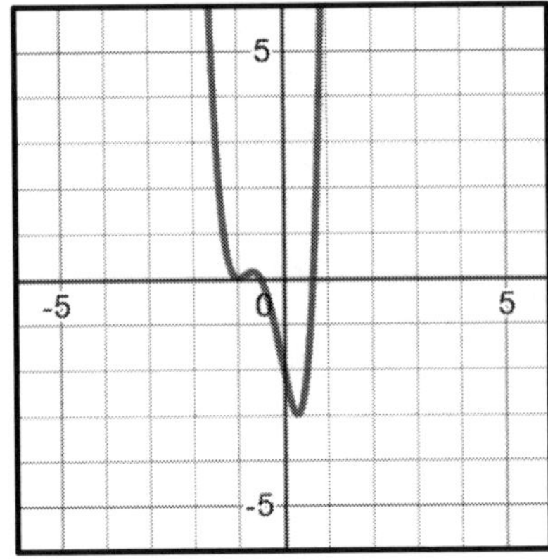

Solution:

(A) degree is 4 and leading coefficient is 6

(B) as $x \to -\infty, f(x) \to \infty$, and as $x \to \infty, f(x) \to \infty$

(C) y-intercept at $(0, -2)$

(D) x-intercepts at $(-1,0)$, $\left(-\frac{1}{2}, 0\right)$, and $\left(\frac{2}{3}, 0\right)$

Explanation of steps:

(A) Multiplying the highest degree terms of all the factors will give us the leading term of the polynomial. *[Multiplying $(2x)(3x)(x)(x)$ results in a first term of $6x^4$, so the degree is 4 and leading coefficient is 6.]*

(B) The end behavior is determined by the degree and leading coefficient. *[Since the degree is even (4) and the leading coefficient is positive (6), the end behavior is the same as a parabola that opens up.]*

(C) The y-intercept can be found by evaluating $f(0)$. The graph will intercept the y-axis at the point $(0, f(0))$. *[$f(0) = (1)(-2)(1)(1) = -2$, so $(0, -2)$ is the y-intercept.]*

(D) Set each factor equal to 0 and solve to find the roots. For any <u>real</u> root a, there will be an x-intercept at $(a, 0)$. *[Setting $2x + 1 = 0$ gives us a root of $-\frac{1}{2}$ and setting $3x - 2 = 0$ gives us a root of $\frac{2}{3}$. Setting $x + 1 = 0$ gives us a double root of -1, one for each repeated factor, $(x + 1)$.]*

PRACTICE PROBLEMS

1. What are the degree and leading coefficient of the function $f(x) = -3x^5 + 2x^2 - 7x + 1$?

   Describe the end behavior of the graph of $f(x)$.

2. What are the degree and leading coefficient of the function $g(x) = 5(x+1)^2(x-1)(x+4)$?

   Describe the end behavior of the graph of $g(x)$.

3. What are the degree and leading coefficient of the function
   $$h(x) = -4(x^2 - 2)(2x - 5)^2$$

   Describe the end behavior of the graph of $h(x)$.

4. What are the degree and leading coefficient of the function
   $$j(x) = x^3(x^2 + 3x - 2)(x - 1)^2$$

   Describe the end behavior of the graph of $j(x)$.

5. State the y-intercept and x-intercepts of the graph of
   $$h(x) = -4(x^2 - 2)(2x - 5)^2$$

6. State the y-intercept and x-intercepts of the graph of
   $$k(x) = 5x(x^2 + 1)(x + 4)^2$$

7. Use the calculator to find the x-intercepts of the graph of $f(x) = -2x^3 - 5x^2 + 2$, rounding coordinates to the *nearest hundredth*.

8. Use the calculator to find the relative maxima and relative minima of the function $f(x) = x^3 - 12^2 + 45x - 49$.

   Over which interval(s) of x is the function increasing or decreasing?

9. Use the calculator to find the relative maxima and relative minima of the function $f(x) = 20(x+3)^2(x+2)^2$.

   Over which interval(s) of x is the function increasing or decreasing?

7.4 Graph Polynomial Functions

Key Terms and Concepts

Even without the use of a calculator to graph a polynomial function, if we can find the x-intercepts and y-intercept algebraically, and we can determine its the end behavior, we can draw a rough sketch.

Example: To sketch $f(x) = \frac{1}{2}(x+3)(x+2)(x-3)$, the x-intercepts are $-3, -2$, and 3, the y-intercept is $f(0) = \frac{1}{2}(3)(2)(-3) = -9$, and with an odd degree (3) and a positive leading coefficient $\left(\frac{1}{2}\right)$, the graph has the end behavior shown to the right.

Putting all of this together, the graph will look like this:

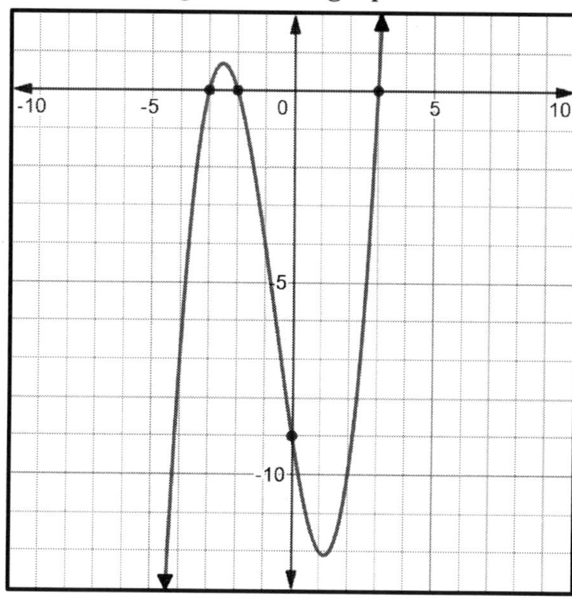

CALCULATOR TIP

If we are given a set of points, we can use the calculator to find the best-fitting quadratic, cubic or quartic equation using the corresponding regression function. The quadratic regression function requires at least three points, the cubic regression function requires at least four points, and the quartic regression function requires at least five points.

Example: To find the best-fitting quartic equation given the points (0,10), (1,2), (2,4), (3,7), (4,8), and (5,10), follow these steps.

1. Press [STAT][1] and enter the x-values as L1 and the y-values as L2.
2. Press [STAT][CALC][7] for quartic regression.
3. On the TI-84 models, select L1 for XList and L2 for YList (using [2nd][LIST]) and optionally select Y1 to store the resulting equation (using [ALPHA][F4]). [On the TI-83, this screen is skipped, so just press [ENTER].]
4. The calculator will show the values of the coefficients as a, b, c, d, and e. For the given points, we get $y = .25x^4 - 3x^3 + 12.25x^2 - 17.5x + 10$.

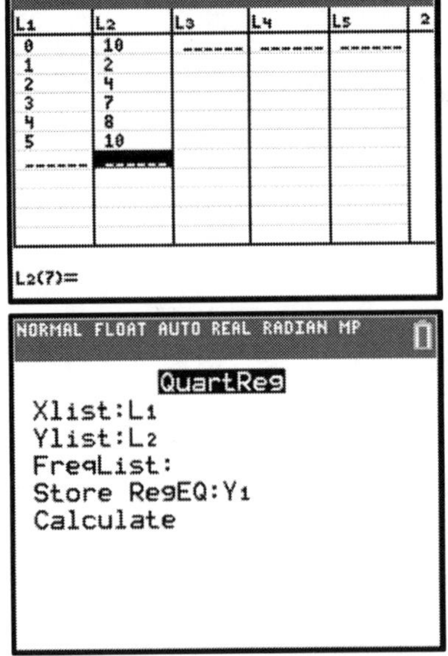

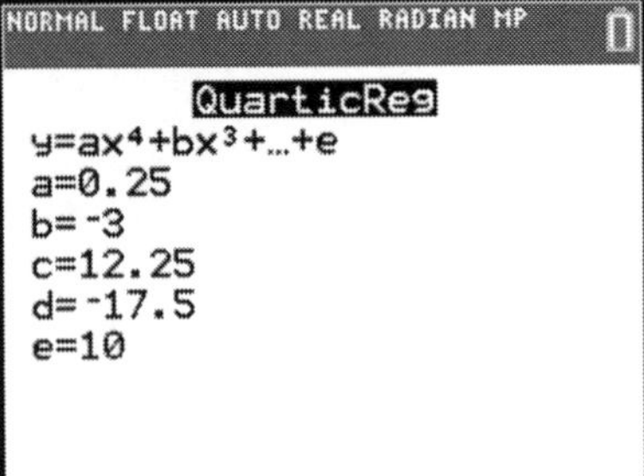

When using the calculator's regression function, a value of zero may be expressed by a close approximation of zero in scientific notation.

Example: If the calculator displays that a value is 9E-12, which is scientific notation for 0.000000000009, it may be exactly zero.

Suppose you are given a set of points at equal intervals for x but you are not told the degree of the polynomial function. You can determine the degree by looking at **finite differences.**

To determine the degree of a polynomial function using finite differences:

1. Write out the values of $f(x)$ for a set of equally spaced values of x.
2. Find the *first differences* by subtracting consecutive values. If the differences are constant, then the polynomial is of the *first degree* (linear).
3. If not, find the *second differences* by subtracting the values found in the previous step. If the differences are constant, then the polynomial is of the *second degree* (quadratic).
4. If not, find the *third differences* by subtracting the values found in the previous step. If the differences are constant, then the polynomial is of the *third degree* (cubic).
5. Continue this process as necessary.

Example: Given the following points of a polynomial function,

$$(-2,-1),(-1,7),(0,5),(1,5),(2,19),(3,59),(4,137),$$

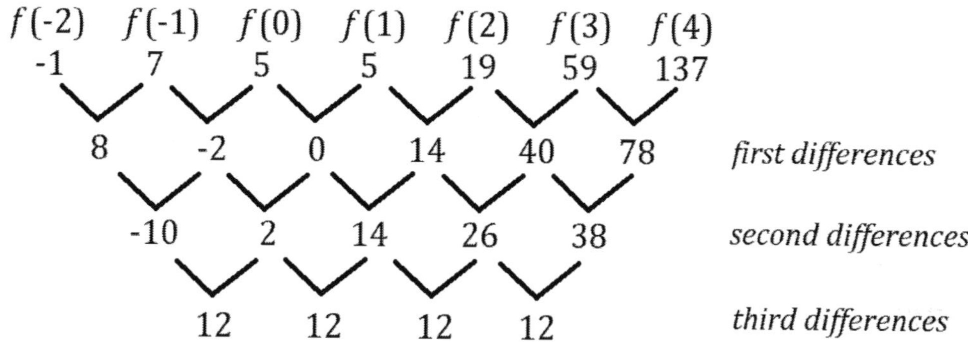

Because the third differences are constant, this is a third degree (cubic) function. Using cubic regression, the equation is $f(x) = 2x^3 + x^2 - 3x + 5$.

Model Problem

The graph of a polynomial function $f(x)$ includes the points $(-3,0)$, $(-2,0)$, $(-1,4)$, $(0,6)$, $(1,0)$, $(2,-20)$, and $(3,-60)$.

a) Use finite differences to determine the degree of $f(x)$.

b) Use the corresponding regression function to find the equation of the function.

Solution:

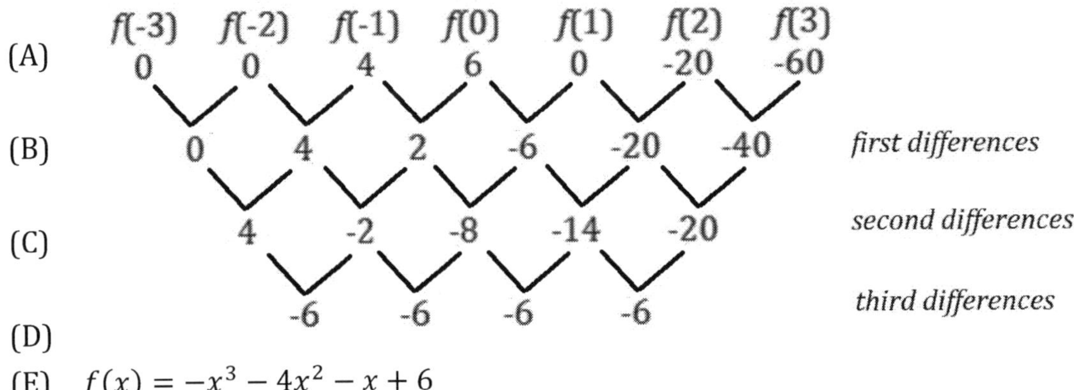

(E) $f(x) = -x^3 - 4x^2 - x + 6$

Explanation of steps:

(A) Write out the values of $f(x)$ for a set of equally spaced values of x.
[The given points have equally spaced values of x: $-3, -2, -1, 0, 1, 2, 3$.]

(B) Find the first differences by subtracting consecutive values. If the differences are constant, then the polynomial is of the first degree.
[The first differences – 0, 4, 2, etc. – are not constant.]

(C) If not, find the second differences by subtracting the values found in the previous step. If the differences are constant, then the polynomial is of the second degree.
[The second differences are not constant.]

(D) If not, find the third differences by subtracting the values found in the previous step. If the differences are constant, then the polynomial is of the third degree.
[The third differences are constant, so the function is third degree, or cubic.]

(E) Enter the points into the calculator and use the appropriate *[cubic]* regression function to find the equation of the function.

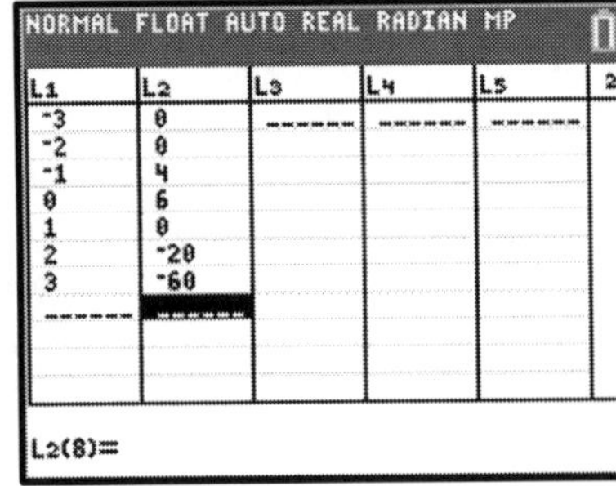

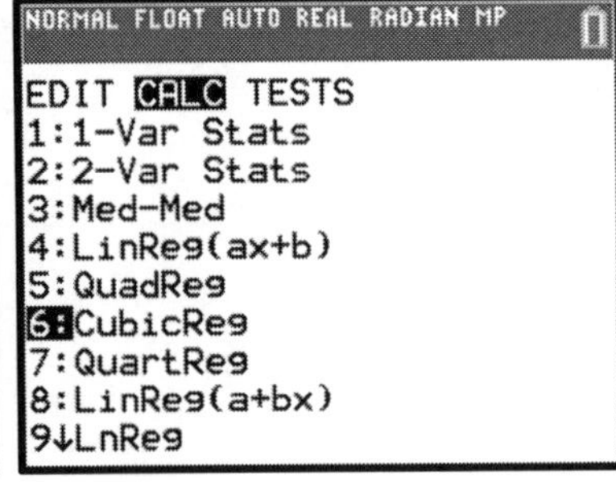

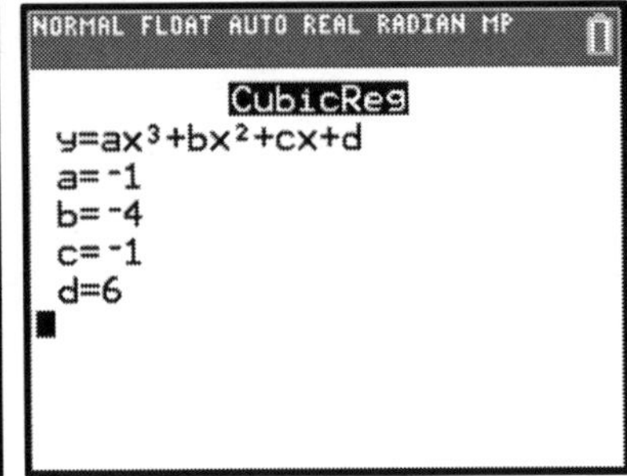

Practice Problems

1. Without your calculator, sketch a graph of the function, $f(x) = x(x+2)(x-3)$ on the grid below. Then, graph it on your calculator to check how close your sketch is to the actual graph.

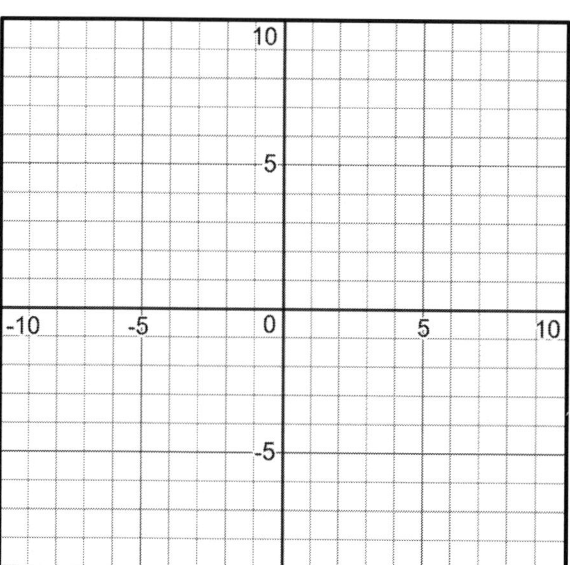

2. Without your calculator, sketch a graph of $g(x) = 2(x+1)^2(x-1)(x+2)$ on the grid below. Then, graph it on your calculator to check how close your sketch is to the actual graph.

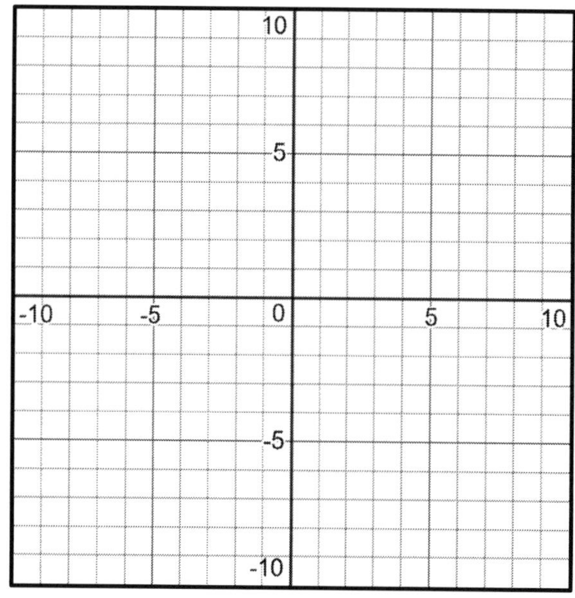

3. Use the calculator's regression function to determine the equation of the cubic function $c(x)$ that includes the points $(-4, 0), (-2, 30), (0, 12)$, and $(2, -6)$.

4. Use the calculator's regression function to determine the equation of the quartic function $q(x)$ that includes the points $(-2, 0), (-1, -2), (0, 0), (1, 0), (2, 40)$.

5. The graph of a polynomial function $f(x)$ includes the points $(-2,-24)$, $(-1,-6)$, $(0,4)$, $(1,6)$, and $(2,0)$. Use finite differences to determine the degree of $f(x)$, then use the calculator's corresponding regression function to find the equation of the function.

6. The graph of a polynomial function $g(x)$ includes $(-2,0)$, $(-1,-15)$, $(0,-16)$, $(1,-15)$, and $(2,0)$. Use finite differences to determine the degree of $g(x)$, then use the calculator's corresponding regression function to find the equation of the function.

7.5 Polynomial Transformations

Key Terms and Concepts

A polynomial function $f(x)$ may be transformed in any of the following ways:

$f(x)+k$	$(x,y)\to(x,y+k)$	vertically shifts the graph up $(k>0)$ or down $(k<0)$
$f(x+k)$	$(x,y)\to(x-k,y)$	horizontally shifts the graph left $(k>0)$ or right $(k<0)$
$-f(x)$	$(x,y)\to(x,-y)$	reflects the graph over the x-axis
$f(-x)$	$(x,y)\to(-x,y)$	reflects the graph over the y-axis
$k\cdot f(x)$	$(x,y)\to(x,ky)$	vertically stretches by a factor of k
$f(kx)$	$(x,y)\to(kx,y)$	horizontally stretches by a factor of $\frac{1}{k}$

For example, here are two **translations** of $f(x)=x^3$.

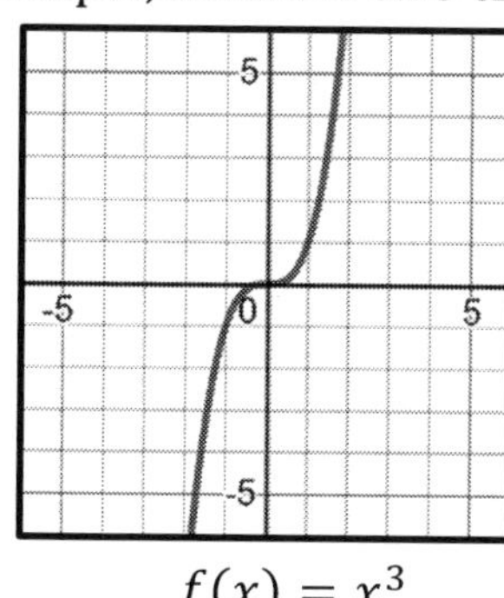

$f(x)=x^3$

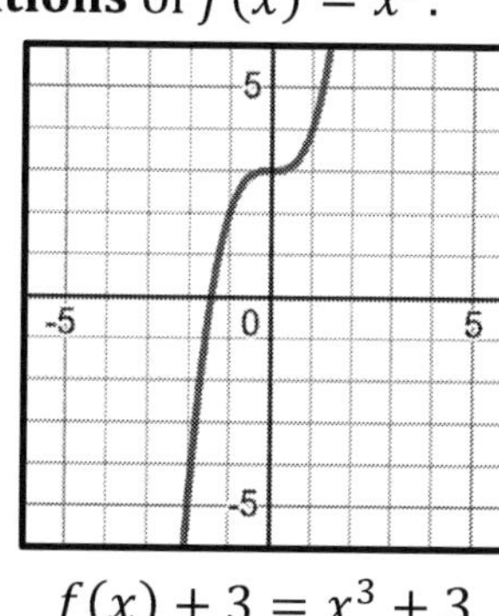

$f(x)+3=x^3+3$

shifted up 3

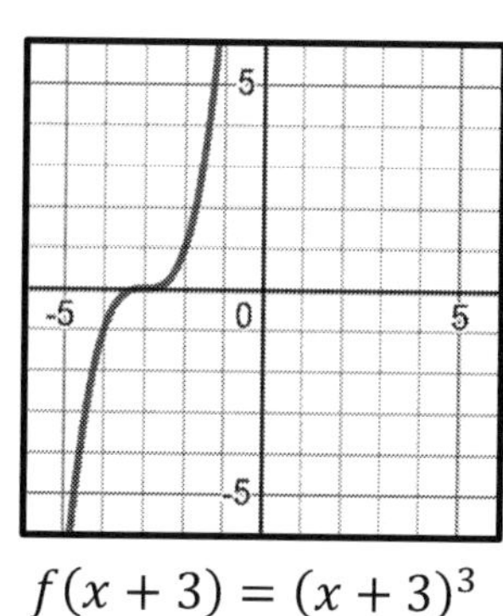

$f(x+3)=(x+3)^3$

shifted left 3

Here are two **reflections** of $f(x)=x^3$. Note that they both result in the same graph.

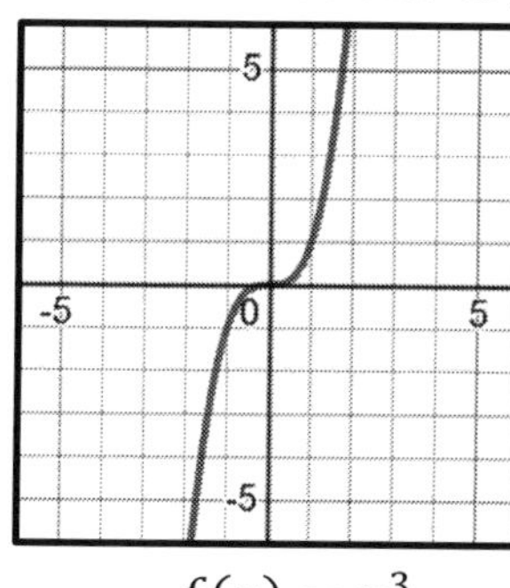

$f(x)=x^3$

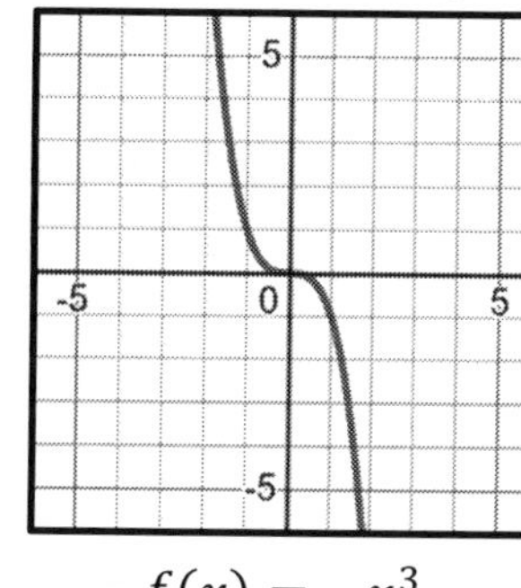

$-f(x)=-x^3$

reflected over x-axis

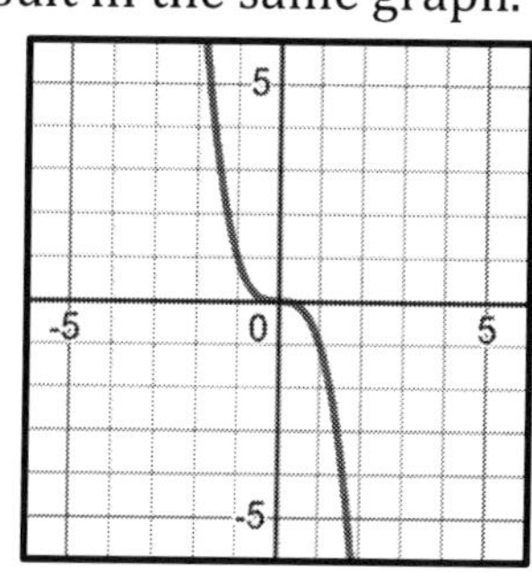

$f(-x)=(-x)^3$

reflected over y-axis

Here are two **stretches** of $f(x) = x^3$.

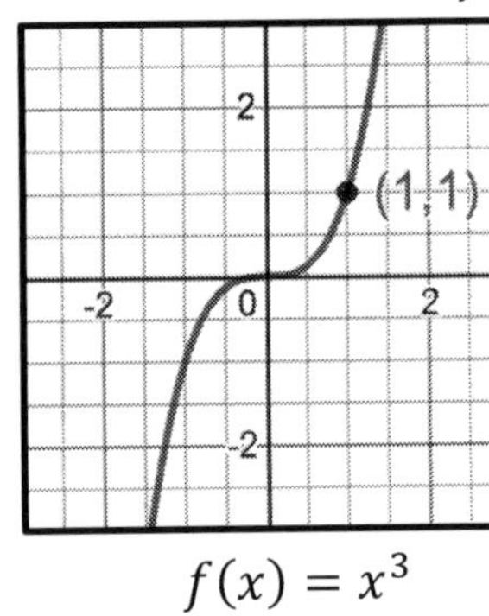

$f(x) = x^3$

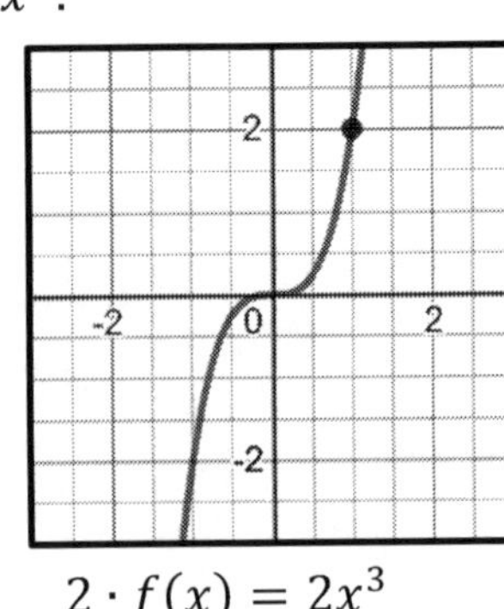

$2 \cdot f(x) = 2x^3$

vertically stretched by 2

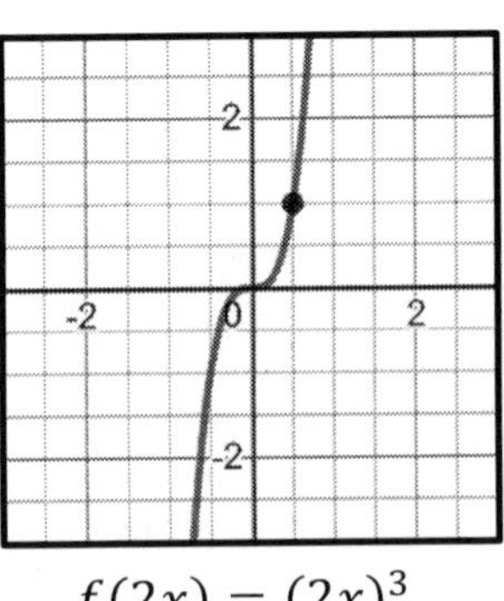

$f(2x) = (2x)^3$

horizontally stretched by $\frac{1}{2}$

The stretched graphs above are zoomed in further to show detail. Note that $(1,1) \to (1,2)$ after a vertical stretch by a factor of 2, and that $(1,1) \to \left(\frac{1}{2}, 1\right)$ after a horizontal stretch by a factor of $\frac{1}{2}$. Note that we will use the word "stretch" to refer to both expanding (by a factor greater than 1) or contracting (by a factor between 0 and 1).

When transforming a function graph, it is usually easiest to select specific points on the original graph and plot their images after the transformation.

Example: The maximum point on the graph of the equation $y = g(x)$ is $(2, -3)$. To find the maximum point on the graph of the equation $y = g(x - 4)$, plot the image of the given point. The mapping rule for $g(x + k)$ is a translation $(x, y) \to (x - k, y)$, so $(2, -3)$ maps to $(6, -3)$.

Of course, we should also graph the image function on the calculator to check that our new graph is as accurate as possible. Most importantly, use the calculator to confirm any intercepts.

MODEL PROBLEM

$g(x)$ is the image of $f(x) = x^3$ after the following sequence of transformations:

1. a vertical stretch by a factor of 2
2. a reflection over the x-axis
3. a translation 4 units up and 3 units right

Write an equation for $g(x)$.

Solution: $g(x) = -2(x-3)^3 + 4$

Explanation of steps:

Start with the original graph *[$y = x^3$]* and determine how each transformation, in order, changes the equation of the graph. *[After a vertical stretch by a factor of 2, $y = 2x^3$. Then, after a reflection over the x-axis, $y = -2x^3$. After a translation 4 units up, $y = -2x^3 + 4$. Finally, after a translation 3 units right, $y = -2(x-3)^3 + 4$.]*

PRACTICE PROBLEMS

1. The graph below represents the equation $y = f(x)$.

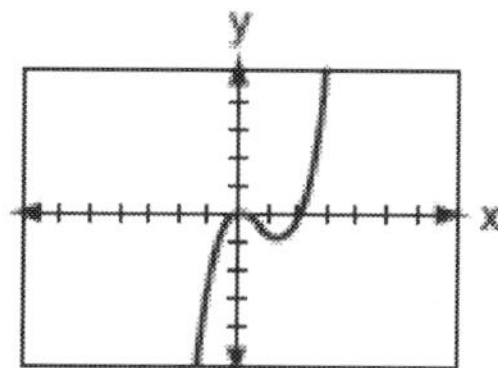

Which graph represents $g(x)$ if $g(x) = -f(x)$?

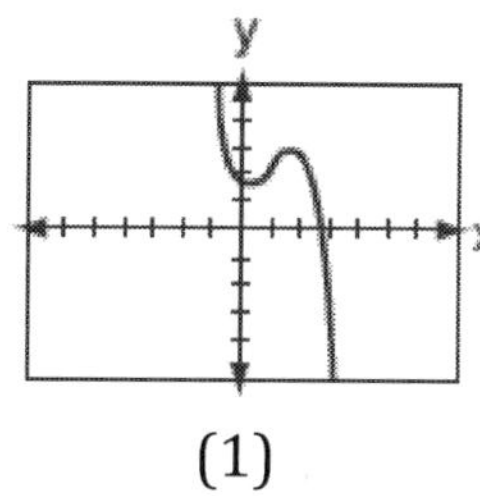

(1)

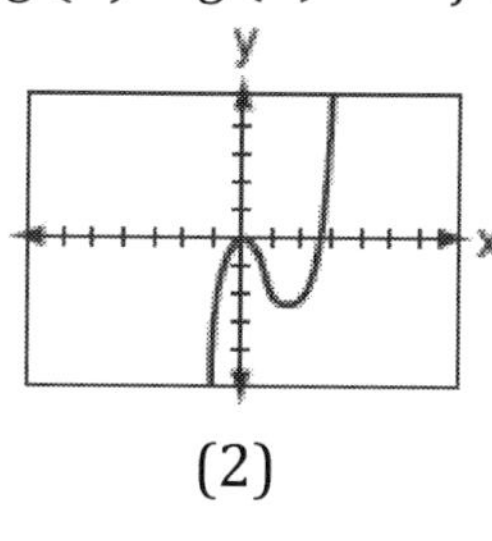

(2)

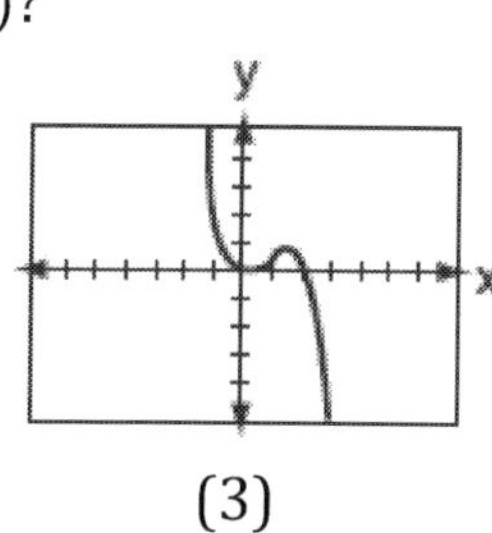

(3)

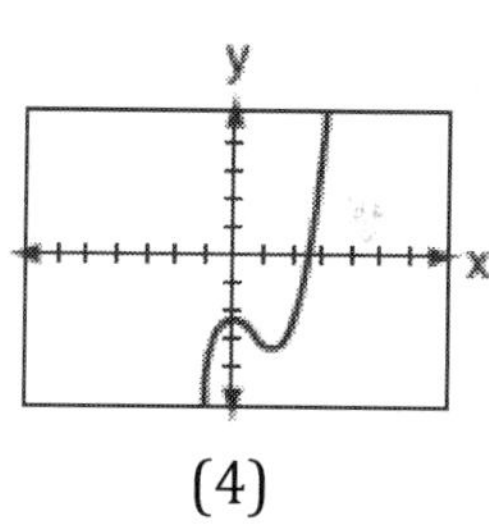

(4)

2. The graph below shows the function $f(x)$.

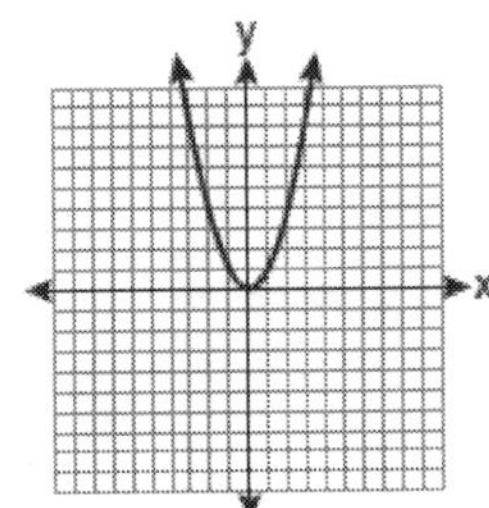

Which graph represents the function $f(x+2)$?

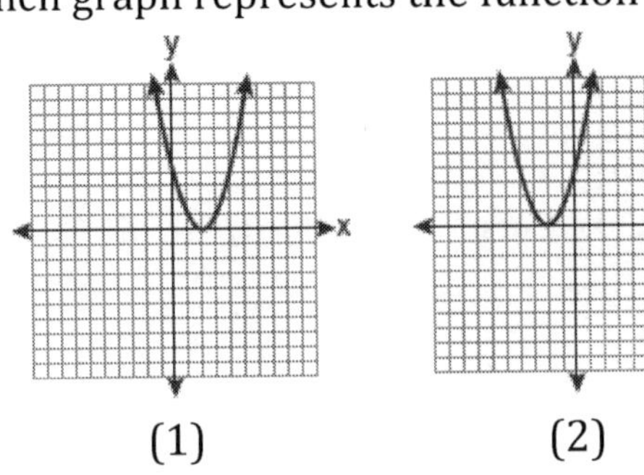

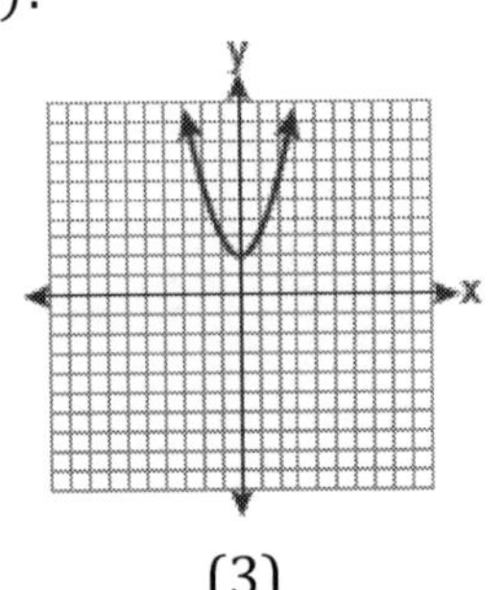

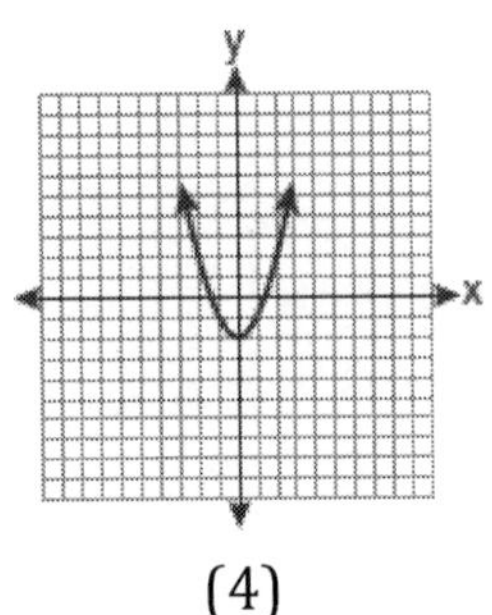

(1) (2) (3) (4)

3. The graph of $y = f(x)$ is shown below.

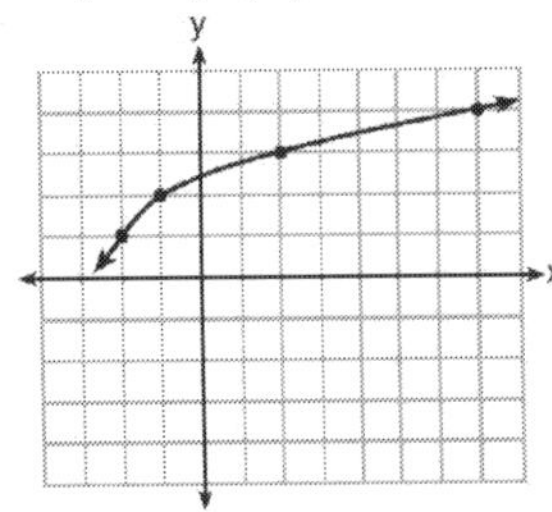

What is the graph of $y = f(x+1) - 2$?

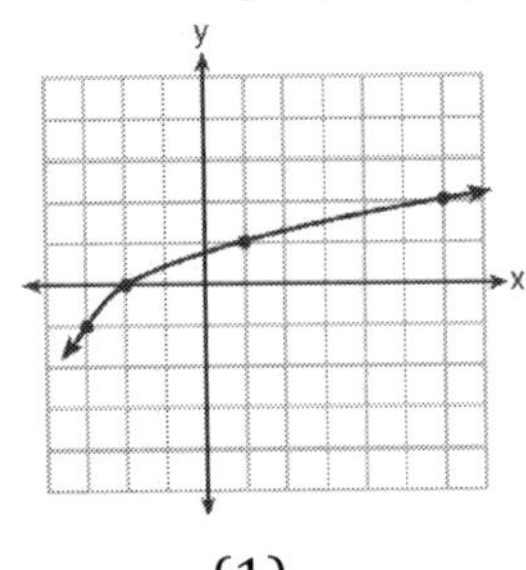

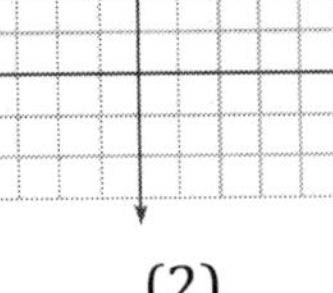

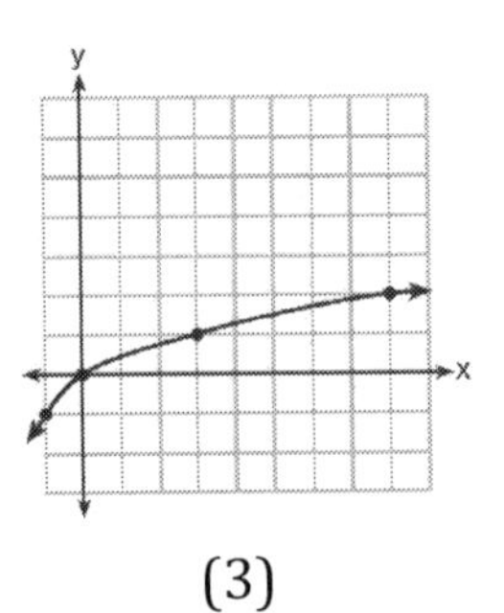

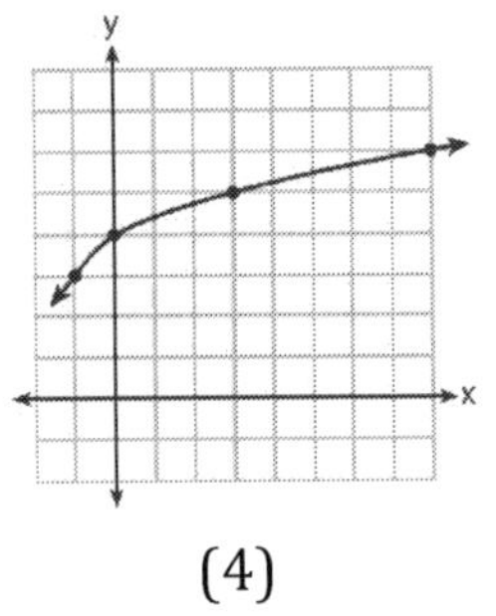

(1) (2) (3) (4)

4. The graph of $b(x)$ is shown below.

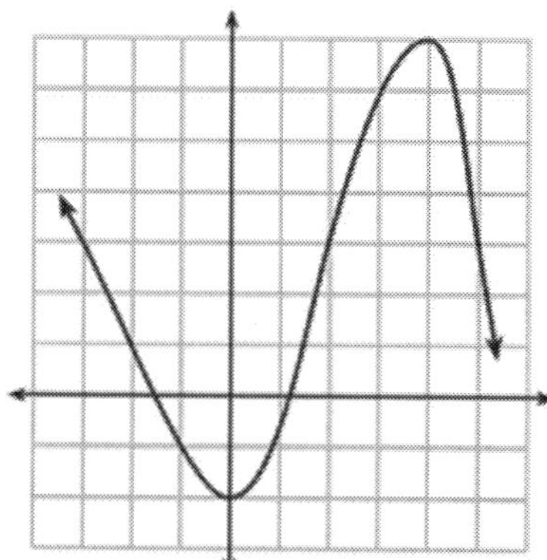

The graph of $b(x)$ is transformed using the equation $c(x) = b(x-2)+3$. Describe how the graph of $b(x)$ changed to form the graph of $c(x)$.

5. Given $f(x) = x(x+4)(x-3)$, write an equation for $g(x)$, a translation of $f(x)$ by 2 units to the right.

6. Given $m(x) = 2x^4 - 3x^3 + x - 5$, write an equation for $n(x)$, the image of $m(x)$ after the following sequence of transformations:
   1. Vertical stretch by a factor of 3
   2. Reflection over the y-axis
   3. Translation of 6 units up

7.6 Systems of Polynomial Functions

Key Terms and Concepts

A **nonlinear system** of equations is a system in which at least one of the equations is not linear. A system of polynomial functions can be solved algebraically by using the *substitution method.* We saw this in Algebra I when we solved quadratic-linear systems.

To solve a system of polynomial functions algebraically:

1. Solve each equation for y so that y is set equal to expressions in terms of x.
2. Since both expressions are equal to y, set them equal to each other, and solve the resulting equation for x.
3. Once you have the value(s) of x, find the corresponding value(s) of y by substituting into one of the original equations.

Alternatively, we can solve a system graphically by finding the points of intersection of the graphs. We can use the "intersect" function of the graphing calculator to find these points.

Calculator Tip

To solve a system graphically using the calculator:

1. Press [Y=] and enter both equations.
2. Press [GRAPH] to graph them.
3. Press [2nd] [CALC] [intersect].
4. Move the blinking cursor onto each curve and press [ENTER] for the "First curve?" and "Second curve?" prompts.
5. For the "Guess?" prompt, use the arrow keys to move the cursor near one of the points of intersection. Then press [ENTER].
6. The coordinates of the closest point of intersection will be shown.
7. If there appear to be additional points of intersection, repeat steps 2 to 5 but move the cursor closer to the next point in response to the "Guess?" prompt.

MODEL PROBLEM 1: *SOLVE A SYSTEM ALGEBRAICALLY*

Given $f(x) = x^3 - 2x + 1$ and $g(x) = x^2 + 1$. Algebraically find the value(s) of x where $f(x) = g(x)$. State the points of intersection of the two functions' graphs.

Solution:

(A) $x^3 - 2x + 1 = x^2 + 1$

(B) $x^3 - x^2 - 2x = 0$

$x(x^2 - x - 2) = 0$

$x(x+1)(x-2) = 0$

$x = \{0, -1, 2\}$

(C) For $x = 0, y = 0^2 + 1 = 1$

For $x = -1, y = (-1)^2 + 1 = 2$

For $x = 2, y = 2^2 + 1 = 5$

$(0,1)$, $(-1,2)$, and $(2,5)$

Explanation of steps:

(A) Set the functions' expressions equal to each other.

(B) Solve the resulting equation for x. *[This equation can be factored.]*

(C) To find the solutions (that is, the coordinates of the points of intersection), substitute each value of x into one of the original expressions.

PRACTICE PROBLEMS

1. Solve algebraically: $y = x^2 + 2x - 1$, $y = 3x + 5$	2. Solve algebraically: $y = x^2 + 7x + 22$, $y + 3x = 1$

3. Solve algebraically: $x^2 + 2x = y + 7$ $y = 2x + 1$	4. Solve algebraically: $y = x^3 - 6x + 1$ $y = -2x + 1$

5. Solve algebraically, in the set of complex numbers: $y = x^2 - 3x + 9$ $y + 2 = 3x$	6. Solve algebraically, in the set of complex numbers: $y = x^3 + 5x^2 - 2$ $y = 7x^2 - 5x + 2$

MODEL PROBLEM 2: *SOLVE A SYSTEM GRAPHICALLY*

Given $f(x) = x^3 - 2x + 1$ and $g(x) = x^2 + 1$. Graphically find the value(s) of x where $f(x) = g(x)$. State the solutions as a set of points.

Solution:

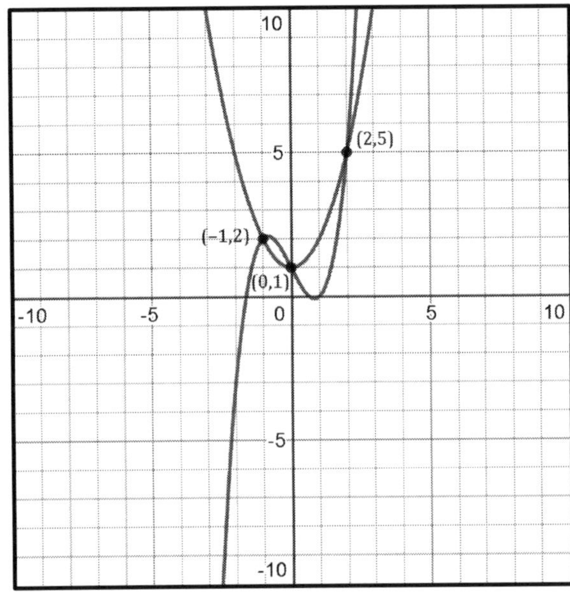

(A)

NORMAL FLOAT AUTO REAL RADIAN MP

Plot1 Plot2 Plot3

\Y1=X³-2X+1

\Y2=X²+1

\Y3=

\Y4=

\Y5=

\Y6=

\Y7=

\Y8=

(B)

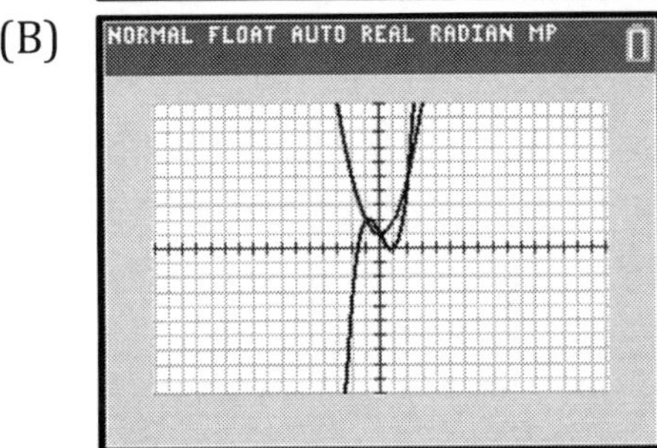

(C)

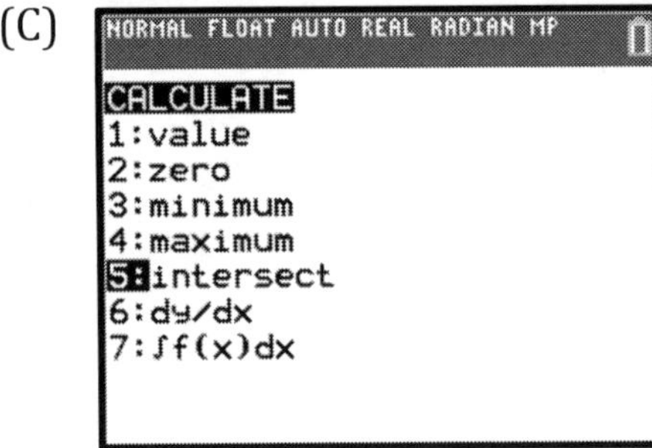

(D) Graphs intersect at $x = \{-1,0,2\}$.
Solutions are $(-1,2)$, $(0,1)$, and $(2,5)$.

Explanation of steps:

(A) On the calculator, enter the two equations.

(B) Press [GRAPH] to graph them.

(C) Press [2nd][CALC][5] for intersect. Press [ENTER] twice for the First Curve and Second Curve prompts. Then, at the "Guess?" prompt, use the arrow keys to move close to the first point of intersection and press [ENTER] again. The coordinates of this point will be shown. *[The leftmost point of intersection is* $(-1,2)$*.]*

(D) Repeat step (C) for each additional point of intersection *[*$(0,1)$ *and* $(2,5)$*]*.

Practice Problems

7. Find all real solutions graphically, and sketch the graphs on the grid below.

$y = x^2 + 2x - 7$

$y = 1$

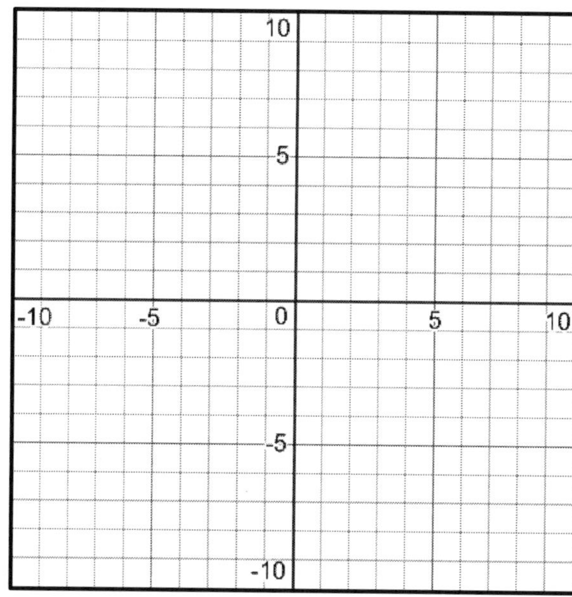

8. Find all real solutions graphically, and sketch the graphs on the grid below.

$y = x^3 - 6x + 1$

$y = -2x + 1$

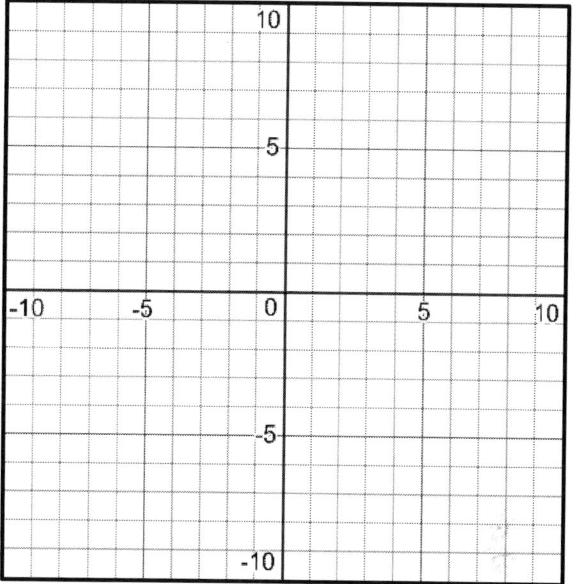

9. Find all real solutions graphically, and sketch the graphs on the grid below.

$y = -x^2 + 4x + 6$

$y = (x - 2)^3 - 2$

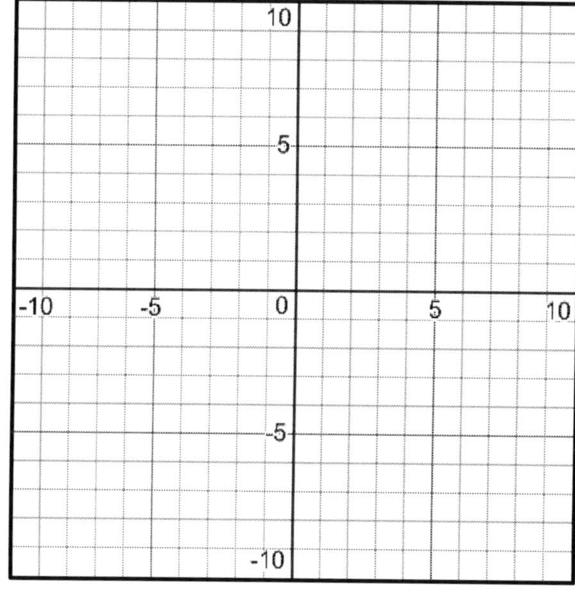

10. Find all real solutions graphically, and sketch the graphs on the grid below.

$y = x^4 + 2x^3$

$y = x^3 + 2x^2$

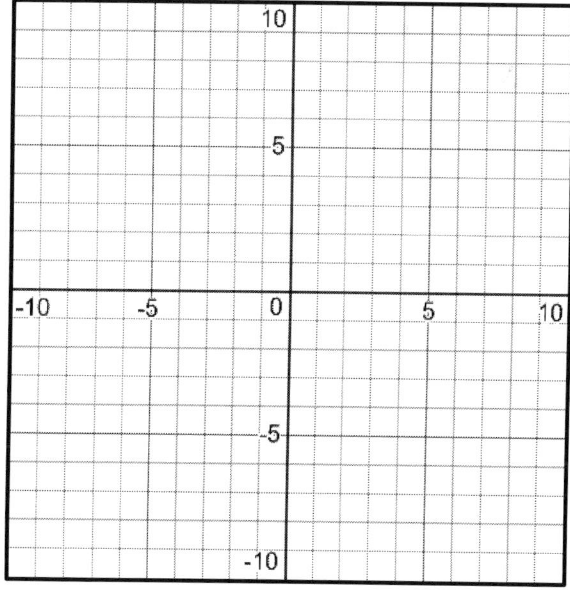

Chapter 8. Radicals and Rational Exponents

8.1 *n*th Roots

Key Terms and Concepts

So far, we've only worked with square roots, but we can also work with cube roots $\left(\sqrt[3]{}\right)$, fourth roots $\left(\sqrt[4]{}\right)$, or any other *n*th roots in algebraic expressions or equations.

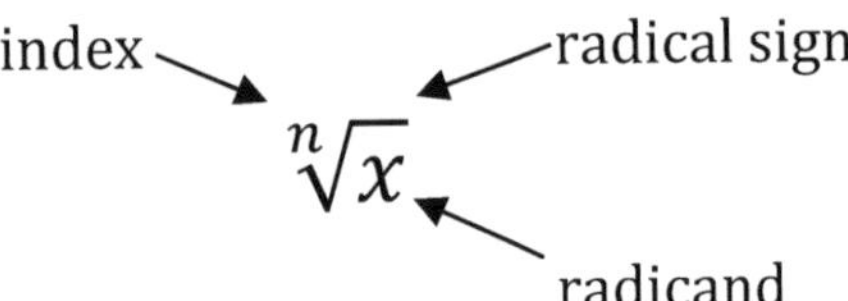

Whenever the index is not written, it is assumed to be 2, meaning the square root.

The **cube root** of a number is a factor whose cube equals that number.

Example: $\sqrt[3]{8} = 2$ because $2^3 = 8$.

In general, the ***n*th root** of a number is a value which, when raised to the power *n*, is equal to the number.

Example: $\sqrt[5]{243} = 3$ because $3^5 = 243$.

Calculator Tip

To calculate a cube root on the calculator, use the MATH function $\sqrt[3]{}$, which we get by pressing [MATH] [4].

Example: To calculate $\sqrt[3]{125}$, enter [MATH][4] [1][2][5]. The answer is 5.

You can calculate the *n*th root for any other index *n* on the calculator by typing the index followed by the MATH function $\sqrt[x]{}$, which we get by pressing [MATH] [5].

[On the TI-84 models, you can also access the $\sqrt[x]{}$ function by pressing [ALPHA][F2][6].]

Example: To calculate $\sqrt[4]{625}$, enter [4][MATH][5] [6][2][5]. The answer is 5.

We can **solve equations involving x^n** by taking the nth root of both sides. It is important to remember, however, that when n is an even number, there are two roots: one positive and one negative. Therefore, write $\pm$.

Example: If $x^4 = 81$, then we can take the fourth root of both sides, $x = \pm\sqrt[4]{81}$, to get $x = \pm 3$. Both $3 \cdot 3 \cdot 3 \cdot 3 = 81$ and $(-3) \cdot (-3) \cdot (-3) \cdot (-3) = 81$.

For an even number n, there are no real nth roots of negative numbers. However, for an odd number n, the nth root of a negative number is real (and negative).

Example: $\sqrt[4]{-81}$ is not a real number. However, $\sqrt[3]{-27} = -3$.

[Note: For this course, you will not be asked to find the complex nth roots of negative numbers where n is an even integer greater than 2.]

It may not be easy for larger numbers, but we can **simplify nth roots** by hand using the same method as we did for simplifying square roots, except that instead of placing pairs of factors in boxes, we *place in boxes as many repeated factors as the index tells us.*

To simplify a radical into simplest radical form:

1. Write the prime factorization of the radicand.
2. Group repeated factors into boxes, using the index to determine how many factors to place in each box.
3. Remove one factor from the radicand for each box of factors inside the radicand.
4. Multiply all factors outside the radicand, and all factors remaining inside the radicand.

Example: To simplify $\sqrt[3]{250}$, the prime factorization of 250 is $2 \cdot 5 \cdot 5 \cdot 5$, so $\sqrt[3]{2 \cdot \boxed{5 \cdot 5 \cdot 5}} = 5\sqrt[3]{2}$. The index 3 tells us to box repeated factors in groups of 3. We still "take out" one factor for every box (a box of three 5s inside the radicand becomes a single 5 outside the radicand).

To simplify nth roots with *variables in the radicand*, you can also box the variables as you do with prime factors:

Example: $\sqrt[3]{y^5} = \sqrt[3]{\boxed{y \cdot y \cdot y} \cdot y \cdot y} = y \cdot \sqrt[3]{y^2}$

One exception needs to be recognized here, due to the fact that the principal root $\sqrt[n]{}$ is defined as always positive:

$$\sqrt[n]{x^n} = \begin{cases} x, & \text{if } n \text{ is odd} \\ |x|, & \text{if } n \text{ is even} \end{cases}$$

Therefore, unless the problem restricts the domains of the variables to positive values, you will need to use an *absolute value symbol around any variable written outside the radical* whenever (a) the radical has an even index and (b) the variable is raised to an odd power.

Example: $\sqrt[4]{x^8y^{12}} = x^2|y^3|$. Since 4 is an even index, we need the absolute value of y^3 because the principle root is positive but y^3 may be negative. We don't need the absolute value symbol for x^2 because it is always nonnegative.

A shortcut method for simplifying nth roots with variables is to divide each variable's exponent by the index. The quotient tells you the exponent to use for the variable outside the radicand and the remainder tells you the exponent to use for the variable inside the radicand.

Example: For $\sqrt[3]{y^{14}}$, dividing 14 by 3 gives us 4 with a remainder of 2, so y^4 is taken out, and y^2 remains inside: $y^4\sqrt[3]{y^2}$.

MODEL PROBLEM

Simplify $\sqrt[4]{64a^5b^4c^3d^{13}}$.

Solution:

$2|abd^3|\sqrt[4]{4ac^3d}$

Explanation of steps:

(A) For the numerical part, use prime factorization to take out factors in multiples specified by the index. *[$64 = 2^6$, so take out a 2, and $2^2 = 4$ remains inside.]*

(B) For each variable, divide the exponent by the index; the quotient tells you the exponent to use outside and the remainder tells you the exponent to use inside. *[For a^5, divide $5 \div 4 = 1R1$, so a^1 comes out and a^1 remains inside. For b^4, divide $4 \div 4 = 1R0$, so b^1 comes out and nothing remains. For c^3, divide $3 \div 4 = 0R3$, so nothing comes out and c^3 remains inside. For d^{13}, divide $13 \div 4 = 3R1$, so d^3 comes out and d^1 remains.]*

(C) For radicals with an even index, place an absolute value symbol around any variables that are taken out with an odd exponent. *[Since the index is 4, we need an absolute value symbol around abd^3 because the exponents of all three of these variables are odd.]*

PRACTICE PROBLEMS

1. Use the calculator to find $\sqrt[3]{1331}$.	2. Use the calculator to find $\sqrt[5]{7776}$.
3. Find $\sqrt[4]{72}$ to the *nearest hundredth.*	4. Simplify $\sqrt[4]{162}$
5. Solve for *x*: $x^2 - 6 = 19$	6. Solve for *x*: $x^5 = 243$
7. Solve for x: $x^6 = 46{,}656$	8. $x^3 = 515$. Find the value of *x* to *the nearest hundredth.*

9. Simplify $\sqrt[3]{64y^6}$	10. Simplify $\sqrt[3]{x^{12}y^{16}}$
11. Simplify $\sqrt[5]{4a^8b^{14}c^5}$	12. Simplify $\sqrt[4]{x^{12}y^{16}}$

8.2 Operations with Radicals

Key Terms and Concepts

To add or subtract *n*th roots, the radicals must have the same index and same radicand. Otherwise, they are *not* like radicals.

Examples: $2\sqrt[3]{6}$ and $-5\sqrt[3]{6}$ are like radicals. Their sum is $-3\sqrt[3]{6}$.

$2\sqrt{5}$ and $2\sqrt[3]{5}$ are *not* like radicals (different indices).

$2\sqrt[3]{2}$ and $2\sqrt[3]{5}$ are *not* like radicals (different radicands).

To combine radicals by multiplication or division, the radicals must have the *same index* but may have different radicands.

To multiply *n*th roots: $\left(\sqrt[n]{a}\right)\left(\sqrt[n]{b}\right) = \sqrt[n]{ab}$

Examples:

(1) $\sqrt[5]{15} \cdot \sqrt[5]{2} = \sqrt[5]{30}$

(2) $\sqrt[3]{4} \cdot \sqrt[3]{6} = \sqrt[3]{24} = \sqrt[3]{8} \cdot \sqrt[3]{3} = 2\sqrt[3]{3}$

(3) $\sqrt[3]{x^2} \cdot \sqrt[3]{x^7} = \sqrt[3]{x^2x^7} = \sqrt[3]{x^9} = x^3$

To divide *n*th roots: $\dfrac{\sqrt[n]{a}}{\sqrt[n]{b}} = \sqrt[n]{\dfrac{a}{b}}$

Examples:

(1) $\dfrac{\sqrt[4]{162}}{\sqrt[4]{2}} = \sqrt[4]{\dfrac{162}{2}} = \sqrt[4]{81} = 3$

(2) $\dfrac{\sqrt{50x^5}}{\sqrt{2x^2}} = \sqrt{\dfrac{50x^5}{2x^2}} = \sqrt{25x^3} = 5|x|\sqrt{x}$

We can **rationalize denominators containing *n*th roots** by multiplying by the appropriate unit fraction. To find the appropriate unit fraction, use this rule:
If the denominator is $\sqrt[n]{b^p}$, then multiply the numerator and denominator by $\sqrt[n]{b^{n-p}}$.

Examples: (a) To simplify $\frac{1}{\sqrt[4]{3}}$, use $\sqrt[4]{3^{4-1}} = \sqrt[4]{3^3}$: $\frac{1}{\sqrt[4]{3}} \cdot \left(\frac{\sqrt[4]{3^3}}{\sqrt[4]{3^3}}\right) = \frac{\sqrt[4]{27}}{3}$

(b) To simplify $\frac{3}{\sqrt[5]{x^2}}$, use $\sqrt[5]{x^{5-2}} = \sqrt[5]{x^3}$: $\frac{3}{\sqrt[5]{x^2}} \cdot \left(\frac{\sqrt[5]{x^3}}{\sqrt[5]{x^3}}\right) = \frac{3\sqrt[5]{x^3}}{x}$

(c) To simplify $\frac{10}{\sqrt[3]{2x^2y}}$, use $\sqrt[3]{2^{3-1}x^{3-2}y^{3-1}} = \sqrt[3]{2^2x^1y^2}$:

$$\frac{10}{\sqrt[3]{2x^2y}} \cdot \left(\frac{\sqrt[3]{4xy^2}}{\sqrt[3]{4xy^2}}\right) = \frac{10\sqrt[3]{4xy^2}}{2xy} = \frac{5\sqrt[3]{4xy^2}}{xy}$$

To simplify a radical with a **fractional radicand**, we can split the numerator and denominator, then rationalize the denominator.

Example: $\sqrt[4]{\frac{2}{3}} = \frac{\sqrt[4]{2}}{\sqrt[4]{3}}$, giving us $\frac{\sqrt[4]{2}}{\sqrt[4]{3}}\left(\frac{\sqrt[4]{3^3}}{\sqrt[4]{3^3}}\right) = \frac{\sqrt[4]{2}\cdot\sqrt[4]{27}}{3} = \frac{\sqrt[4]{54}}{3}$

An expression involving a radical is in **simplest radical form** when

(a) the radicands have no perfect *n*th powers greater than 1 as factors (for index *n*),
(b) there are no radicals in the denominator of a fraction, and
(c) there are no radicands containing fractions.

Model Problem

Write the product $(3\sqrt[3]{4})(2\sqrt[3]{12})$ in simplest form.

Solution:

$6\sqrt[3]{48} = 12\sqrt[3]{6}$

Explanation of steps:

To multiply radicals with the same index, multiply the coefficients *[*$3 \times 2 = 6$*]* and multiply the radicands *[*$4 \times 12 = 48$*]*. Then simplify, if possible *[*$\sqrt[3]{48} = \sqrt[3]{\boxed{2\cdot 2\cdot 2}\cdot 2\cdot 3} = 2\sqrt[3]{6}$*]*.

Practice Problems

1. Write the sum of $5\sqrt[4]{2}$ and $4\sqrt[4]{2}$ in simplest form.	2. Write the product of $5\sqrt[3]{2}$ and $2\sqrt[3]{5}$ in simplest form.
3. Write $\sqrt[3]{128}$ divided by $\sqrt[3]{2}$ in simplest form.	4. Express $\left(\sqrt[4]{8}\right)\left(3\sqrt[4]{6}\right) - \left(2\sqrt[4]{3}\right)$ in simplest form.
5. Rationalize the denominator of $\frac{3}{\sqrt[3]{9}}$.	6. Rationalize the denominator of $\frac{6\sqrt[4]{8}}{\sqrt[4]{16}}$.

8.3 Solve Equations with Radicals

KEY TERMS AND CONCEPTS

We can **solve equations involving $\sqrt[n]{x}$** by isolating the radical and then raising both sides to the nth power.

Example: If $\sqrt[4]{x}+7=10$, first isolate the radical by subtracting 7 from both sides, giving us $\sqrt[4]{x}=3$. Then raise both sides to the fourth power, $\left(\sqrt[4]{x}\right)^4=3^4$, giving us $x=81$.

To solve an equation involving the variable under a radical:

1. Isolate the term with the radical.
2. Raise both sides to the nth power to eliminate the radical.
3. Solve.

There may be any algebraic expression inside the radical sign.

Example:

$\sqrt{2x-1}+5=8$	
$\sqrt{2x-1}=3$	[subtract 5 from both sides]
$\left(\sqrt{2x-1}\right)^2=3^2$	[square both sides]
$2x-1=9$	
$2x=10$	[solve]
$x=5$	

If we raise both sides an equation to an even power, it's possible that one or more of the roots of the resulting equation may be *extraneous*. This is because raising both sides of an equation to an even power is not a reversible operation.

Example: If $x=2$, then $x^2=2^2$ is true. But if $x^2=2^2$, then it doesn't necessarily mean that $x=2$, since x could equal –2.

Therefore, it is important to test each root in the original equation to determine whether it leads to a false statement, meaning that the root is extraneous.

Example: Solve $\sqrt{2x+15}=x$.

$\left(\sqrt{2x+15}\right)^2=x^2$

$2x+15=x^2$

$x^2-2x-15=0$

$(x+3)(x-5)=0$

$\{-3, 5\}$ (with -3 crossed out)

-3 is an extraneous root because $\sqrt{2(-3)+15}\neq -3$

If there are *more than one radical term* in an equation, then unless we can combine like radicals, we'll need to isolate and eliminate each radical one at a time.

Example: Solve $\sqrt{x-5}+\sqrt{x}=5$.

$\sqrt{x-5}=5-\sqrt{x}$ *[Isolate $\sqrt{x-5}$.]*

$\left(\sqrt{x-5}\right)^2=\left(5-\sqrt{x}\right)^2$ *[Square both sides.]*

$x-5=25-10\sqrt{x}+x$

$10\sqrt{x}=30$ *[Isolate $10\sqrt{x}$.]*

$\sqrt{x}=3$ *[Divide both sides by 10.]*

$\left(\sqrt{x}\right)^2=3^2$ *[Square both sides.]*

$x=9$

MODEL PROBLEM 1: *SIMPLE RADICANDS*

Solve for x: $2\sqrt[3]{x}+5=21$

Solution:

(A) $2\sqrt[3]{x}=16$

$\sqrt[3]{x}=8$

(B) $\left(\sqrt[3]{x}\right)^3=8^3$

(C) $x=512$

Explanation of steps:

(A) Isolate the radical. *[Subtract 5 from both sides, then divide both sides by 2.]*

(B) To find x, raise both sides to the power indicated by the index *[3]*.

(C) Simplify.

PRACTICE PROBLEMS

1. Solve for x: $\sqrt[3]{x} = 7$	2. Solve for x: $3\sqrt[4]{x} - 5 = 13$

MODEL PROBLEM 2: *LEADING TO A LINEAR EQUATION*

Solve for x: $\sqrt{2x-1} + 2 = 5$

Solution:

$\sqrt{2x-1} + 2 = 5$

(A) $\sqrt{2x-1} = 3$

(B) $\left(\sqrt{2x-1}\right)^2 = 3^2$

$2x - 1 = 9$

(C) $2x = 10$

$x = 5$

Explanation of steps:

(A) Isolate the term with the radical *[subtract 2 from both sides]*.

(B) Square both sides.

(C) Solve.

PRACTICE PROBLEMS

3. Solve: $\sqrt{x-4} = 7$	4. Solve: $\sqrt{2x-1} + 2 = 5$

5. Solve: $\sqrt[3]{2x+3}-5=-2$	6. Given $\sqrt{x-a}=b$ and $x>a$, solve for x in terms of a and b.

MODEL PROBLEM 3: *LEADING TO A QUADRATIC EQUATION*

Solve for x: $\sqrt{3x+16}=x+2$

Solution:

(A) $\sqrt{3x+16}=x+2$

(B) $\left(\sqrt{3x+16}\right)^2=(x+2)^2$

$3x+16=x^2+4x+4$

(C) $x^2+x-12=0$

$(x+4)(x-3)=0$

(D) $\{\cancel{-4},3\}$

Explanation of steps:

(A) Isolate the term with the radical *[the radical is already isolated here].*

(B) Square both sides.

(C) Solve.

(D) Test each root in the original equation. *[$\sqrt{3(-4)+16}=(-4)+2$ is untrue because $\sqrt{4}\neq-2$, so the root -4 is extraneous. The only solution is $x=3$.]*

PRACTICE PROBLEMS

7. Solve: $\sqrt{x^2-3x+3}=1$	8. Solve: $\sqrt{x+3}-3=x$

9. Solve: $\sqrt{2x-4} = x-2$	10. Solve: $x = 2\sqrt{2x-3}$
11. Solve: $\sqrt{9x+10} = x$	12. Solve: $\sqrt{5x+29} = x+3$
13. Solve: $\sqrt[3]{x^3-2x^2-5} = x-1$	14. Solve: $\sqrt[4]{15x^4+81} = 2x$

MODEL PROBLEM 4: *MORE THAN ONE RADICAL TERM*

Solve for x: $\sqrt{x+2} + \sqrt{x-3} = 5$

Solution:

(A) $\sqrt{x+2} = 5 - \sqrt{x-3}$

(B) $\left(\sqrt{x+2}\right)^2 = \left(5 - \sqrt{x-3}\right)^2$

$x + 2 = 25 - 10\sqrt{x-3} +$

$x - 3$

(C) $10\sqrt{x-3} = 20$

$\sqrt{x-3} = 2$

(D) $\left(\sqrt{x-3}\right)^2 = 2^2$

(E) $x - 3 = 4$

$x = 7$

Explanation of steps:

(A) We cannot combine like radicals, so isolate one of the radical terms. *[Isolate $\sqrt{x+2}$.]*

(B) Square both sides.

(C) Isolate the other radical term. *[Isolate $10\sqrt{x-3}$, then divide both sides by 10 to simplify the equation.]*

(D) Square both sides.

(E) Solve the linear equation.

PRACTICE PROBLEMS

15. Solve: $\sqrt{x+4} - \sqrt{x-3} = 1$	16. Solve: $\sqrt{x+4} = \sqrt{5x} - 2$

17. Solve: $\sqrt{x+3}+1=\sqrt{-2x}$	18. Solve: $\sqrt{x+1}-\sqrt{x-4}=\sqrt{x-7}$

8.4 Graphs of Radical Functions

KEY TERMS AND CONCEPTS

A **square root function** is a function that has the independent variable, x, in the radicand of a square root symbol.

Examples: $y = \sqrt{x} + 2$ or $y = \sqrt{x - 3}$

The simplest square root function (the *parent function*), $y = \sqrt{x}$, can be graphed as follows:

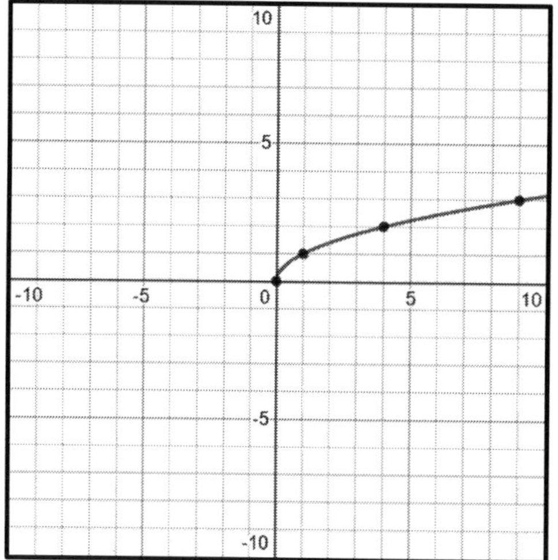

x	y
0	0
1	1
4	2
9	3

Note that the radicand cannot be negative. This is because the square root of a negative number is imaginary and the coordinate plane is limited to real numbers only.

CALCULATOR TIP

To enter a square root on the calculator, press [2nd] [√‾], followed by the radicand.

Be sure to end the radicand by pressing [ )] on the TI-83 or by pressing [▸] on the TI-84.

A **cube root function** is similar to a square root function except that a cube root symbol is used.

Examples: $y = \sqrt[3]{x} + 2$ or $y = \sqrt[3]{x - 3}$

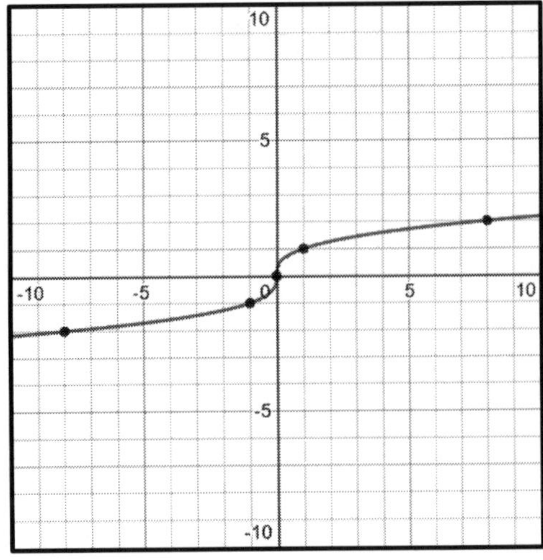

The simplest cube root function (the *parent function*), $y = \sqrt[3]{x}$, can be graphed as follows:

x	y
-8	−2
−1	−1
0	0
1	1
8	2

Not that the domain of a cube root function is *not* restricted to non-negative values of x. The cube root of a negative number is a real, negative number.

Example: $\sqrt[3]{-8} = -2$ because $(-2)(-2)(-2) = -8$.

CALCULATOR TIP

To enter a cube root on the calculator, press [MATH][4], followed by the radicand.

Be sure to end the radicand by pressing [)] on the TI-83 or by pressing [▸] on the TI-84.

Below are some additional graphs of nth root parent functions. As you can see, the even root function graphs look very much alike, as do the odd root function graphs.

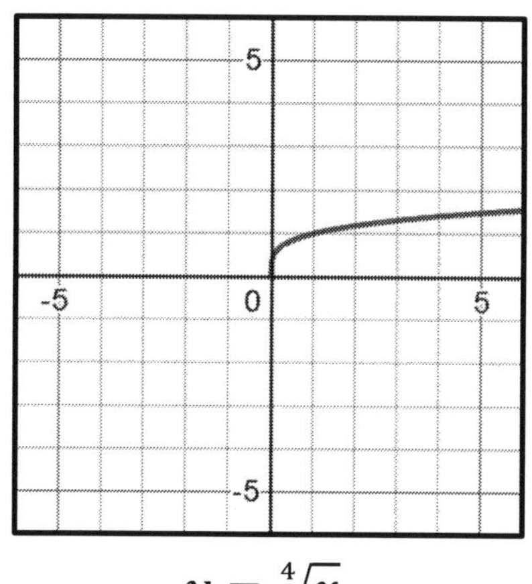

$y = \sqrt[4]{x}$

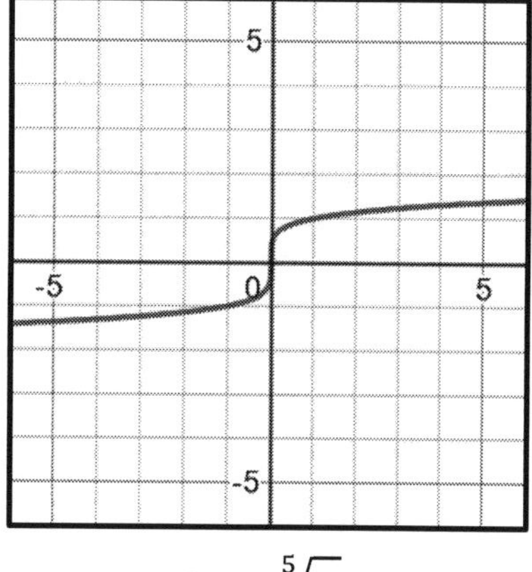

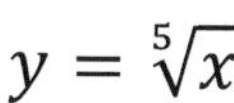
$y = \sqrt[5]{x}$

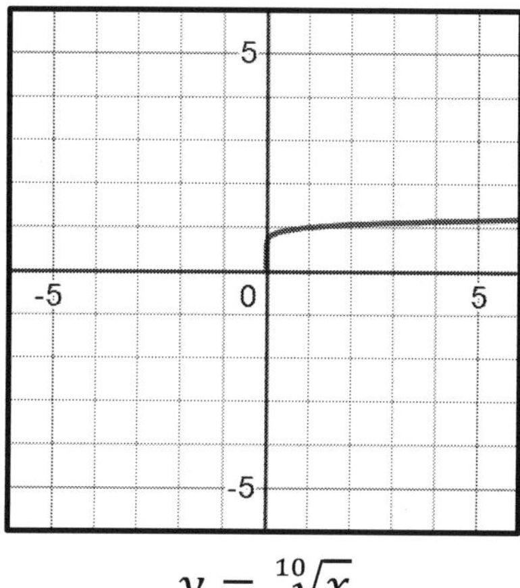

$y = \sqrt[10]{x}$

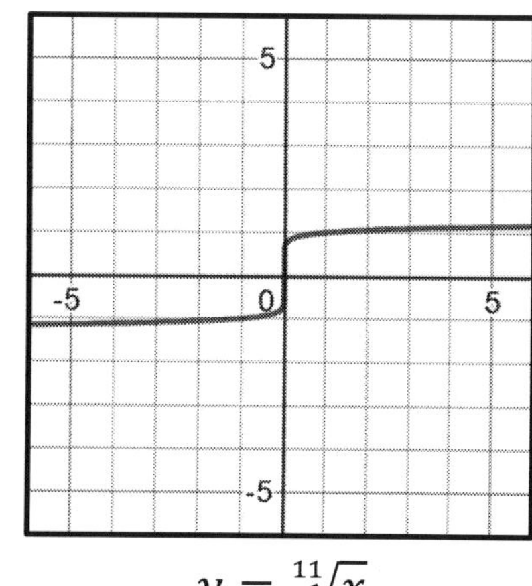

$y = \sqrt[11]{x}$

A radical function $f(x)$ may be transformed in any of the following ways:

$f(x)+k$	$(x,y) \to (x, y+k)$	vertically shifts the graph up $(k>0)$ or down $(k<0)$
$f(x+k)$	$(x,y) \to (x-k, y)$	horizontally shifts the graph left $(k>0)$ or right $(k<0)$
$-f(x)$	$(x,y) \to (x,-y)$	reflects the graph over the x-axis
$f(-x)$	$(x,y) \to (-x,y)$	reflects the graph over the y-axis
$k \cdot f(x)$	$(x,y) \to (x,ky)$	vertically stretches by a factor of k
$f(kx)$	$(x,y) \to (kx,y)$	horizontally stretches by a factor of $\frac{1}{k}$

For example, here are two **translations** of $f(x) = \sqrt{x}$.

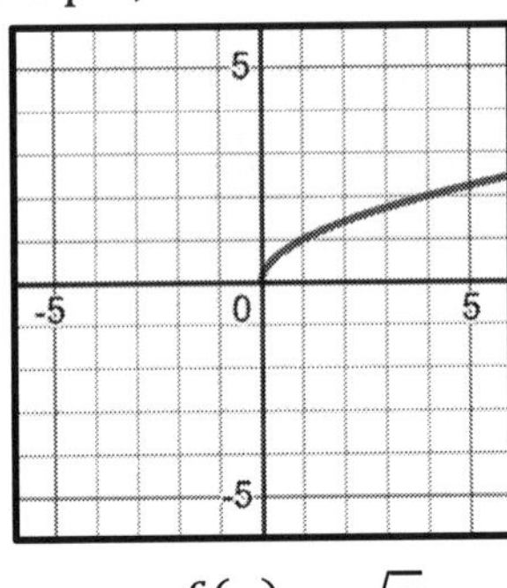

$f(x) = \sqrt{x}$

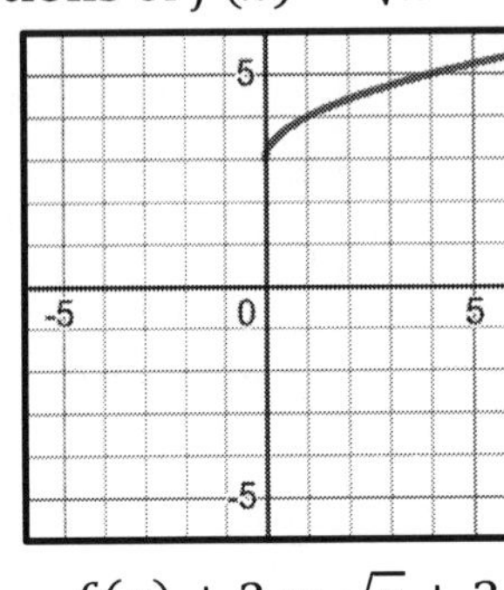

$f(x) + 3 = \sqrt{x} + 3$

shifted up 3

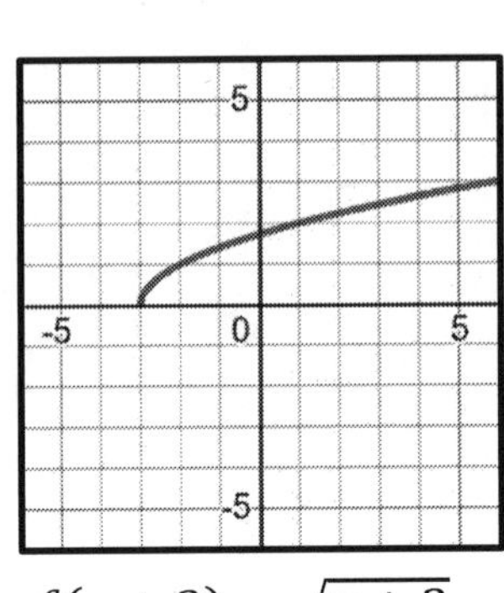

$f(x+3) = \sqrt{x+3}$

shifted left 3

Here are two **reflections** of $f(x) = \sqrt{x}$.

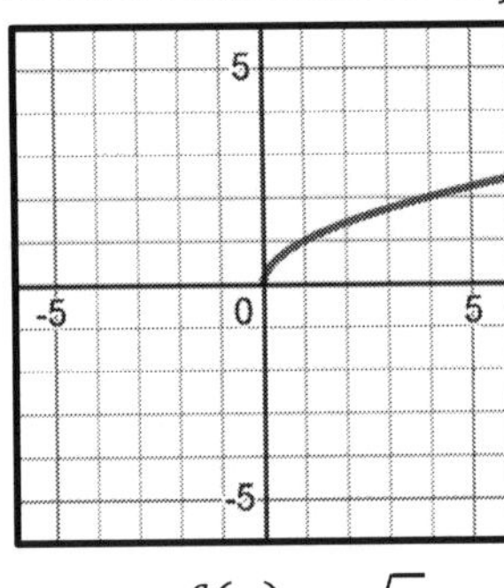

$f(x) = \sqrt{x}$

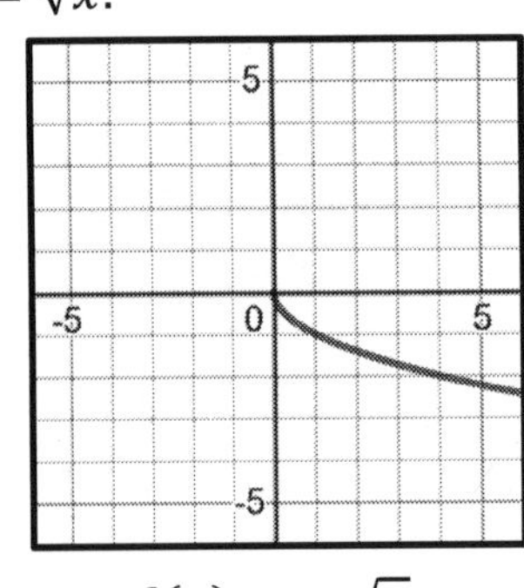

$-f(x) = -\sqrt{x}$

reflected over x-axis

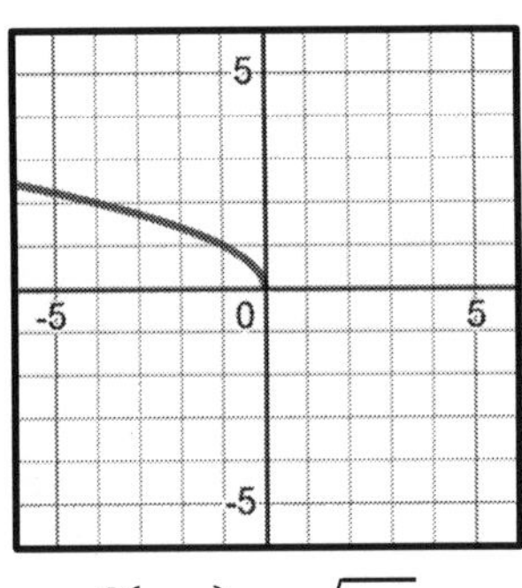

$f(-x) = \sqrt{-x}$

reflected over y-axis

Here are two **stretches** of $f(x) = \sqrt{x}$.

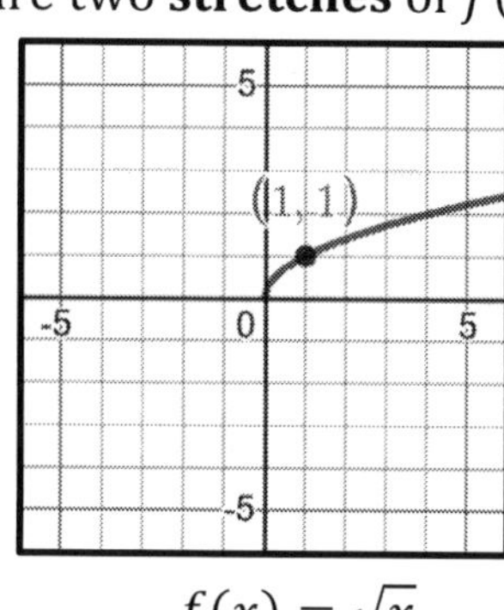

$f(x) = \sqrt{x}$

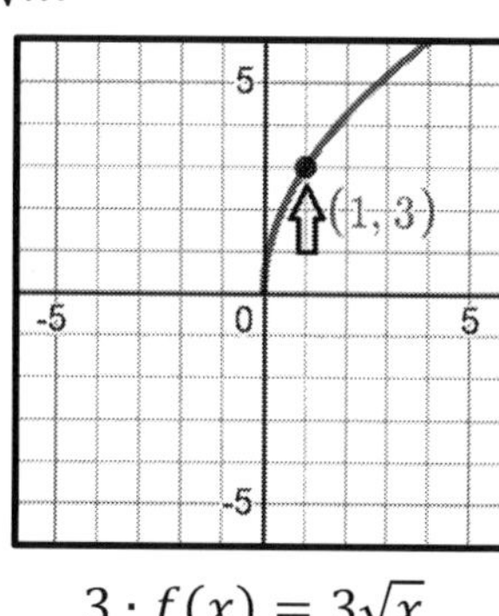

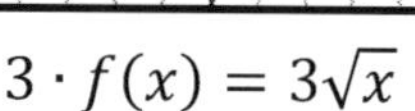

$3 \cdot f(x) = 3\sqrt{x}$

vertically stretched by 3

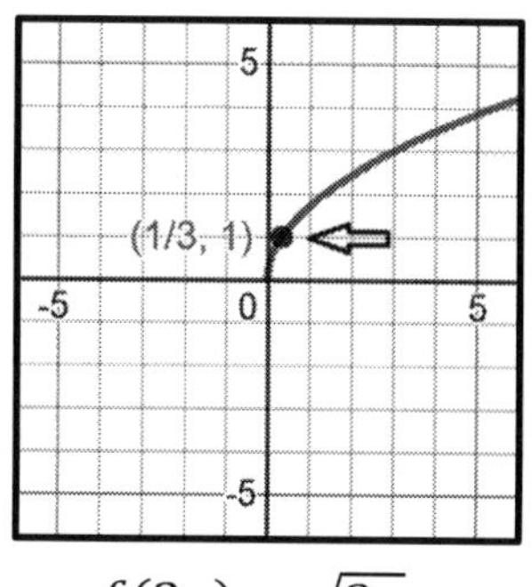

$f(3x) = \sqrt{3x}$

horizontally stretched by $\frac{1}{3}$

Model Problem

What are the real solutions to the system of equations $y = \sqrt{x}$ and $y = \sqrt[3]{x}$?

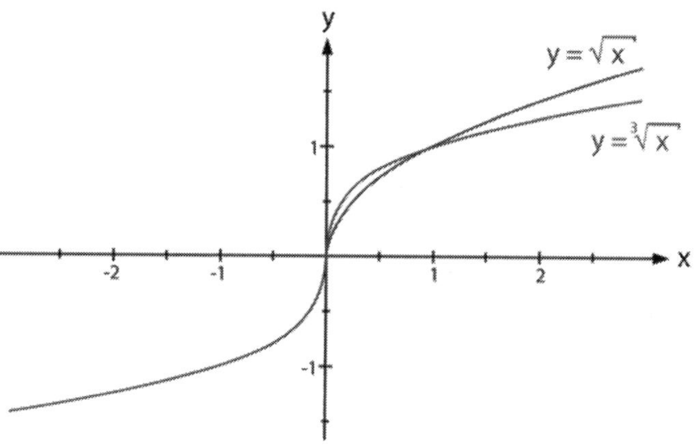

Solution: (0,0) and (1,1)

Explanation:

Graph both functions on the same plane and look for intersections. *[These functions intersect at only two points,* (0,0) *and* (1,1)*. These points of intersection show that* $\sqrt{0} = \sqrt[3]{0} = 0$ *and* $\sqrt{1} = \sqrt[3]{1} = 1$*.]*

PRACTICE PROBLEMS

1. Which of the following graphs represent the function $y = \sqrt{x + 1}$?

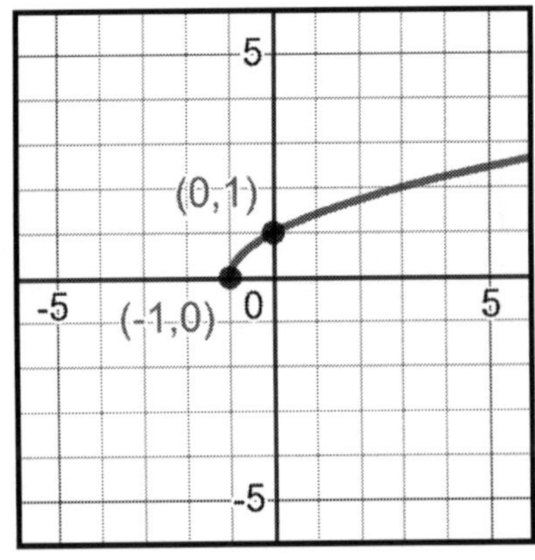

(1)

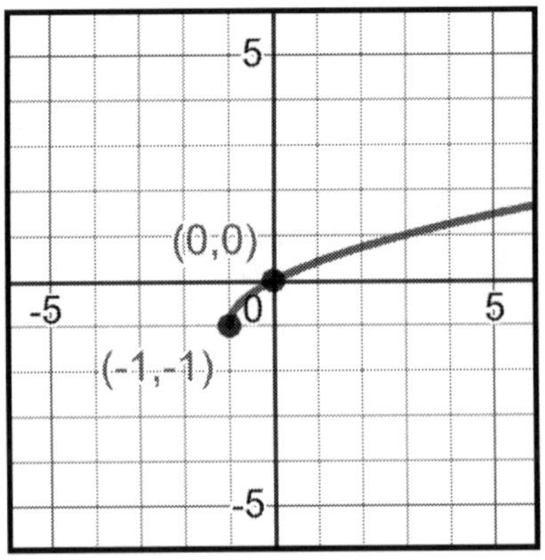

(3)

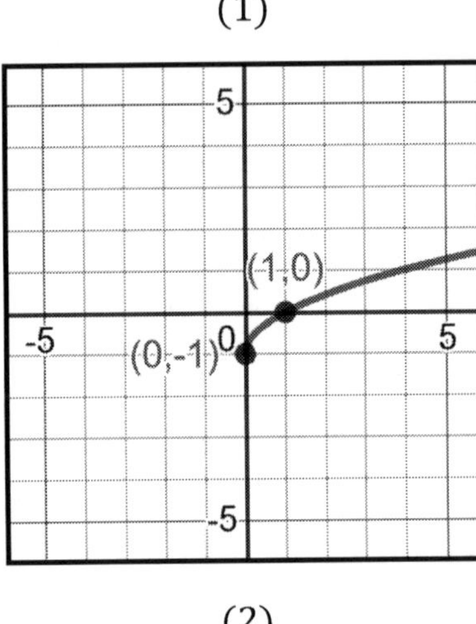

(2)

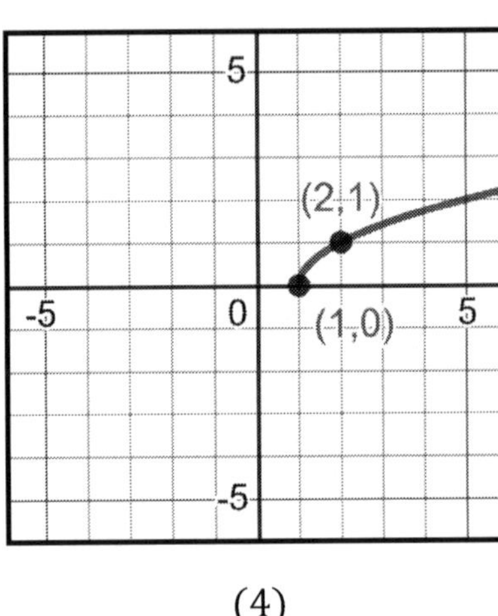

(4)

2. Which of the following graphs represent the function $y = \sqrt[3]{x} + 3$?

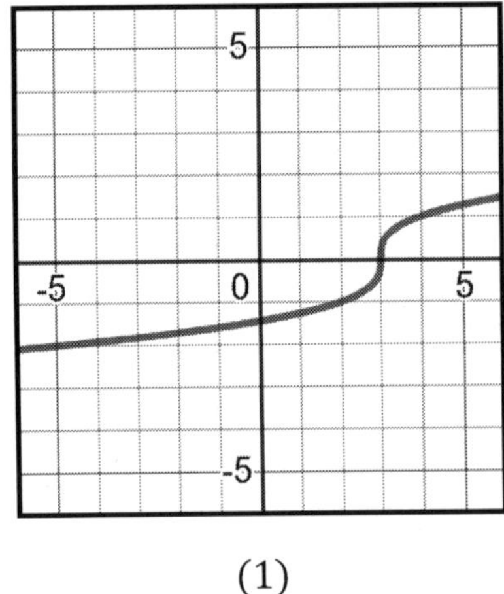

(1)

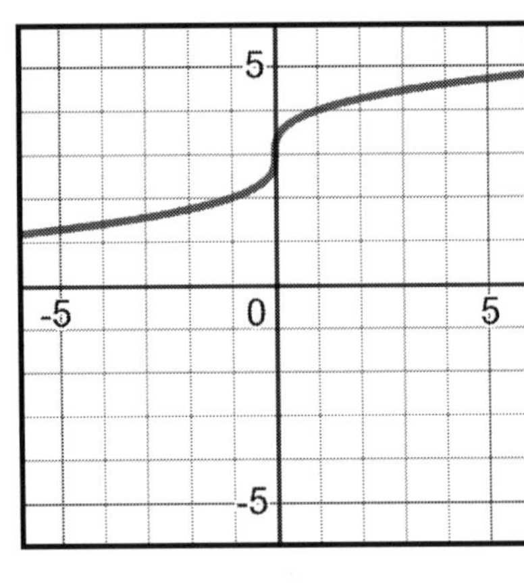

(2)

3. Two functions are graphed below: $f(x) = \sqrt{x}$ is graphed as a dashed line and $g(x)$, the image of $f(x)$ after a sequence of transformations, is graphed as a solid line.

   a) What sequence of transformations maps $f(x)$ to $g(x)$?

   b) Write an equation for $g(x)$.

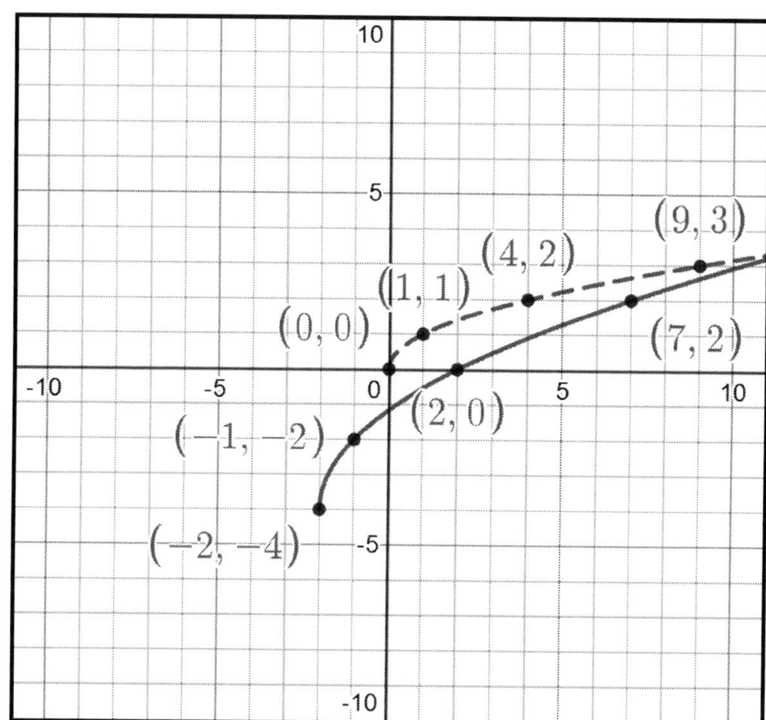

4. Williams High School relocated into a larger building and immediately began recruiting students to increase enrollment. The number of students enrolled, y, is modeled by the function $y = 90\sqrt{3x} + 400$, where x is the number of months the new school building has been open.

   a) Construct a table of values.
   b) Sketch the function on the grid.
   c) Find the number of students enrolled exactly 3 months after the building opened.

   d) After how many months will 940 students be enrolled?

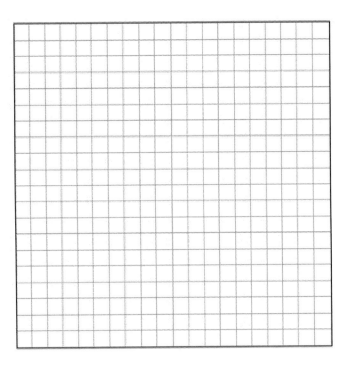

8.5 Negative Exponents

Key Terms and Concepts

We know that when we divide expressions with the same bases, we can subtract the exponents.

Example: $\frac{x^5}{x^3} = x^{5-3} = x^2$, which we can confirm by $\frac{\cancel{x} \cdot \cancel{x} \cdot \cancel{x} \cdot x \cdot x}{\cancel{x} \cdot \cancel{x} \cdot \cancel{x}} = x^2$

This rule still holds when the denominator has a greater exponent.

Example: $\frac{x^3}{x^5} = x^{3-5} = x^{-2}$. Since $\frac{\cancel{x} \cdot \cancel{x} \cdot \cancel{x}}{\cancel{x} \cdot \cancel{x} \cdot \cancel{x} \cdot x \cdot x} = \frac{1}{x^2}$, we get $x^{-2} = \frac{1}{x^2}$.

Also, since $\frac{1}{x^{-2}} = 1 \div x^{-2} = 1 \div \frac{1}{x^2} = 1 \times \frac{x^2}{1} = x^2$, we see that $\frac{1}{x^{-2}} = x^2$.

This leads to two general rules regarding **negative exponents**:

$$x^{-y} = \frac{1}{x^y} \quad \text{and} \quad \frac{1}{x^{-y}} = x^y$$

In other words, when we have a base raised to a negative exponent in the numerator, we can change it to a positive exponent by moving it to the denominator. Likewise, when we have a base raised to a negative exponent in the denominator, we can change it to a positive exponent by moving it to the numerator.

Examples: $5x^{-4} = \frac{5}{x^4}$ $\quad \frac{1}{a^2b^{-3}} = \frac{b^3}{a^c}$ $\quad \frac{r^6m^{-3}}{p^2n^{-6}} = \frac{r^6n^6}{p^2m^3}$

Also recall that when any value is raised to a **zero exponent**, the result is 1, and any value raised to the **first power** is the value itself.

Examples: $12x^0 = 12(1) = 12$ $\quad -2y^1 = -2y$

Whenever we **simplify expressions**, any negative exponents, as well as exponents of zero or one, should be eliminated.

The following **properties of exponents** may also be applied when simplifying expressions:

$a^m a^n = a^{m+n}$ $\frac{a^m}{a^n} = a^{m-n}$ $(a^m)^n = a^{mn}$

$(ab)^n = a^n b^n$ $\left(\frac{a}{b}\right)^n = \frac{a^n}{b^n}$

MODEL PROBLEM

Rewrite $3x^{-1}$ without using a negative exponent.

Solution:

(A) (B)

$$3x^{-1} = \frac{3}{x^1} = \frac{3}{x}$$

Explanation of steps:

(A) Whenever we have a base to a negative exponent in the numerator, we can change it to a positive exponent by moving it to the denominator.
$\left[x^{-1} \rightarrow \frac{1}{x^1}\right]$

(B) Anything to the first power is itself $[x^1 = x]$.

PRACTICE PROBLEMS

1. Rewrite p^{-7} using only positive exponents.	2. Rewrite $(5x)^{-2}$ using only positive exponents.
3. Simplify the expression $3^0 + 3^{-2}$.	4. If $f(x) = x^{-x} - x^0 + 2^x$, find $f(2)$.

5. Simplify: $-2m^3n^{-5}$	6. Simplify: $\frac{5}{x^{-3}}$
7. Simplify: $-\frac{3b}{5c^{-1}}$	8. Simplify: $\left(\frac{2x}{5y}\right)^{-3}$
9. Simplify: $\left(\frac{3}{4}\right)^2 \cdot \left(\frac{1}{4}\right)^{-2}$	10. Simplify: $\frac{a^2b^{-3}}{a^{-4}b^2}$
11. If $a = 3$ and $b = -2$, what is the value of the expression $\frac{a^{-2}}{b^{-3}}$?	12. Simplify: $(5^{-2}a^3b^{-4})^{-1}$

13. Simplify: $\frac{x^{-1}y^2}{x^2y^{-4}}$	14. Simplify: $\frac{x^{-1}y^4}{3x^{-5}y^{-1}}$
15. Simplify: $\frac{12x^{-5}y^5}{24x^{-3}y^{-2}}$	16. Simplify: $\frac{3x^{-4}y^5}{(2x^3y^{-7})^{-2}}$

8.6 Rational Exponents

Key Terms and Concepts

To raise a value to the one-half power is equivalent to taking the square root of the value. That is, $x^{\frac{1}{2}} = \sqrt{x}$. Why is this true? Consider that $\left(x^{\frac{1}{2}}\right)^2 = x^{\frac{1}{2}} \cdot x^{\frac{1}{2}} = x^{\left(\frac{1}{2}+\frac{1}{2}\right)} = x^1 = x$. So, if x is the square of $x^{\frac{1}{2}}$, then $x^{\frac{1}{2}}$ is the square root of x, or $x^{\frac{1}{2}} = \sqrt{x}$.

Similarly, raising a value to the one-third power is the same as taking its cube root: $x^{\frac{1}{3}} = \sqrt[3]{x}$. In fact, for any positive integer n, $\boxed{x^{\frac{1}{n}} = \sqrt[n]{x}}$.

If the numerator of a fractional exponent is greater than one, treat it as a normal power. In other words, $x^{\frac{p}{n}}$ is the nth root of x, raised to the p power, or $\boxed{x^{\frac{p}{n}} = \left(\sqrt[n]{x}\right)^p}$. This works because $x^{\frac{p}{n}} = x^{\frac{1}{n}\cdot p} = \left(\sqrt[n]{x}\right)^p$.

Example: $27^{\frac{2}{3}} = \left(\sqrt[3]{27}\right)^2 = 3^2 = 9$.

Note that the power can be applied either inside or outside the radical. In other words, $x^{\frac{p}{n}}$ could have been rewritten as $\sqrt[n]{x^p}$. After all, we could have used $x^{\frac{p}{n}} = x^{p\cdot\frac{1}{n}} = \sqrt[n]{x^p}$ instead. Therefore, $\boxed{\sqrt[n]{x^p} = \left(\sqrt[n]{x}\right)^p}$.

Examples: (1) $\left(\sqrt{x}\right)^3 = \sqrt{x^3}$. Both can be represented by $x^{\frac{3}{2}}$.

(2) $\left(\sqrt[3]{2}\right)^2 = \sqrt[3]{2^2} = \sqrt[3]{4}$. Therefore, $2^{\frac{2}{3}} = \sqrt[3]{4}$.

Note that for the above rule, the *entire radicand* is raised to a power.

Example: $\sqrt[3]{x^2+5}$ is *not* equivalent to $\left(\sqrt[3]{x+5}\right)^2$!

However, $\sqrt[3]{(x+5)^2} = \left(\sqrt[3]{x+5}\right)^2$ is true.

The **laws of exponents** apply to fractional exponents.

$x^a x^b = x^{a+b}$ $\quad\quad$ $\frac{x^a}{x^b} = x^{a-b}$ $\quad\quad$ $(x^a)^b = x^{ab}$ $\quad\quad$ $x^{-a} = \frac{1}{x^a}$

Examples: (1) $x^{\frac{1}{3}} \cdot x^{\frac{1}{3}} = x^{\left(\frac{2}{3}+\frac{1}{3}\right)} = x^1 = x$

(2) $\frac{x^{\frac{2}{3}}}{x^{\frac{1}{3}}} = x^{\left(\frac{2}{3}-\frac{1}{3}\right)} = x^{\frac{1}{3}} = \sqrt[3]{x}$

(3) $\left(x^{\frac{2}{3}}\right)^{\frac{1}{3}} = x^{\left(\frac{2}{3}\cdot\frac{1}{3}\right)} = x^{\frac{2}{9}} = \sqrt[9]{x^2}$

(4) $x^{-\frac{2}{3}} = \frac{1}{x^{\frac{2}{3}}}$ for $x \neq 0$

To simplify expressions with fractional exponents, it is often helpful to represent the fractional exponents as equivalent fractions using the least common denominator.

Example: To simplify $\frac{\sqrt[4]{x^3}}{\sqrt[3]{x^2}}$, we can rewrite the expression using fractional exponents.

$\frac{\sqrt[4]{x^3}}{\sqrt[3]{x^2}} = \frac{x^{\frac{3}{4}}}{x^{\frac{2}{3}}}$. Then, we can rewrite the exponents using fractions with a common denominator, which will allow us to simplify:

$$\frac{x^{\frac{3}{4}}}{x^{\frac{2}{3}}} = \frac{x^{\frac{9}{12}}}{x^{\frac{8}{12}}} = x^{\frac{1}{12}} = \sqrt[12]{x}$$

CALCULATOR TIP

Operations involving fractions can be done on the calculator.

Example: $\frac{3}{4}-\frac{2}{3}$ can be entered as [3][÷][4][-][2][÷][3] [MATH][1].

[On the TI-84 models, fractions can also be entered by pressing [ALPHA][F1][1].

Some TI-84 models also allow fractions to be entered by pressing [ALPHA][$\frac{\square}{\square}$].*]*

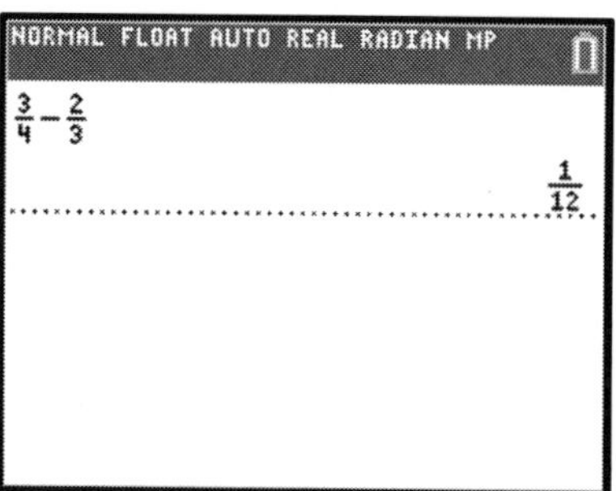

TI-83 Model *TI-84 Model*

To solve an equation with a variable raised to a fractional exponent, we should isolate the term with the fractional exponent and then raise both sides to its reciprocal power.

Example: To solve $x^{\frac{3}{5}}=8$, we can raise both sides to the $\frac{5}{3}$ power (the reciprocal of $\frac{3}{5}$).

$\left(x^{\frac{3}{5}}\right)^{\frac{5}{3}}=8^{\frac{5}{3}}$ gives us $x=32$. This works because $\left(x^{\frac{3}{5}}\right)^{\frac{5}{3}}=x^{\left(\frac{3}{5}\cdot\frac{5}{3}\right)}=x^1=x$.

When solving equations of this type, we need to be careful to include a $\pm$ symbol whenever we take an even root of both sides; that is, whenever we raise both sides to a fractional exponent having an *even number in its denominator.*

Example: To solve $x^{\frac{4}{3}}=1{,}296$, we get $\left(x^{\frac{4}{3}}\right)^{\frac{3}{4}}=\pm 1{,}296^{\frac{3}{4}}$, or $x=\pm 216$.

MODEL PROBLEM 1: ***SIMPLIFY EXPRESSIONS***

Simplify: $\left(\dfrac{x^3}{x^{\frac{8}{3}}}\right)^{-2}$

Solution:

(A) (B) (C)

$$\left(\frac{x^3}{x^{\frac{8}{3}}}\right)^{-2} = \left(x^{\frac{1}{3}}\right)^{-2} = x^{-\frac{2}{3}} = \frac{1}{x^{\frac{2}{3}}} = \frac{1}{\sqrt[3]{x^2}} \qquad \frac{1}{\sqrt[3]{x^2}} \cdot \left(\frac{\sqrt[3]{x}}{\sqrt[3]{x}}\right) = \frac{\sqrt[3]{x}}{x}$$

Explanation of steps:

(A) Use the law of exponents to simplify an expression involving rational exponents. *[When dividing the same bases, subtract exponents:* $3 - \frac{8}{3} = \frac{1}{3}$. *When raising a power to a power, multiply exponents:* $\frac{1}{3} \cdot (-2) = -\frac{2}{3}$. *When raising a base to a negative exponent, write the reciprocal with a positive exponent:* $x^{-\frac{2}{3}}$ *is the reciprocal of* $x^{\frac{2}{3}}$.*]*

(B) Change terms with rational exponents into radicals using $x^{\frac{p}{n}} = \sqrt[n]{x^p}$.

(C) Rationalize any denominators with radicals.

Practice Problems

1. Rewrite $\left(\sqrt[5]{x}\right)^4$ using a fractional exponent.	2. Rewrite $\sqrt[4]{x^7}$ using a fractional exponent.
3. Rewrite $\frac{3}{\sqrt[5]{x}}$ using a fractional exponent.	4. Rewrite $x^{\frac{3}{2}}$ as a radical.
5. Rewrite $x^{-\frac{2}{3}}$ as a radical. Then rationalize the denominator.	6. Find the exact value of $\left(\frac{27}{64}\right)^{-\frac{2}{3}}$ written as a fraction in simplest form.
7. Simplify: $\left(\frac{x^2}{x^{\frac{1}{9}}}\right)^3$	8. Simplify: $\left(\frac{27x^4}{xy^{-\frac{2}{3}}}\right)^{\frac{1}{3}}$

9. Simplify: $(9x^2y^6)^{-\frac{1}{2}}$

10. Simplify: $\left(x^{\frac{1}{2}}y^{-\frac{2}{3}}\right)^{-6}$

11. Simplify: $\left(\frac{x^{-5}}{x^{-9}}\right)^{\frac{1}{2}}$

12. Simplify: $\frac{(m^6)^{-\frac{2}{3}}}{m^2}$

13. Use fractional exponents to simplify $\frac{\sqrt[3]{x^2}}{\sqrt[6]{x}}$ into a single radical.

14. Use fractional exponents to simplify $\frac{\sqrt{x^5}+x^3}{\sqrt[3]{x^7}}$ into the sum of two radicals.

MODEL PROBLEM 2: *SOLVE EQUATIONS*

Solve for x: $4x^{\frac{3}{4}} + 5 = 113$

Solution:

(A) $4x^{\frac{3}{4}} = 108$

$x^{\frac{3}{4}} = 27$

(B) $\left(x^{\frac{3}{4}}\right)^{\frac{4}{3}} = 27^{\frac{4}{3}}$

(C) $x = \left(\sqrt[3]{27}\right)^4 = 3^4 = 81$

Explanation of steps:

(A) Isolate the variable with the fractional exponent. *[Subtract 5 from both sides, then divide both sides by 4.]*

(B) Raise both sided to the reciprocal power. This will eliminate the exponent of the variable.

(C) Evaluate, either by simplifying the expression or by using the calculator.

PRACTICE PROBLEMS

15. Solve for x: $2x^{\frac{3}{4}} + 5 = 133$	16. Solve for x: $x^{\frac{7}{10}} \cdot \sqrt{x} = 729$
17. Solve for y in terms of x: $\sqrt[3]{y^2} = x^{\frac{5}{6}}$	18. Solve for b in terms of V and a: $V = \sqrt{a^4 b^{\frac{1}{3}}}$

Chapter 9. Rational Functions

9.1 Undefined Expressions

Key Terms and Concepts

When we divide $6 \div 3$ we get 2, because when we multiply 2 times 3, the product is 6. However, if we try to divide $6 \div 0$ there is no solution, since there are no values that when multiplied by 0 would give us a product of 6. The expression $6 \div 0$, or $\frac{6}{0}$, is undefined.

In fact, any expression involving division is **undefined** when the divisor (denominator) is equal to zero.

A **rational expression** is an expression involving algebraic fractions with polynomials in the numerators and denominators.

A rational expression is undefined whenever its denominator is zero. Therefore, to determine when an expression is undefined, set the denominator equal to zero and solve.

Example: $\frac{x^2+1}{x-5}$ is undefined when $x - 5 = 0$, or when $x = 5$. Therefore, if the rational expression $\frac{x^2+1}{x-5}$ is included in a problem, it should be accompanied by a statement that $x \neq 5$.

Model Problem

For which set of values of *x* is the algebraic expression $\frac{x^2-9}{x^2-4x}$ undefined?

Solution:

(A) $x^2 - 4x = 0$

(B) $x(x - 4) = 0$

$\{0,4\}$

Explanation of steps:

(A) The expression is undefined when the denominator equals zero.

(B) Solve and state the solutions.

PRACTICE PROBLEMS

1. For which value of x is the expression $\frac{x-7}{x+2}$ undefined?	2. For which value of x is the expression $\frac{3x}{3x+1}$ undefined?
3. For which values of x is the expression $\frac{2x+3}{x^2-4}$ undefined?	4. For which values of x is the expression $\frac{x^2-16}{x^2-4x-12}$ undefined?
5. For which values of x is the function $f(x)=\frac{x-7}{9-x^2}$ undefined?	6. For which values of x is the function $g(x)=\frac{x^2+x-6}{x^2+5x-6}$ undefined?
7. For which values of x is the function $h(x)=\frac{x-3}{x+2}+\frac{x}{x-1}$ undefined?	8. In the domain of real numbers, for which values of x is the expression $\frac{x-2}{2x^2+1}$ undefined?

9.2 Simplify Rational Expressions

KEY TERMS AND CONCEPTS

When **dividing a polynomial by a monomial**, we may be able to simplify by dividing each term of the polynomial in the numerator by the monomial in the denominator. For each new fraction, divide the coefficients and subtract exponents of the same base. The result should have as many terms as the original polynomial.

Example: $$\frac{21a^2b - 3ab}{3ab} = \frac{21a^2b}{3ab} - \frac{3ab}{3ab} = 7a - 1 \quad (\text{for } a \neq 0, b \neq 0)$$

We could also have simplified this algebraic fraction by **factoring** and then cancelling out (dividing) the common monomial factors in both the numerator and denominator.

Example: $$\frac{21a^2b - 3ab}{3ab} = \frac{\cancel{3ab}(7a - 1)}{\cancel{3ab}} = 7a - 1 \quad (\text{for } a \neq 0, b \neq 0)$$

When **dividing a polynomial by a polynomial**, we may be able simplify the rational expression by factoring. **Factor the numerator and denominator completely and cancel any common factors in both.**

Example: $$\frac{x^2 - 2x}{x^2 - 4} = \frac{x\cancel{(x-2)}}{(x+2)\cancel{(x-2)}} = \frac{x}{x+2} \quad (\text{for } x \neq \pm 2)$$

Be sure to **cancel only factors** and not individual terms of a polynomial. As an analogy, think of plus or minus signs as bonds that may not be broken.

Example: $$\frac{\cancel{x^2} - 2x}{\cancel{x^2} - 4}$$ This cancelling is ***INCORRECT!***

Steps for Simplifying Fractional Expressions with Polynomials in the Denominator:

1. **Factor** both the numerator and denominator **completely**.
2. If there are **monomial factors**, divide if possible. Cancel out any **polynomial factors** that are common to both the numerator and denominator.
3. Express in standard form.

Example: To simplify $\frac{12x^2 + 12x}{3x^2 - 6x - 9}$,

$\frac{12x(x+1)}{3(x+1)(x-3)} =$ *[factor completely]*

$\frac{^{4}\cancel{12}x\cancel{(x+1)}}{\cancel{3}\cancel{(x+1)}(x-3)} =$ *[divide monomial factors and cancel polynomial factors]*

$\frac{4x}{x-3}$ *[express in standard form]*

When simplifying, it may be helpful to remember that $a - b = -(b - a)$.

Example: $\frac{2x-4}{2-x} = \frac{2(x-2)}{2-x} = \frac{2\cancel{(x-2)}}{-\cancel{(x-2)}} = \frac{2}{-1} = -2$

Sometimes, the simplest form may not always be the easiest form with which to work. It may be helpful, at times, to rewrite a rational expression in a different, less simple form. The ability to recognize different ways to rewrite expressions is an important skill.

Example: We can view the rational expression $\frac{x+4}{x+3}$ as $\frac{(x+3)+1}{(x+3)}$ which is $1 + \frac{1}{x+3}$.

Model Problem

Simplify: $\dfrac{6x^4 - 6x^2}{2x^2 - 2x}$ (for $x \neq 0$ and $x \neq 1$)

Solution:

$$\frac{6x^4 - 6x^2}{2x^2 - 2x} \overset{(A)}{=} \frac{6x^2(x^2 - 1)}{2x(x-1)} \overset{(B)}{=} \frac{6x^2(x+1)\cancel{(x-1)}}{2x\cancel{(x-1)}} \overset{(C)}{=} 3x(x+1) = 3x^2 + 3x$$

Explanation of steps:

(A) Factor both the numerator and denominator completely.
[First, the GCF of each is factored out. Then, since the numerator still contains a difference of two squares, $x^2 - 1$, we factor it further.]

(B) If there are monomial factors, divide if possible $\left[\frac{6x^2}{2x} = 3x\right]$. Cancel out any polynomial factors that are common to both the numerator and denominator *[$(x - 1)$ is common to both].*

(C) Express in standard form. *[Multiply to eliminate the parentheses.]*

Practice Problems

1. For all values for which the fraction is defined, write $\dfrac{x^2 + 2x}{x}$ in simplest form.	2. For all values for which the fraction is defined, simplify $\dfrac{6x^3 + 9x^2 + 3x}{3x}$.
3. For all values for which the fraction is defined, simplify $\dfrac{9x^4 - 27x^6}{3x^3}$.	4. For all values for which the fraction is defined, simplify $\dfrac{2x^6 - 18x^4 + 2x^2}{2x^2}$.

5. For all values for which the fraction is defined, simplify $\frac{45a^4b^3 - 90a^3b}{15a^2b}$.

6. For all values for which the fraction is defined, simplify $\frac{2x^2 - 12x}{x - 6}$.

7. For all values for which the fraction is defined, simplify $\frac{25x - 125}{x^2 - 25}$.

8. For all values for which the fraction is defined, simplify $\frac{x^2 - 2x - 15}{x^2 + 3x}$.

9. For all values for which the fraction is defined, simplify $\frac{x^2 + 6x + 5}{x^2 - 25}$.

10. For all values for which the fraction is defined, simplify $\frac{9x^2 - 15xy}{9x^2 - 25y^2}$.

11. For all values for which the fraction is defined, simplify $\frac{x^2 - 2x - 15}{x^2 + 3x}$.	12. For all values for which the fraction is defined, simplify $\frac{x^2 - x - 6}{x^2 - 5x + 6}$.
13. For all values for which the fraction is defined, simplify $\frac{2x^2 + 10x - 28}{4x + 28}$.	14. For all values for which the fraction is defined, simplify $\frac{3 - x}{2x - 6}$.
15. For all values for which the fraction is defined, simplify $\frac{y - x}{x^2 - y^2}$.	16. For all values for which the fraction is defined, simplify $\frac{x^2 - 9x}{45x - 5x^2}$.

17. For all values for which the fraction is defined, simplify $\frac{3y^2 - 12y}{4y^2 - y^3}$.	18. For all values for which the fraction is defined, simplify $\frac{x^2y^2 - 9}{3 - xy}$.
19. The area of a rectangle is represented by $x^2 + 8x + 15$. If the length of the rectangle is represented by $x + 5$, express the width of the rectangle as a binomial.	20. An isosceles triangle has two sides of length $4x + 6$. The perimeter of the triangle is $14x + 21$. What is the ratio of the base to the perimeter?

9.3 Multiply and Divide Rational Expressions

Key Terms and Concepts

When multiplying numeric fractions like $\frac{9}{20} \cdot \frac{5}{6}$, we can simply factor, cancel common factors, and then multiply the remaining factors in the numerator and in the denominator.

Example: $\frac{9}{20} \cdot \frac{5}{6} = \frac{\cancel{3} \cdot 3}{2 \cdot 2 \cdot \cancel{5}} \cdot \frac{\cancel{5}}{2 \cdot \cancel{3}} = \frac{3}{8}$

To multiply algebraic fractions, the same process works: **factor, cancel common factors,** and **multiply across**.

Example: $\frac{5x}{x^2 - 4} \cdot \frac{x-2}{10} = \frac{\cancel{5}x}{(x+2)\cancel{(x-2)}} \cdot \frac{\cancel{x-2}}{2 \cdot \cancel{5}} = \frac{x}{2(x+2)} = \frac{x}{2x+4}$

Reminder: Only cancel common factors, not terms.

To divide algebraic fractions, simply **flip the second fraction** (that is, invert it into its reciprocal) and then **multiply**.

Example: $\frac{2x+6}{8xy} \div \frac{x+3}{2y^2} = \frac{2x+6}{8xy} \cdot \frac{2y^2}{x+3} = \frac{\cancel{2}\cancel{(x+3)}}{\cancel{2} \cdot \cancel{2} \cdot 2 \cdot x \cdot \cancel{y}} \cdot \frac{\cancel{2} \cdot \cancel{y} \cdot y}{\cancel{x+3}} = \frac{y}{2x}$

(Assume all rational expressions are defined for their respective domains.)

Model Problem

Express the product of $\frac{x^2 - x - 6}{3x^2 - 9x}$ and $\frac{12}{x+2}$ in simplest form, for all values for which the fractions are defined.

Solution:

$$\frac{x^2-x-6}{3x^2-9x} \cdot \frac{12}{x+2} \overset{(A)}{=} \frac{\cancel{(x-3)}\cancel{(x+2)}}{\cancel{3}x\cancel{(x-3)}} \cdot \frac{2 \cdot 2 \cdot \cancel{3}}{\cancel{x+2}} \overset{(B)}{} \overset{(C)}{=} \frac{4}{x}$$

Explanation of steps:

(A) Factor each numerator and denominator completely.

(B) Cancel factors common to both a numerator and denominator.

(C) Multiply the remaining factors in the numerators and in the denominators.

Practice Problems

1. Perform the indicated operation and express the result in simplest terms, for all values for which the fractions are defined: $\frac{7x^2}{3} \cdot \frac{9}{14x}$	2. Perform the indicated operation and express the result in simplest terms, for all values for which the fractions are defined: $\frac{4x^2}{7y^2} \cdot \frac{21y^3}{20x^4}$
3. Perform the indicated operation and express the result in simplest terms, for all values for which the fractions are defined: $\frac{x^2-1}{x} \cdot \frac{4x^2}{x+1}$	4. Perform the indicated operation and express the result in simplest terms, for all values for which the fractions are defined: $\frac{4x}{x-1} \cdot \frac{x^2-1}{3x+3}$
5. What is the product of $\frac{x^2-1}{x+1}$ and $\frac{x+3}{3x-3}$ expressed in simplest form, for all values for which the fractions are defined?	6. Express the product of $\frac{x+2}{2}$ and $\frac{4x+20}{x^2+6x+8}$ in simplest form, for all values for which the fractions are defined.

7. Perform the indicated operation and express the result in simplest terms, for all values for which the fractions are defined: $$\frac{x+2}{3x+3} \cdot \frac{x^2+5x+4}{2x+4}$$	8. Perform the indicated operation and express the result in simplest terms, for all values for which the fractions are defined: $$\frac{x^2-9}{x^2+9x+18} \cdot \frac{x}{x^2-3x}$$
9. Perform the indicated operation and express the result in simplest terms, for all values for which the fractions are defined: $$\frac{x}{x+3} \div \frac{3x}{x^2-9}$$	10. What is the quotient of $\frac{x}{x+4}$ divided by $\frac{2x}{x^2-16}$, for all values for which the fractions are defined?
11. Perform the indicated operation and express the result in simplest terms, for all values for which the fractions are defined: $$\frac{9x^2}{x^2+12x+36} \div \frac{12x}{x^2+6x}$$	12. Perform the indicated operation and express the result in simplest terms, for all values for which the fractions are defined: $$\frac{3x+6}{4x+12} \div \frac{x^2-4}{x+3}$$

13. Express in simplest form for all values for which the fractions are defined: $$\frac{2x^2 - 8x - 42}{6x^2} \div \frac{x^2 - 9}{x^2 - 3x}$$	14. Express in simplest form for all values for which the fractions are defined: $$\frac{3x^2 + 9x}{x^2 + 5x + 6} \div \frac{x^2 - 9}{x^2 - x - 6}$$
15. Express in simplest form for all values for which the fractions are defined: $$\frac{x^2 + 9x + 14}{x^2 - 49} \div \frac{3x + 6}{x^2 + x - 56}$$	16. If the length of a rectangular garden is represented by $\frac{2x - 6}{2x + 4}$ and its width by $\frac{x^2 + 2x}{x^2 + 2x - 15}$, write an expression for the area of the garden in simplest form, for all values for which the fractions are defined.

17. Perform the indicated operation and express the result in simplest terms, for all values for which the fractions are defined: $$\frac{x^2+2x-15}{x^2-4x-45} \div \frac{x^2+x-12}{x^2-5x-36}$$	18. Perform the indicated operations and express the result in simplest terms, for all values for which the fractions are defined: $$\frac{x^2+4x+3}{2x^2-x-10} \cdot \frac{2x^2+4x^3}{x^2+3x} \cdot \frac{x^2+4x+4}{x^2+3x+2}$$

9.4 Add and Subtract Rational Expressions

Key Terms and Concepts

To add or subtract two fractions that have the **same denominator**, we simply add or subtract the numerators and keep the denominators unchanged, and then simplify (reduce) if possible.

Example: $\frac{7}{10}-\frac{3}{10}=\frac{4}{10}=\frac{\cancel{2}\cdot 2}{\cancel{2}\cdot 5}=\frac{2}{5}$

We can follow the same procedure when adding or subtracting algebraic fractions with the same denominators.

Examples: (1) $\frac{5x}{6y}-\frac{x}{6y}=\frac{\overset{2}{\cancel{4}}x}{{}_{3}\cancel{6}y}=\frac{2x}{3y}$

(2) $\frac{2x}{x+5}+\frac{x+1}{x+5}=\frac{3x+1}{x+5}$

When subtracting, be careful to subtract the entire numerator of the second fraction.

Example: $\frac{x}{5}-\frac{x-1}{5}=\frac{x-(x-1)}{5}=\frac{x-x+1}{5}=\frac{1}{5}$

To add or subtract fractions with different denominators, we need to find the **least common multiple (LCM)** of the denominators, also known as the **least common denominator (LCD)**. We then need to change each one into an equivalent fraction that has the LCM as the denominator before adding. To do so, we multiply each term by a **fractional form of one** including any factors missing from its denominator, so that each term ends up with the LCM as its denominator.

Example: To add $\frac{3}{10}+\frac{7}{15}$ we use the LCM of 10 and 15, which is 30.

So, $\frac{3}{10}+\frac{7}{15}=\frac{3}{10}\left(\frac{3}{3}\right)+\frac{7}{15}\left(\frac{2}{2}\right)=\frac{9}{30}+\frac{14}{30}=\frac{23}{30}$

The same process works with the addition or subtraction of algebraic fractions. The LCM of two or more algebraic expressions is the smallest (or simplest) expression that is divisible by each of the expressions.

To find the LCM of two or more algebraic expressions:

1. Factor each expression completely. For monomial factors, write the prime factorization of the coefficients.
2. For each monomial factor, take each numeric or variable factor to the highest power that it appears in any one expression
3. Include any polynomial factors, each to the highest power that it appears.
4. The LCM is the product of all the factors taken in steps 2 to 3.

Examples: (a) To find the LCM of $54x^2y^2$ and $180xy^3z$, write the prime factorizations of the coefficients, as in $2 \cdot \underline{3^3} \cdot \underline{x^2}y^2$ and $\underline{2^2} \cdot 3^2 \cdot \underline{5} \cdot x\underline{y^3z}$. Then take each factor to the highest power it appears in either term (as shown by the underlined factors): $2^2 \cdot 3^3 \cdot 5 \cdot x^2y^3z$, which gives us $540x^2y^3z$.

(b) To find the LCM of $x^2 - y^2$ and $(x+y)^2$, first factor $x^2 - y^2$ into $(x+y)(x-y)$. Then take each binomial factor to its highest power, giving us $(x+y)^2(x-y)$.

Once we know the LCM, we can multiply each term by a **fractional form of one** including any factors missing from its denominator, so that each term ends up with the LCM as its denominator. Once the denominators are the same, we can add or subtract numerators.

Examples: (a) $\dfrac{5x}{6} + \dfrac{1}{2x}$ LCM of 6 and $2x$ is $6x$.

$$\text{So, } \frac{5x}{6}\left(\frac{x}{x}\right) + \frac{1}{2x}\left(\frac{3}{3}\right) = \frac{5x^2}{6x} + \frac{3}{6x} = \frac{5x^2+3}{6x}$$

(b) $\dfrac{2}{x} + \dfrac{3}{x^2}$ LCM of x and x^2 is x^2.

$$\text{So, } \frac{2}{x}\left(\frac{x}{x}\right) + \frac{3}{x^2} = \frac{2x}{x^2} + \frac{3}{x^2} = \frac{2x+3}{x^2}$$

If any of the denominators are polynomials, factor them completely to find the LCM.

Examples: (a) $\frac{4a}{15b} - \frac{b}{20a+5} = \frac{4a}{3 \cdot 5b} - \frac{b}{5(4a+1)}$ LCM is $15b(4a+1)$

$$\frac{4a}{15b}\left(\frac{4a+1}{4a+1}\right) - \frac{b}{5(4a+1)}\left(\frac{3b}{3b}\right) =$$

$$\frac{16a^2+4a}{15b(4a+1)} - \frac{3b^2}{15b(4a+1)} = \frac{16a^2+4a-3b^2}{60ab+15b}$$

(b) $\frac{3x}{2x+2} + \frac{4x}{x-2} = \frac{3x}{2(x+1)} + \frac{4x}{x-2}$ LCM is $2(x+1)(x-2)$

$$\frac{3x}{2(x+1)}\left(\frac{x-2}{x-2}\right) + \frac{4x}{x-2}\left(\frac{2(x+1)}{2(x+1)}\right) =$$

$$\frac{3x^2-6x}{2(x+1)(x-2)} + \frac{8x^2+8x}{2(x+1)(x-2)} = \frac{11x^2+2x}{2x^2-2x-4}$$

These steps are summarized below.

To add or subtract algebraic fractions:

1. Factor all numerators and denominators completely. Reduce if possible.
2. Find the LCM of the denominators.
3. For each fraction, determine which factor(s) of the LCM are missing from the denominator and multiply by the fractional form of one using these factor(s).
4. Add (or subtract) the numerators, and place the result over the common denominator. Factor completely and reduce if possible.

MODEL PROBLEM 1: *SAME DENOMINATORS*

Express the sum $\frac{x^2 - 2x}{6x + 6} + \frac{x^2 + 4x}{6x + 6}$ in simplest form.

Solution:

$$\frac{x^2 - 2x}{6x + 6} + \frac{x^2 + 4x}{6x + 6} \overset{(A)}{=} \frac{2x^2 + 2x}{6x + 6} \overset{(B)}{=} \frac{\cancel{2}x\cancel{(x+1)}}{\cancel{2} \cdot 3\cancel{(x+1)}} = \frac{x}{3}$$

Explanation of steps:

(A) Since the denominators are the same, combine the fractions by adding the numerators.

(B) Simplify by factoring the numerator and denominator and cancelling common factors.

PRACTICE PROBLEMS

1. Express the sum in simplest form: $\frac{3}{x^2 + 1} + \frac{5}{x^2 + 1}$	2. What is the sum of $\frac{3x^2}{x - 2}$ and $\frac{x^2}{x - 2}$, for $x \neq 2$?
3. Find the sum of $\frac{-x + 7}{2x + 4}$ and $\frac{2x + 5}{2x + 4}$ for $x \neq -2$.	4. What is $\frac{2 + x}{5x} - \frac{x - 2}{5x}$, for $x \neq 0$, expressed in simplest form?

5. What is the sum of $\frac{2y}{y+5}$ and $\frac{10}{y+5}$ in simplest form, for $y \neq -5$?	6. Express the sum in simplest form: $\frac{x^2}{x+1} + \frac{6x+5}{x+1}$, for $x \neq -1$

Model Problem 2: *Finding the LCM*

What is the LCM of $24xy^3$ and $40x^2y^2$?

Solution:

(A) $2^3 \cdot 3 \cdot x \cdot y^3$ and $2^3 \cdot 5 \cdot x^2 \cdot y^2$

(B) LCM is $2^3 \cdot 3 \cdot 5 \cdot x^2 \cdot y^3$

(C) LCM is $120x^2y^3$

Explanation of steps:

(A) Write the prime factorizations of the coefficients. *[*$24 = 2^3 \cdot 3$ *and* $40 = 2^3 \cdot 5$*.]*

(B) Take each factor to the highest power that it appears in either expression.

(C) The LCM is the product of all the factors.

Practice Problems

7. Find the LCM of $84a^3b$ and $27a^2b^2$.	8. Find the LCM of $x^2 - 4$ and $x + 2$.

9. Find the LCM of $2x^2, x^2 + x$, and $3x^3 + 3x$.	10. Find the LCM of $x^2 + 3x - 4$, $x^2 - 2x + 1$, and $x^2 + 6x + 8$.

MODEL PROBLEM 3: *DIFFERENT MONOMIAL DENOMINATORS*

What is $\frac{6x}{5x} - \frac{x+5}{x}$ in simplest form, for $x \neq 0$?

Solution:

$$\frac{6x}{5x} - \frac{x+5}{x} \overset{(A)}{=} \frac{6x}{5x} - \frac{x+5}{x}\left(\frac{5}{5}\right) = \frac{6x}{5x} - \frac{5x+25}{5x} \overset{(B)}{=} \frac{6x-5x-25}{5x} \overset{(C)}{=} \frac{x-25}{5x}$$

Explanation of steps:

(A) Find the LCM of the denominators *[5x]*, and change each fraction so that it has the same common denominator by multiplying by a fractional form of one. *[Only the second fraction needs to be changed by multiplying it by $\frac{5}{5}$.]*

(B) Combine into one fraction by adding or subtracting the numerators and keeping the same common denominator.

(C) Simplify if possible.

Practice Problems

11. Express the sum in simplest form: $\frac{5x}{6}+\frac{x}{4}$	12. Express the difference in simplest form: $\frac{3}{x}-\frac{2}{5}$
13. Express the difference in simplest form: $\frac{3}{7n}-\frac{7}{3n}$	14. Express the sum in simplest form: $\frac{a}{x}+\frac{b}{2x}$
15. Express in simplest form: $\frac{a}{b}-\frac{1}{3}$	16. Express in simplest form: $\frac{7}{12x}-\frac{y}{6x^2}$

MODEL PROBLEM 4: *DIFFERENT POLYNOMIAL DENOMINATORS*

What is $\frac{3x+1}{x^2-1} - \frac{1}{x+1}$ in simplest form, for $x \neq \pm 1$?

Solution:

(A) $\frac{3x+1}{x^2-1} - \frac{1}{x+1} = \frac{3x+1}{(x+1)(x-1)} - \frac{1}{x+1}$ (B) LCM is $(x+1)(x-1)$

(C) $= \frac{3x+1}{(x+1)(x-1)} - \frac{1}{x+1}\left(\frac{x-1}{x-1}\right) = \frac{3x+1}{(x+1)(x-1)} - \frac{x-1}{(x+1)(x-1)}$

(D) $= \frac{2x+2}{(x+1)(x-1)} = \frac{2\cancel{(x+1)}}{\cancel{(x+1)}(x-1)} = \frac{2}{x-1}$

Explanation of steps:

(A) Factor all numerators and denominators completely *[*$x^2-1=(x+1)(x-1)$*]*. Reduce if possible.

(B) Find the LCM of the denominators *[*$(x+1)(x-1)$*]*.

(C) For each fraction, determine what factor (or factors) is missing from the denominator and multiply by the fractional form of one using this factor. *[The second fraction needs* $(x-1)$ *in the denominator, so multiply the second fraction by* $\left(\frac{x-1}{x-1}\right)$*.]*

(D) Add (or subtract) the numerators, and place the result over the common denominator *[*$(3x+1)-(x-1)=2x+2$*]*. Factor completely and reduce if possible.

Practice Problems

17. Express in simplest form, for all values for which the fractions are defined: $\frac{6}{y-5}-\frac{y+5}{y^2-25}$	18. Express in simplest form, for all values for which the fractions are defined: $\frac{5}{x^2+5x}+\frac{x}{x+5}$
19. Express in simplest form, for all values for which the fractions are defined: $\frac{4x}{x^2-1}-\frac{3x}{2x+2}$	20. Express in simplest form, for all values for which the fractions are defined: $\frac{6x^2+6x}{x^2-x-2}+\frac{x^2+2x+1}{3x^2-6x}$

21. What is the sum of $(y-5)+\frac{3}{y+2}$?

22. Express in simplest form, for all values for which the fractions are defined:

$$\frac{3}{a-1}+\frac{3}{1-a}$$

23. What is the sum of $\frac{3x}{2x-6}+\frac{9}{6-2x}$, for all values for which the fractions are defined?

24. What is $\frac{x}{x-1}-\frac{1}{2-2x}$ expressed as a single fraction, for all values for which the fractions are defined?

9.5 Simplify Complex Fractions

Key Terms and Concepts

A **complex fraction** is a fraction that contains a fraction in its numerator or denominator. To simplify a complex fraction with only *single fractions* in the numerator or denominator, it is usually easiest to rewrite it as a division.

Example: $\dfrac{a^5}{\frac{a^4}{b^2}} = a^5 \div \dfrac{a^4}{b^2} = a^5 \cdot \dfrac{b^2}{a^4} = ab^2$ (for $a \neq 0$ and $b \neq 0$)

However, for complex fractions with multiple terms or fractions in the numerator or denominator, this might not be as easy to do. For these expressions, multiply the numerator and denominator by the LCM of all the denominators, and simplify if possible.

Example: For the fraction $\dfrac{x}{\frac{1}{x}+\frac{1}{y}}$, the LCM of the denominators is xy. Therefore, we can simplify the fraction by multiplying the numerator and denominator by xy.

$\dfrac{x}{\frac{1}{x}+\frac{1}{y}}\left(\dfrac{xy}{xy}\right) = \dfrac{x^2y}{y+x}$, for $x \neq 0$ and $y \neq 0$.

MODEL PROBLEM

Express in simplest form, for $x \neq 0$:

$$\frac{\frac{1}{2}+\frac{1}{x}}{\frac{1}{4}-\frac{1}{x^2}}$$

Solution:

(A) (B) (C)

$$\frac{\frac{1}{2}+\frac{1}{x}}{\frac{1}{4}-\frac{1}{x^2}}\left(\frac{4x^2}{4x^2}\right)=\frac{\frac{1}{2}(4x^2)+\frac{1}{x}(4x^2)}{\frac{1}{4}(4x^2)-\frac{1}{x^2}(4x^2)}=\frac{2x^2+4x}{x^2-4}=\frac{2x\cancel{(x+2)}}{\cancel{(x+2)}(x-2)}=\frac{2x}{x-2}$$

Explanation of steps:

(A) Find the LCM of all the denominators *[LCM of 2, x, 4, and* x^2 *is* $4x^2$*]* and multiply the complex fraction by a fractional form of one using this LCM.

(B) Multiply by distributing the LCM. By multiplying each fraction by the LCM, the fractions will disappear and the result will no longer be complex. *[Distribute* $4x^2$ *by each fraction.]*

(C) Factor completely, and cancel common factors.

PRACTICE PROBLEMS

1. Express in simplest form, for all values for which the fraction is defined: $\frac{x^2}{\frac{1}{x}}$	2. Express in simplest form, for all values for which the fraction is defined: $\frac{\frac{2x}{y}}{\frac{4x}{y^2}}$

3. Express in simplest form, for all values for which the fraction is defined:

$$\frac{\frac{1}{x}+\frac{1}{y}}{\frac{1}{x}-\frac{1}{y}}$$

4. Express in simplest form, for all values for which the fraction is defined:

$$\frac{1-\frac{1}{x^2}}{1+\frac{1}{x}}$$

5. Express in simplest form, for all values for which the fraction is defined:

$$\frac{6-\frac{x}{x-1}}{4-\frac{x}{x+1}}$$

6. Express in simplest form, for all values for which the fraction is defined:

$$1-\frac{1}{1-\frac{1}{x-5}}$$

9.6 Solve Rational Equations

Key Terms and Concepts

When an equation involves fractional expressions, we can solve it more easily by **eliminating the fractions.** We can accomplish this by **multiplying each term of the equation by the LCM of the denominators (LCD).** This will eliminate the denominator of each term since it will divide evenly into the LCM.

Example: To solve $\frac{6}{x}+\frac{x-3}{2x}=2$, we can multiply all terms by the LCM, $2x$.

$$2x\left(\frac{6}{x}\right)+2x\left(\frac{x-3}{2x}\right)=2x(2)$$

$$12+(x-3)=4x$$

$$9+x=4x$$

$$9=3x$$

$$3=x$$

When any of the fractions have **polynomial denominators**, factor the denominators in order to find the LCM.

Example: For $\frac{2}{x+1}-\frac{1}{x}=\frac{3x}{x^2+x}$, we can factor the denominator on the right side of the equation: $x^2+x=x(x+1)$. We can now easily see that the LCM is $x(x+1)$ and can multiply both sides of the equation (i.e., each term) by the LCM.

$$x(x+1)\left(\frac{2}{x+1}\right)-x(x+1)\left(\frac{1}{x}\right)=x(x+1)\left(\frac{3x}{x(x+1)}\right)$$

$$2x-(x+1)=3x$$

$$2x-x-1=3x$$

$$-1=2x$$

$$-\tfrac{1}{2}=x$$

Note: To show that we are multiplying both sides of the equation by the LCM, it may be easier to place the whole equation in square brackets and write the LCM only once.

Example: When solving the problem above, the first step could have been written as $x(x+1)\left[\frac{2}{x+1}-\frac{1}{x}=\frac{3x}{x(x+1)}\right]$. This is acceptable shorthand notation.

When working with rational equations, it is important to check that the solutions are valid. In particular, substitute each solution for the variable in each denominator to make sure that the denominator would not be zero. If any solution (root) would result in an undefined fraction (division by zero), then that solution should be rejected. Any such solutions are called **extraneous roots**.

MODEL PROBLEM 1: *MONOMIAL DENOMINATORS*

Solve for x: $\frac{x+1}{4} - \frac{2x}{3} = \frac{x}{12}$

Solution:

(A) $12\left(\frac{x+1}{4}\right) - 12\left(\frac{2x}{3}\right) = 12\left(\frac{x}{12}\right)$

(B) $3(x+1) - 4(2x) = x$

(C) $3x + 3 - 8x = x$

$-5x + 3 = x$

$3 = 6x$

$\frac{1}{2} = x$

Explanation of steps:

(A) Find the LCM *[12]*. Multiply each term by the LCM.

(B) Each denominator will divide evenly into the LCM, thereby eliminating the denominator.

(C) Solve the resulting equation.

PRACTICE PROBLEMS

1. Solve for x: $\frac{x}{16} + \frac{1}{4} = \frac{1}{2}$	2. What is the value of x in the equation $\frac{x}{2} + \frac{x}{6} = 2$?

3. Solve for x: $\frac{3}{5}x + \frac{2}{5} = 4$	4. What is the value of x in the equation $\frac{3}{4}x + 2 = \frac{5}{4}x - 6$?
5. Solve for x: $\frac{3}{4}x = \frac{1}{3}x + 5$	6. Solve for n: $\frac{5}{n} - \frac{1}{2} = \frac{3}{6n}$
7. Solve for x: $\frac{2x}{5} + \frac{1}{3} = \frac{7x - 2}{15}$	8. Solve for x: $\frac{2}{x} - 3 = \frac{26}{x}$

9. Solve for x: $\frac{2x}{3} + \frac{x}{6} = 5$	10. Solve for x: $\frac{1}{7} + \frac{2x}{3} = \frac{15x - 3}{21}$
11. Solve for x: $\frac{x}{3} + \frac{x+1}{2} = x$	12. Solve for x: $\frac{8}{3x} - \frac{x-1}{12} = \frac{1}{6x}$
13. Solve for x: $\frac{3}{4}(x + 3) = 9$	14. Solve for x: $\frac{3}{5}(x + 2) = x - 4$

15. Solve for m: $\frac{m}{5}+\frac{3(m-1)}{2}=2(m-3)$	16. Solve for x: $1-\frac{6}{x^2}=\frac{1}{x}$

MODEL PROBLEM 2: ***POLYNOMIAL DENOMINATORS***

Solve for x: $\frac{12}{x^2-16}-\frac{24}{x-4}=3$

Solution:

(A) $\frac{12}{(x+4)(x-4)}-\frac{24}{x-4}=3$

(B) LCM is $(x+4)(x-4)$.

(C) $(x+4)(x-4)\left[\frac{12}{(x+4)(x-4)}-\frac{24}{x-4}=3\right]$

$12-24(x+4)=3(x+4)(x-4)$

(D) $12-24x-96=3(x^2-16)$

$12-24x-96=3x^2-48$

$3x^2+24x+36=0$

$3(x^2+8x+12)=0$

$3(x+6)(x+2)=0$

$\{-6,-2\}$

Explanation of steps:

(A) Factor the denominators. *[$x^2-16=(x+4)(x-4)$]*

(B) Find the LCM of the denominators.

(C) Multiply the whole equation (i.e., all terms on both sides) by the LCM.

(D) Solve the resulting equation. Check for any extraneous roots. *[Substituting -6 or -2 for x would not result in a zero for any of the denominators here.]*

Practice Problems

17. Solve for x: $\frac{4x}{x+2} - \frac{12}{x} = 1$	18. Solve for x: $\frac{2}{3x} + \frac{4}{x} = \frac{7}{x+1}$
19. Solve for x: $\frac{3}{x+3} + \frac{2}{x-4} = \frac{4}{3}$	20. Solve for x: $\frac{x}{x+5} + \frac{9}{x-5} = \frac{50}{x^2-25}$

21. Solve for x: $\frac{x}{x-4} - \frac{1}{x+3} = \frac{28}{x^2 - x - 12}$	22. Solve for x: $\frac{4}{x} - \frac{3}{x+1} = 7$

Model Problem 3: *Complex Roots*

Solve for x: $\frac{x+8}{5} + \frac{x+5}{x} = 1$

Solution:

(A) $5x\left[\frac{x+8}{5} + \frac{x+5}{x} = 1\right]$

$x(x+8) + 5(x+5) = 5x$

(B) $x^2 + 8x + 5x + 25 = 5x$

$x^2 + 8x + 25 = 0$

(C) $x^2 + 8x = -25$

$x^2 + 8x + 16 = -9$

$(x+4)^2 = -9$

$x + 4 = \pm\sqrt{-9}$

(D) $x = -4 \pm 3i$

Explanation of steps:

(A) Eliminate denominators by multiplying the equation by the LCM.

(B) Simplify.

(C) Solve by competing the square.

(D) Express the square root of a negative number as an imaginary number. Write the solution in $a \pm bi$ form.

Practice Problems

23. Solve for x: $\frac{x+3}{3}+\frac{x+3}{x}=2$	24. Solve for x: $x+\frac{5}{x}=2$
25. Solve for x: $x=2-\frac{8}{x}$	26. Solve for x: $2x+\frac{3}{x}=-2$

27. Solve for x: $\frac{x}{8}+\frac{8}{9x}=0$	28. Solve for x: $2+\frac{5}{x^2}=\frac{6}{x}$

9.7 Model Rational Expressions and Equations

Key Terms and Concepts

Many real-life situations can be expressed algebraically using rational expressions or equations. Remember that an algebraic fraction represents division. We also see algebraic fractions when we represent a rate using the word "per," as in miles per hour (mph). Algebraic proportions are expressed as rational equations, as well.

Examples: (1) The average of x and y can be represented as $\frac{x+y}{2}$.

(2) A vehicle traveling 100 miles in h hours travels at a speed of $\frac{100}{h}$ mph.

(3) In a scale diagram, 5 inches represents x feet and 8 inches represents $x+6$ feet. We can find x by solving the equation, $\frac{5}{x}=\frac{8}{x+6}$.

Work rate problems will often lead to rational equations. **Work rate** problems involve two or more people working together to complete a single job. For two people, the equation is:

$$\frac{1}{t_1}+\frac{1}{t_2}=\frac{1}{t_b}$$

t_1 = the time taken by the first person to complete the job
t_2 = the time taken by the second person to complete the job
t_b = the time it takes for them working together to complete the job

Another type of problem that will often lead to a rational equation is a **distance problem**. Applying the formula $T=\frac{D}{R}$ or $R=\frac{D}{T}$, where D is distance, R is rate, and T is time, may lead to an equation involving algebraic fractions. A frequently added wrinkle in these problems is the effect of wind or water current. A boat traveling at x mph in still water would travel at $x+4$ mph when traveling with a 4 mph current, or $x-4$ mph when traveling against a 4 mph current. Wind speed has the same effect on airplanes.

When an equation consists of one fraction on each side of the equal sign, the equation could be solved by cross-multiplying. If there are more than one term on either side of the rational equation, we can solve by multiplying the whole equation by the LCM.

MODEL PROBLEM 1: *TRANSLATE WORD PROBLEMS*

Two fractions are formed when 3 is divided by a number and 16 is divided by the square of the number. The sum of these two fractions is equal to 5 divided by the number. Find the number.

Solution:

(A) $\frac{3}{x}+\frac{16}{x^2}=\frac{5}{x}$

(B) $x^2\left[\frac{3}{x}+\frac{16}{x^2}=\frac{5}{x}\right]$

(C) $3x+16=5x$

$2x=16$

$x=8$

Explanation of steps:

(A) Translate the word problem into an equation.

(B) Find the LCM of the denominators and multiply the entire equation by this LCM.

(C) Solve.

PRACTICE PROBLEMS

1. The numerator of a fraction is 3 less than its denominator. If 7 is added to both the numerator and denominator, the resulting fraction is equivalent to 3/4. What is the original fraction?	2. The sum of a number and 64 times its reciprocal is 16. Find the number.

3. When 5 is divided by a number, the result is 3 more than 7 divided by twice the number. What is the number?

4. Given two consecutive even integers, when 6 is divided by the smaller number and 7 is divided by the larger number, the sum of these two fractions is 1. Find the two numbers.

5. The table below shows the number of pets each student in a class has. The average number of pets per student in this class is 2. What is the value of k?

Number of Pets	0	1	2	3	4	5
Number of Students	4	6	10	0	k	2

6. The diagram below represents a parallel electrical circuit.

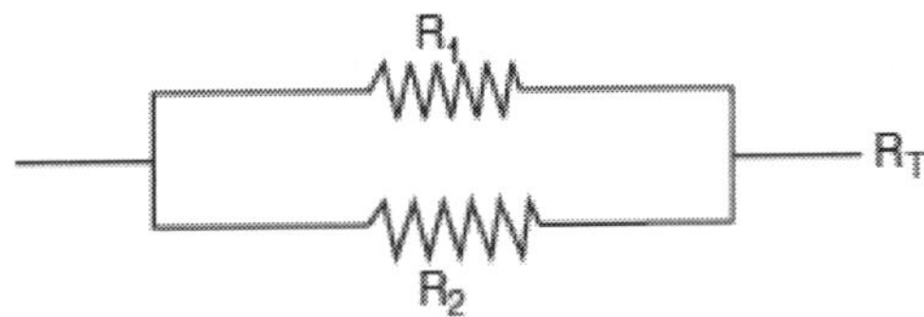

In a parallel circuit, the reciprocal of the total resistance is found by adding the reciprocals of each resistance, using the formula, $\frac{1}{R_1}+\frac{1}{R_2}=\frac{1}{R_T}$.

If $R_1 = x$, $R_2 = x + 3$, and the total resistance $R_T = 2.25$ ohms, find the positive value of R_1 to the *nearest tenth of an ohm.*

MODEL PROBLEM 2: ***WORK RATE PROBLEMS***

Andrew can clean the garage in 3 hours, but it takes Bobby 4 hours to do the same job. How long would it take them to clean the garage if they worked together?

Solution:

(A) $\frac{1}{3}+\frac{1}{4}=\frac{1}{x}$

(B) $12x\left[\frac{1}{3}+\frac{1}{4}=\frac{1}{x}\right]$

$4x + 3x = 12$

(C) $7x = 12$

$x = \frac{12}{7} \approx 1.7$ hrs

Explanation of steps:

(A) Use the work rate equation, $\frac{1}{t_1}+\frac{1}{t_2}=\frac{1}{t_b}$, substituting for known values.

(B) Multiply both sides of the equation by the LCM to eliminate fractions.

(C) Solve the resulting equation and state the solution.

Practice Problems

7. Kent can paint a certain room in 6 hours, but Kendra needs 4 hours to paint the same room. How long does it take them to paint the room if they work together?	8. Byron and Bruce create wedding favors. Byron take 20 minutes to create a favor and Bruce takes 30 minutes to create a favor. If they work together, how long will it take to create 50 favors?

9. Chris can lay bricks twice as fast as Danny. Working together, it takes them 5 hours to build a certain brick wall. How long would it have taken Chris working alone?	10. Two machines cover jars in a factory. When both machines are working, it takes 6 hours to cover 1,000 jars. When working separately, the faster machine takes 5 hours less than the other to cover 1,000 jars. How long does it take the faster machine to cover 1,000 jars on its own?

MODEL PROBLEM 3: *DISTANCE PROBLEMS*

Keandre drives 300 miles to visit his grandparents. After his visit, he drove back home at an average of 10 mph slower, increasing his travel time for the return trip by one hour. What was his average speed on his drive to his grandparents?

Solution:

(A)

	D	***R***	***T***
Going	300	x	$\frac{300}{x}$
Returning	300	$x-10$	$\frac{300}{x-10}$

(B) $\frac{300}{x}+1=\frac{300}{x-10}$

(C) $x(x-10)\left[\frac{300}{x}+1=\frac{300}{x-10}\right]$

$300(x-10)+x(x-10)=300x$

(D) $300x-3000+x^2-10x=300x$

$x^2-10x-3000=0$

$(x+50)(x-60)=0$

~~$x=-50$~~ or $x=60$

(E) Average speed was 60 mph.

Explanation of steps:

(A) In any $D=RT$ word problem, create a table of expressions.
[Distance D is 300 miles each way; let x be the rate on the first trip, so the return trip is 10 mph less; in each case, $T=\frac{D}{R}$.]

(B) Set up an equation.
[The travel time of the return trip is 1 hour longer than the first trip.]

(C) For rational equations, multiply all terms in the equation by the LCM.
[LCM is $x(x-10)$.]

(D) Solve the resulting equation.

(E) State the solution.

Practice Problems

11. Jacob travels 40 miles on his moped in the same amount of time that it takes him to travel 15 miles on his bicycle. If his moped's rate is 20 mph faster than his bicycle's, find the average rate for each.	12. Cameron can row 5 miles per hour in still water. It takes him as long to row 4 miles upstream as 16 miles downstream. How fast is the current?

13. A jet flies 1656 miles with a 12mph tailwind in half the time it takes to fly 3168 miles against the same wind. How fast would the jet fly without any wind?	14. A car travels a distance of 120 miles at an average speed that is 100 mph slower than that of a train. The train covers the same distance in 75 minutes less time. Find the speed of the train.

9.8 Graphs of Rational Functions

Key Terms and Concepts

A rational function is a function that is defined by a rational expression, $f(x) = \frac{N(x)}{D(x)}$, where $N(x)$ and $D(x)$ are polynomials. They belong to a family of functions whose parent function, $f(x) = \frac{1}{x}$, is called the **reciprocal function**.

Example: $f(x) = \frac{1}{x}$, graphed to the right, is called a **hyperbola** and is made up of two symmetrical branches. The domain and range are all nonzero real numbers.

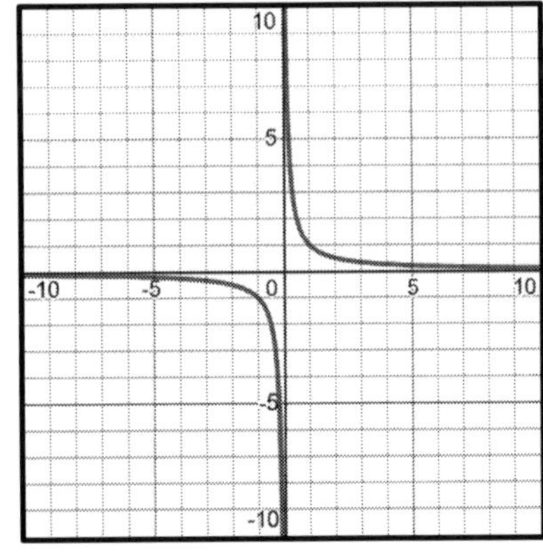

The graph of a rational function will often include asymptotes. **Asymptotes** are lines that the curve of a graph gets closer and closer to but never actually meets.

Example: Below is a graph of $f(x) = \frac{x-3}{x-4}$. It has a vertical asymptote at $x = 4$ and a horizontal asymptote at $y = 1$. Both are shown as dashed lines in the graph.

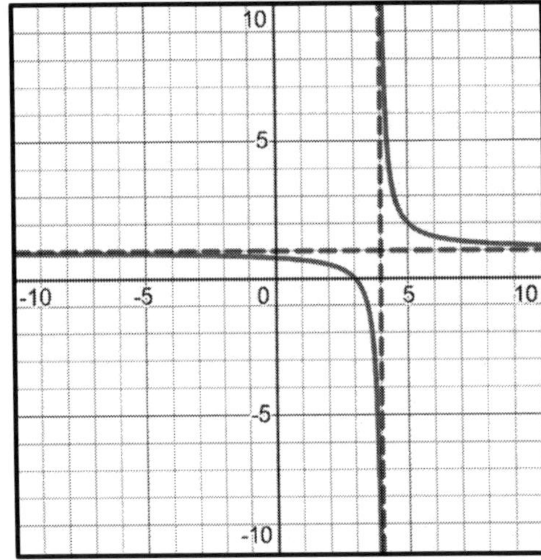

More formally, the vertical line $x = a$ is a **vertical asymptote** of $f(x)$ if $f(x) \to \infty$ or $f(x) \to -\infty$ as $x \to a$ either from the left or from the right. Also, the horizontal line $y = b$ is a **horizontal asymptote** of $f(x)$ if $f(x) \to b$ as $x \to \infty$ or $x \to -\infty$.

CALCULATOR TIP

When entering a rational function to be graphed on the calculator, be sure to use parentheses around the numerator or denominator if it contains more than one term.

Example: To graph $f(x) = \dfrac{x-3}{x-4}$ on the calculator, use parentheses around both the numerator and denominator, as shown below.

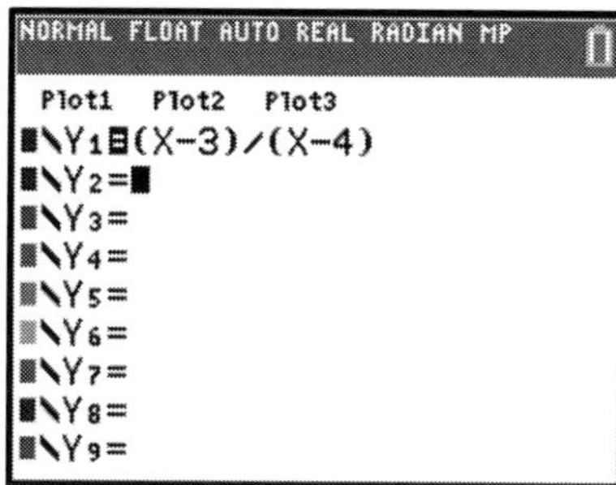

Note that the TI-83 does not always render asymptotes very clearly; the graph looks more like a heartbeat on an ECG where the vertical asymptote should be. The TI-84 does a better job of rendering graphs with asymptotes.

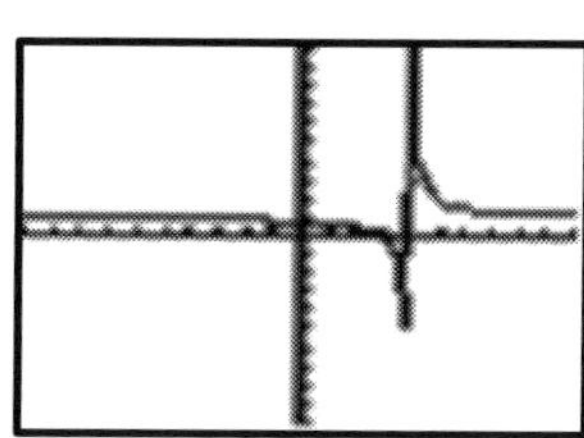
TI-83 Graph

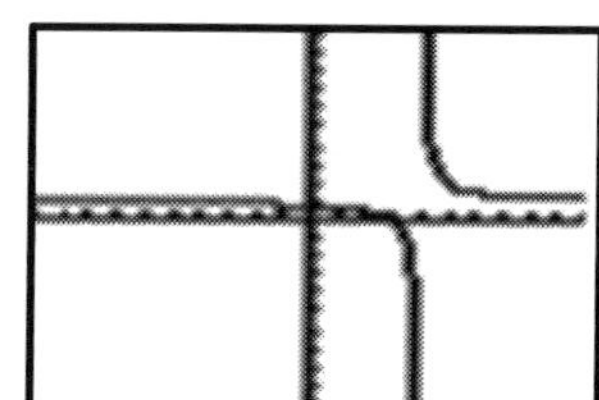
TI-84 Graph

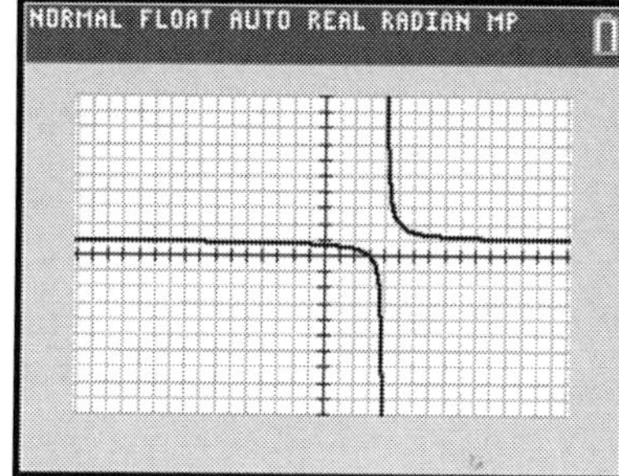

TI-84 CE Graph

Because a rational function never touches an asymptote, the value represented by that asymptote is excluded from the domain (for a vertical asymptote) and *may be* excluded from the range (for a horizontal asymptote).

Example: For the function $f(x) = \dfrac{x-3}{x-4}$ graphed above, the domain excludes $x = 4$ and the range excludes $y = 1$. Note that we should already know that $x = 4$ is excluded from the domain because it would result in a denominator of zero.

If a rational function's graph is a transformation of the reciprocal function $f(x) = \frac{1}{x}$, then it is possible to find its asymptotes by rewriting the function of its hyperbolic image in the form $f(x) = \dfrac{a}{x-h} + k$. If expressed in this form, then the image function will have a vertical asymptote at $x = h$ and a horizontal asymptote at $y = k$.

Examples: (a) The function graphed above, $f(x) = \dfrac{x-3}{x-4}$ can be rewritten as $f(x) = \dfrac{x-4+1}{x-4} = \dfrac{x-4}{x-4} + \dfrac{1}{x-4} = 1 + \dfrac{1}{x-4}$, so $f(x) = \dfrac{1}{x-4} + 1$. Written in this form, we can see that the asymptotes will be at $x = 4$ and $y = 1$.

(b) The function $g(x) = \dfrac{5x-2}{x-1}$ can be rewritten by dividing $5x - 2$ by $x - 1$, which gives us $5 + \dfrac{3}{x-1}$. Therefore, the graph of $g(x)$ will be a hyperbola with asymptotes at $x = 1$ and $y = 5$.

However, not all rational functions are simple transformations of the hyperbola, $f(x) = \frac{1}{x}$. There are more general methods for finding the asymptotes that will work for all rational functions.

To find the **vertical asymptotes** for a rational function of the form $y = \frac{f(x)}{g(x)}$, where $f(x)$ and $g(x)$ are polynomials and the fraction is expressed in simplest form, **set the denominator equal to zero and solve**.

Example: For $g(x) = \frac{x}{x^2 - 4}$, we can solve for $x^2 - 4 = 0$. This gives is $x = \pm 2$. Therefore, the graph of this function, shown below, would have vertical asymptotes at both $x = -2$ and at $x = 2$.

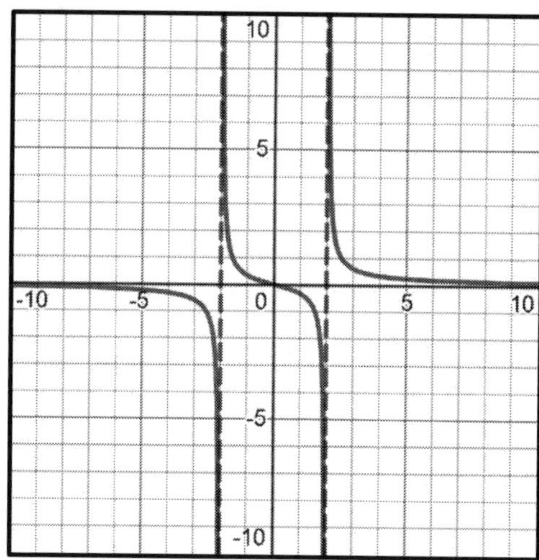

Note that, when finding the vertical asymptotes, the fraction first *needs to be expressed in simplest form before* setting the denominator equal to zero and solving.

Example: You might think $f(x) = \frac{3x + 6}{x^2 - 4}$ would have vertical asymptotes at both $x = -2$ and at $x = 2$, since it also has a denominator of $x^2 - 4$. However, it has only one vertical asymptote, as shown by the graph below.
We can see why if we simplify the expression:

$$f(x) = \frac{3x + 6}{x^2 - 4} = \frac{3\cancel{(x+2)}}{\cancel{(x+2)}(x-2)} = \frac{3}{x-2}$$

In simplest form, the denominator is now $x - 2$, so there is only one vertical asymptote at $x = 2$. We do, however, denote on the graph that $x = -2$ is not in the function's domain by graphing an open circle on the curve where $x = -2$. This is because the function is still undefined when $x = 2$ or $x = -2$.

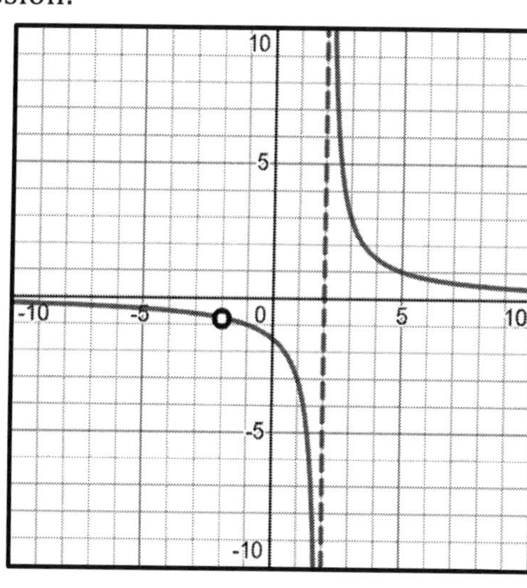

Some rational functions will not have a vertical asymptote.

Example: The graph of $y = \dfrac{1}{x^2 + 1}$, shown below is defined for all real values of x.

We can see that this is true because, if we set the denominator equal to zero and solve, the solutions are imaginary: $x = \pm i$.

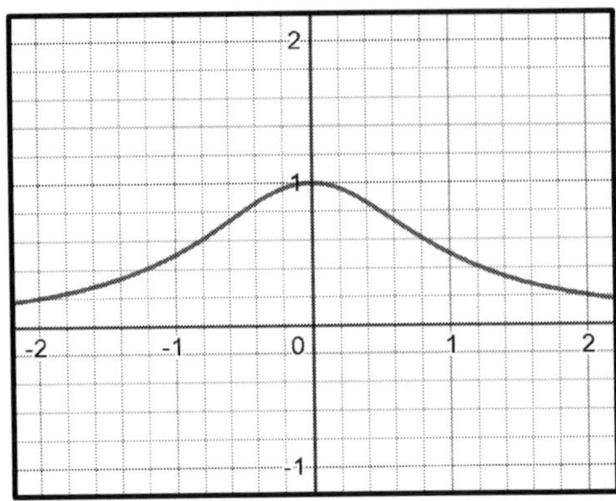

To find the **horizontal asymptotes** for a rational function of the form $y = \dfrac{f(x)}{g(x)}$, where $f(x)$ and $g(x)$ are polynomials, we have to look at the degrees of the polynomials in both the numerator and the denominator. If we let n be the degree of the numerator and d be the degree of the denominator,

- If $n > d$, then the graph will *not* have a horizontal asymptote.
- If $n < d$, then the graph will have a horizontal asymptote at $y = 0$ (the x-axis).
- If $n = d$, then there will be a horizontal asymptote at $y = \dfrac{a}{b}$, where a is the leading coefficient of the numerator and b is the leading coefficient of the denominator.

Example: For the graph of $f(x) = \dfrac{x^2 + 1}{x^2 - 4}$ shown to the right, there is a horizontal asymptote at $y = 1$. Since the degrees of the numerator and denominator are the same (both have a degree of 2), then the asymptote can be found by dividing the leading coefficients: $y = \frac{1}{1} = 1$.

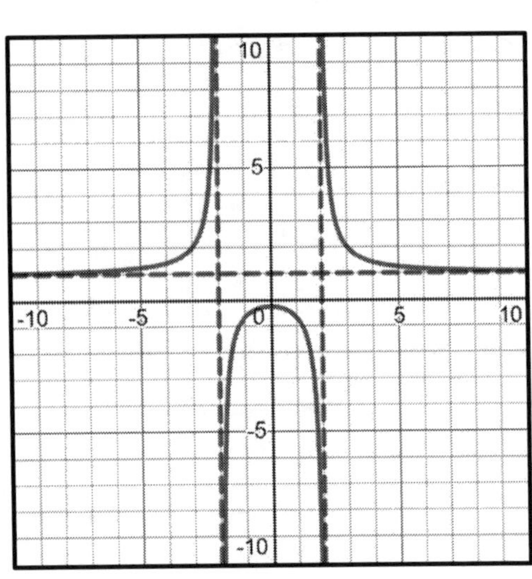

It's important to mention that, although a line may act as a horizontal asymptote along parts of a graph, the graph may actually cross the asymptote at other parts of the graph.

Example: For the graph of $g(x) = \dfrac{x}{x^2 - 4}$ shown below, the degree of the numerator, 1, is smaller than the degree of the denominator, 2, so there is a horizontal asymptote at $y = 0$. Notice that the graph approaches but never touches $y = 0$ as $x \to -\infty$ and as $x \to +\infty$. However, in the middle section of the graph (at $x = 0$) the graph actually crosses $y = 0$ and includes $(0,0)$.

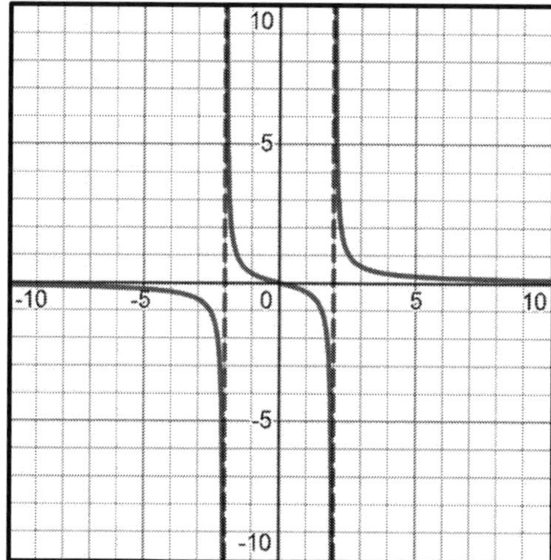

Note that adding a constant can transform a function and change its horizontal asymptote.

Example: Here are the graphs of $f(x) = \dfrac{x-3}{x-4}$ and $h(x) = \dfrac{x-3}{x-4} + 4$.

Note that by adding the constant 4, the horizontal asymptote has changed from $y = 1$ to $y = 5$.

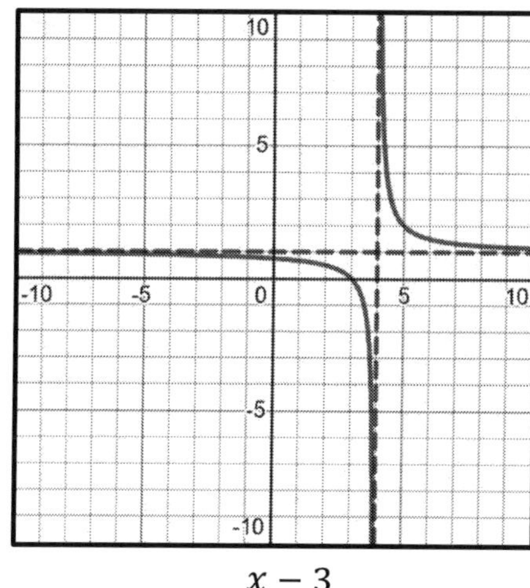

$f(x) = \dfrac{x-3}{x-4}$

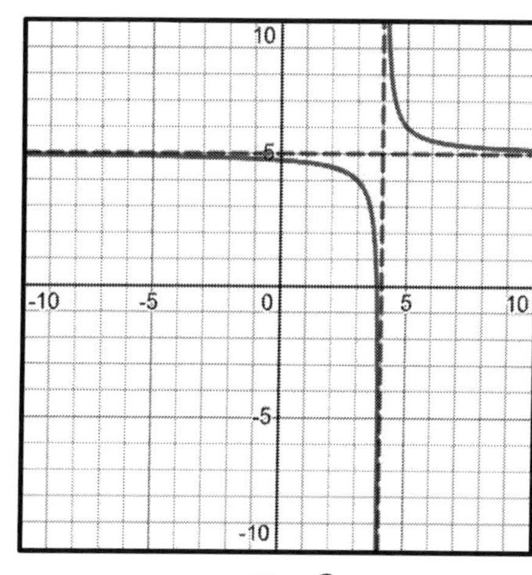

$h(x) = \dfrac{x-3}{x-4} + 4$

To find the **x-intercepts** of a rational function, set the expression equal to zero. If the expression is a single fraction, then it will equal zero when the numerator equals zero.

Example: The x-intercepts of $f(x) = \dfrac{4x^2 - 16}{x^2 - 2}$ occur where the numerator equals zero.

So, solve for $4x^2 - 16 = 0$ to get the x-intercepts at $x = \pm 2$.

To find the **y-intercept** of a rational function (if it exists), find the value of $f(0)$.

Example: To find the y-intercept of $f(x) = \dfrac{4x^2 - 16}{x^2 - 2}$, evaluate $f(0) = \dfrac{4(0)^2 - 16}{(0)^2 - 2} = 8$.

So, the y-intercept is 8.

Below is the graph of $f(x) = \dfrac{4x^2 - 16}{x^2 - 2}$, with the intercepts labelled. Note that there are vertical asymptotes at $x^2 - 2 = 0$, or $x = \pm\sqrt{2}$, and a horizontal asymptote at $y = 4$.

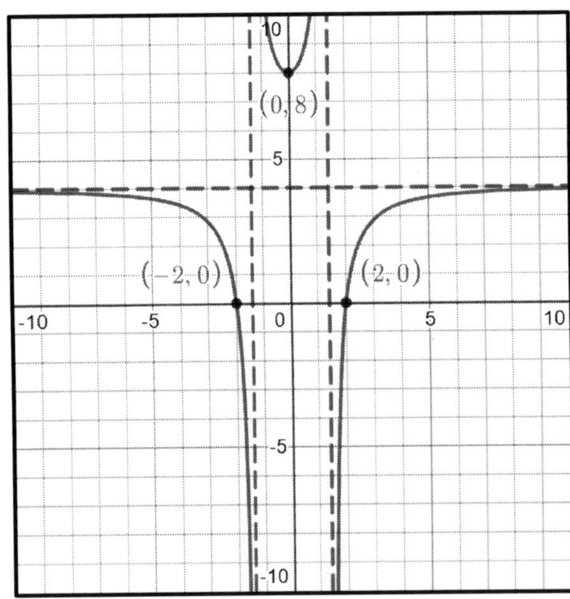

MODEL PROBLEM

For the graph of $f(x) = \frac{x^2+1}{2x^4-8x^2}$,

a) Find the vertical asymptotes, if any exist.
b) Find the horizontal asymptote, if one exists.
c) Find the x-intercepts, if any exist.
d) Find the y-intercept, if it exists.

Solution:

(A) $2x^4 - 8x^2 = 0$
$2x^2(x^2 - 4) = 0$
$x = \{-2, 0, 2\}$
Vertical asymptotes at
$x = -2, x = 0$, and $x = 2$

(B) Horizontal asymptote at
$y = 0$

(C) $x^2 + 1 = 0$
$x = \pm i$ (no real solutions)
No x-intercepts

(D) $f(0) = \frac{(0)^2+1}{2(0)^4 - 8(0)^2} = \frac{1}{0}$
$f(0)$ is *undefined*, so there is no y-intercept.

Explanation of steps:

(A) First, be sure that the rational expression is in simplest form. *[The numerator and denominator of this fraction have no common factors.]* Then set the denominator equal to zero and solve. The solutions tell where the vertical asymptotes appear.

(B) To find the horizontal asymptote, look at the degrees of the polynomials in the numerator and denominator. *[Since the degree of the numerator, 2, is less than the degree of the denominator, 4, the asymptote is $y = 0$.]*

(C) To find the x-intercepts for a single fraction, set the numerator equal to 0 and solve. *[Only real solutions will show as x-intercepts.]*

(D) To find the y-intercept, evaluate $f(0)$. *[If $f(0)$ is undefined, there is no y-intercept.]*

A graph of this function is shown below.

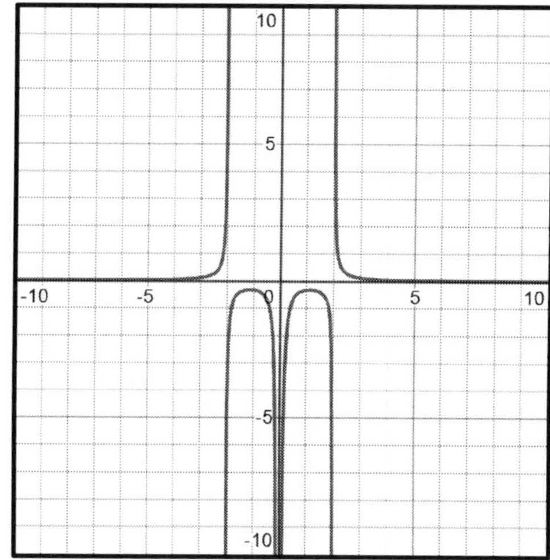

PRACTICE PROBLEMS

1. Find the vertical asymptote(s) of the graph of the function, $f(x) = \frac{x^2 - 9}{(x+2)(x+9)}$.	2. Find all the asymptotes for the function, $g(x) = \frac{2}{x+1} + 3$.
3. Find the horizontal asymptote, if any, of the graph of the function, $f(x) = \frac{x^2 - 9}{(x+2)(x+9)}$.	4. Find the horizontal asymptote, if any, of the graph of the function, $g(x) = \frac{x-1}{x^2-4}$.
5. Find the horizontal asymptote, if any, of the graph of the function, $h(x) = \frac{x^2 + 3x}{2x+1}$.	6. Find all the asymptotes for the function $f(x) = \frac{4x}{2x+1}$.

7. Which of the following functions could be represented by the graph to the right?

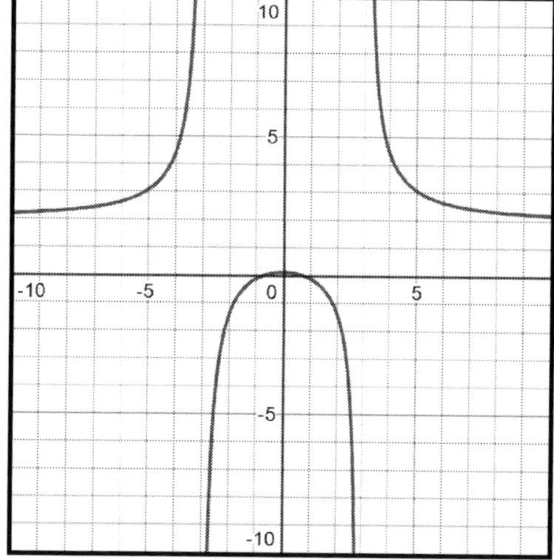

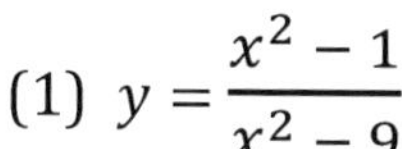

(1) $y = \dfrac{x^2 - 1}{x^2 - 9}$ (2) $y = \dfrac{2x^2 - 1}{x^2 - 9}$

(3) $y = \dfrac{x^2 - 1}{x^2 + 9}$ (4) $y = \dfrac{2x^2 - 1}{x^2 + 9}$

8. Find the x-intercepts and y-intercept, if any, of the graph of $f(x) = \dfrac{4x}{2x + 1}$.

9. Find the x-intercepts and y-intercept, if any, of the graph of $g(x) = \dfrac{x - 1}{x^2 - 4}$.

10. Find the x-intercepts and y-intercept, if any, of the graph of

$$h(x) = \frac{x^2 + 11x + 18}{x^2 + 3x}.$$

11. Find the x-intercepts and y-intercept, if any, of the graph of

$$j(x) = \frac{x^2}{2x + 1} - 1$$

Chapter 10. Exponential Functions

10.1 Solve Simple Exponential Equations

KEY TERMS AND CONCEPTS

Exponential equations are those that have a variable in an exponent. Equations with variables in the exponents may be solved if we can set both sides of the equation to the **same base** raised to different powers.

If both sides already have the same base, then their powers must be equal, and so a new equation can be set up in which the powers are set equal to each other.

Examples: (a) $5^x = 5^3$ can be easily solved. Since they have the same base, the powers must be equal, so $x = 3$.

(b) $10^{1-x} = 10^4$ can be solved by setting the powers equal to each other, $1 - x = 4$, and then solving for x, so $x = -3$.

If the two sides do not have the same base, it may be possible that they can be changed into equivalent expressions that do.

Example: $3^{2x-1} = 27$ is solved by changing 27 to 3^3, resulting in $3^{2x-1} = 3^3$ or $2x - 1 = 3$, so $x = 2$.

MODEL PROBLEM 1: *ISOLATING THE POWER*

Solve $5(2)^x = 160$.

Solution:

(A) $\frac{5(2)^x}{5} = \frac{160}{5}$

$2^x = 32$

(B) $2^x = 2^5$

(C) $x = 5$

Explanation of steps:

(A) Isolate the power *[divide both sides by 5]*.

(B) If possible, change both sides to exponential expressions with the same base *[32 = 2^5]*.

(C) Once the bases are equal, set the powers equal.

Practice Problems

1. Solve for *x*: $10(4)^x = 640$	2. Solve for *x*: $5(3)^x = 405$
3. Solve for *x*: $\frac{1}{5}(5)^x = 5$	4. Solve for *x*: $7\left(\frac{1}{2}\right)^x = \frac{7}{8}$
5. Solve for *x*: $2^{x+1} = 8$	6. Solve for *x*: $3^{2x-2} = 81$
7. Solve for *x*: $3^{x-3} = 1$	8. Solve for *x*: $4^{3x+5} = 16$

9. Solve for x: $3^{x+1} - 5 = 22$	10. Solve for x: $4^4 = 2^{3x-1}$

MODEL PROBLEM 2: *EXPONENTS ON BOTH SIDES*

Solve $7^x = 49^{x+5}$.

Solution:

$7^x = 49^{x+5}$

(A) $7^x = (7^2)^{x+5}$

(B) $7^x = 7^{2(x+5)}$

(C) $x = 2(x+5)$

(D) $x = 2x + 10$

$-x = 10$

$x = -10$

Explanation of steps:

(A) If possible, change both sides to exponential expressions with the same base *[*$49 = 7^2$*]*.

(B) Use the rules for exponents *[when raising a power to a power, multiply the exponents]*.

(C) Once the bases are equal, set the powers equal.

(D) Solve the resulting equation.

Practice Problems

11. Solve for *x*: $5^{2x} = 5^{x+4}$	12. Solve for *x*: $2^{x} = 4^{x+1}$
13. Solve for *x*: $4^{x} = 2^{3x+1}$	14. Solve for *x*: $4^{3x} = 2^{x+5}$
15. Solve for *x*: $3^{x-5} = 9^{x-3}$	16. Solve for *x*: $8^{x-2} = 2^{x}$
17. Solve for *x*: $2 = 2^{2x+1}$	18. Solve for *x*: $8^{x-2} = 4^{x}$

10.2 Rewrite Exponential Expressions

Key Terms and Concepts

We can use the properties of exponents to rewrite expressions for exponential functions.

Examples: $f(x) = 2^{3x}$ can be rewritten as $f(x) = 8^x$ since $2^{3x} = (2^3)^x$.

$g(x) = 3^{x+2}$ can be rewritten as $g(x) = 9(3)^x$ since $3^{x+2} = 3^x \cdot 3^2$.

Model Problem

If $f(x) = 2(0.5)^{3x}$, which of the following is an equivalent function?

(1) $g(x) = 8(0.125)^x$ (3) $g(x) = 2(0.125)^x$

(2) $g(x) = 6(1.5)^x$ (4) $g(x) = 2(1.5)^x$

Solution: (3)

Explanation of steps:

Only the quantity in parentheses *[*(0.5)*]* is being raised to a power. We can simplify by evaluating this quantity raised to the exponent's coefficient.

[$2(0.5)^{3x} = 2(0.5^3)^x = 2(0.125)^x$*]*

Practice Problems

1. Rewrite 5^{2x} as an equivalent expression with only x as the exponent.	2. Rewrite $10(1.1)^{5x}$ as an equivalent expression with only x as the exponent.

3. Rewrite 2^{3x+2} as an equivalent expression with only x as the exponent.	4. Rewrite $4(3)^{x+1}$ as an equivalent expression with only x as the exponent.
5. Rewrite 3^{2x-3} as an equivalent expression with only x as the exponent.	6. Rewrite $\frac{3^{5x+1}}{9^x}$ as an equivalent expression with only x as the exponent.
7. Rewrite $5\left(4^{\frac{x}{2}+2}\right) \cdot 3^{3x}$ as an equivalent expression with only x as the exponent.	8. Given $x = 4y + 5$, find the value of $\frac{2^x}{16^y}$.

9. If $4^x = k^{3x}$, what is the value of k?	10. Given $2^{x+3} - 2^x = k \cdot 2^x$, find the value of k.

11. If $\frac{1}{4}(2^x) = k\left(b^{\frac{x}{4}-2}\right)$, find the values of k and b.

10.3 Graphs of Exponential Functions

KEY TERMS AND CONCEPTS

An **exponential function** is a function in which x appears as an exponent in the equation. An exponential function has the form $f(x) = ab^x$, where $a \neq 0$, $b \neq 1$ and $b > 0$.

Examples: $f(x) = 5^x$ or $g(x) = -3(0.5)^x$

An exponential function can be graphed using a table or a graphing calculator.

Example: We can graph $f(x) = 2^x$ as follows, using $y = f(x)$. Note that the domain is $(-\infty, \infty)$, the range is $(0, \infty)$, the y-intercept is $(0,1)$, and the x-axis is a horizontal asymptote.

x	2^x	$y = f(x)$	(x, y)
-1	2^{-1}	$\frac{1}{2}$	$\left(-1, \frac{1}{2}\right)$
0	2^0	1	(0,1)
1	2^1	2	(1,2)
2	2^2	4	(2,4)
3	2^3	8	(3,8)

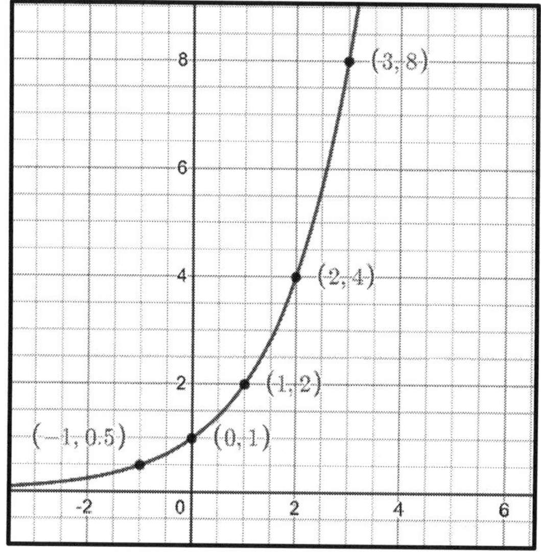

As we saw in Algebra I, if the x-values in a set of points are equally spaced and the y-values have a common ratio, then it is an exponential function (or a geometric sequence).

Example: The sequence of y-values above, $\frac{1}{2}, 1, 2, 4, 8, \ldots$, has a common ratio of 2.

 CALCULATOR TIP

On the calculator, we would enter: [Y=] [2] [^] [X,T,Θ,n] [GRAPH]

Generally, the graph of an exponential function $f(x) = ab^x$ will be shaped like one of these:

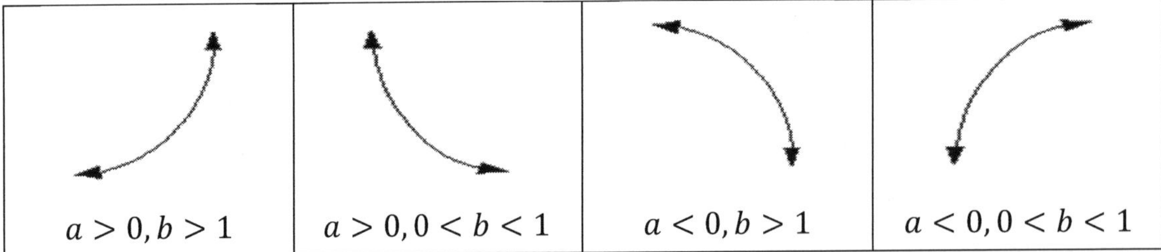

Example: The graph of the function $y = 3\left(\frac{1}{2}\right)^x$, where $a = 3$ and $b = \frac{1}{2}$, looks like this.

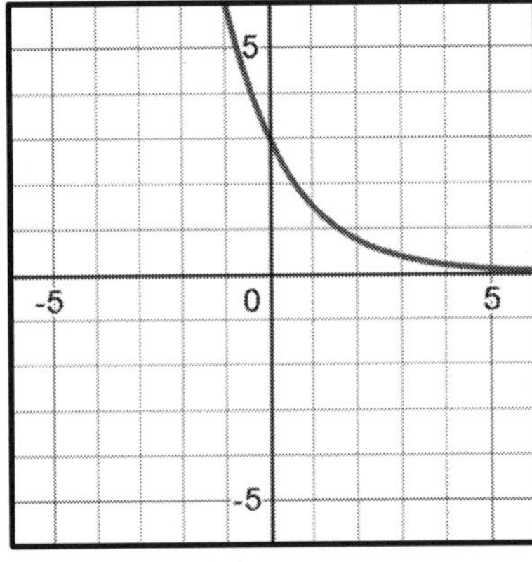

In the function $y = ab^x$, the **constant *a*** will tell us where the curve crosses the y-axis. Therefore, a is the y-intercept. (To find the y-intercept, substitute 0 for x. Since $b^0 = 1$, this gives us $y = a$.)

Example: For the function $y = 3\left(\frac{1}{2}\right)^x$, where $a = 3$ and $b = \frac{1}{2}$, the graph would intersect the y-axis at (0,3), as shown above.

When ***a* is negative**, the graph is reflected over the x-axis and is therefore negative over the entire domain.

An exponential function $y = ab^x$ will **never actually touch** (intersect) the x-axis. Neither a nor b is equal to zero, so ab^x (and therefore y) will never equal zero. This function has an asymptote at the x-axis.

An exponential function $f(x)$ may be transformed in any of the following ways:

$f(x) + k$	$(x, y) \rightarrow (x, y + k)$	vertically shifts the graph up ($k > 0$) or down ($k < 0$)
$f(x + k)$	$(x, y) \rightarrow (x - k, y)$	horizontally shifts the graph left ($k > 0$) or right ($k < 0$)
$-f(x)$	$(x, y) \rightarrow (x, -y)$	reflects the graph over the x-axis
$f(-x)$	$(x, y) \rightarrow (-x, y)$	reflects the graph over the y-axis
$k \cdot f(x)$	$(x, y) \rightarrow (x, ky)$	vertically stretches by a factor of k
$f(kx)$	$(x, y) \rightarrow (kx, y)$	horizontally stretches by a factor of $\frac{1}{k}$

For example, here are two **translations** of $f(x) = 2^x$.

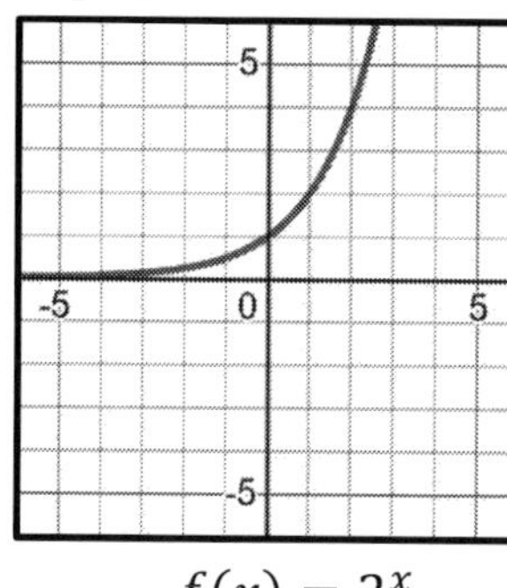

$f(x) = 2^x$

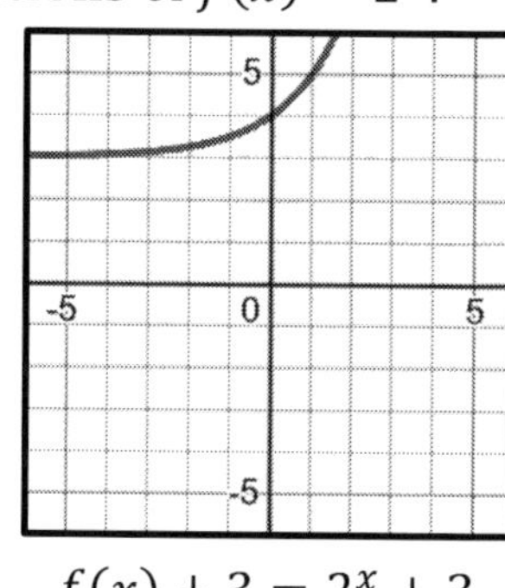

$f(x) + 3 = 2^x + 3$

shifted up 3

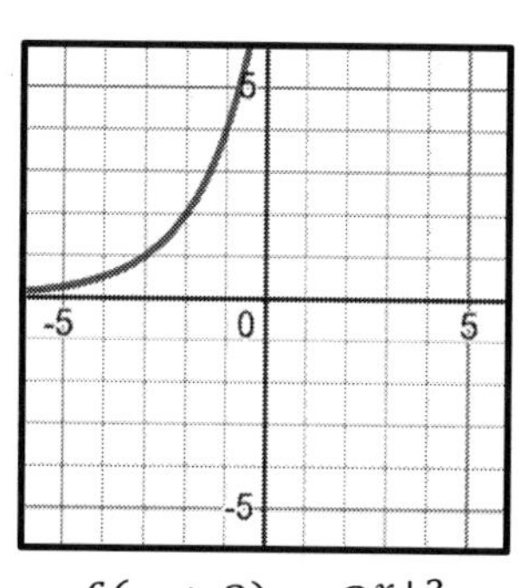

$f(x + 3) = 2^{x+3}$

shifted left 3

Here are two **reflections** of $f(x) = 2^x$.

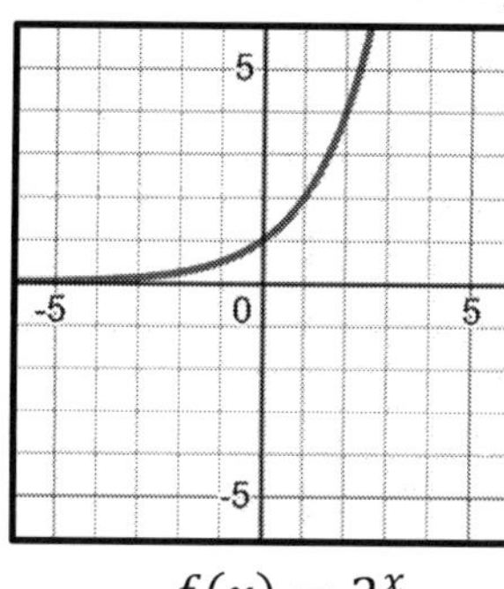

$f(x) = 2^x$

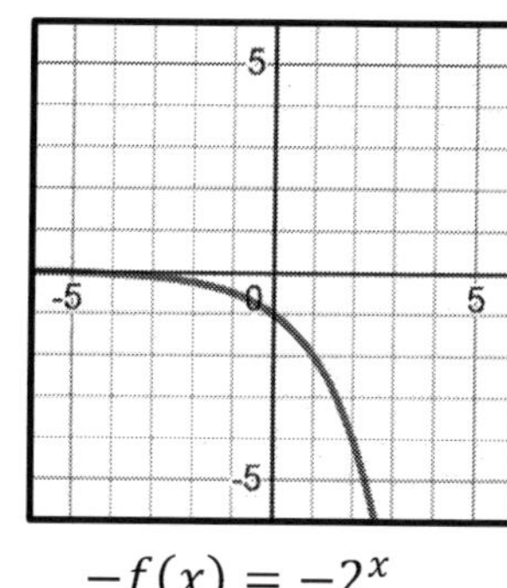

$-f(x) = -2^x$

reflected over x-axis

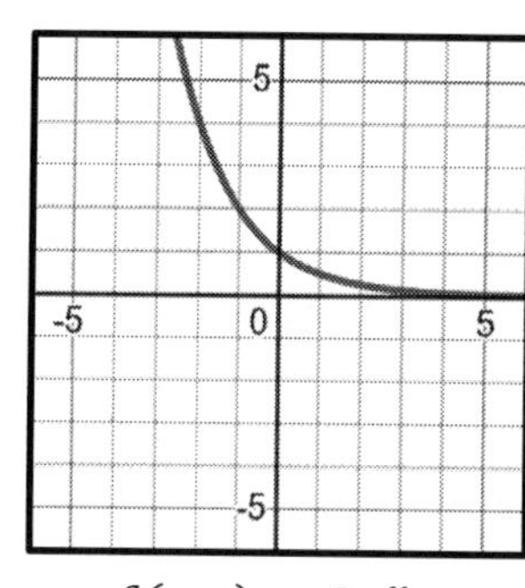

$f(-x) = 2^{-x}$

reflected over y-axis

Here are two **stretches** of $f(x) = 2^x$.

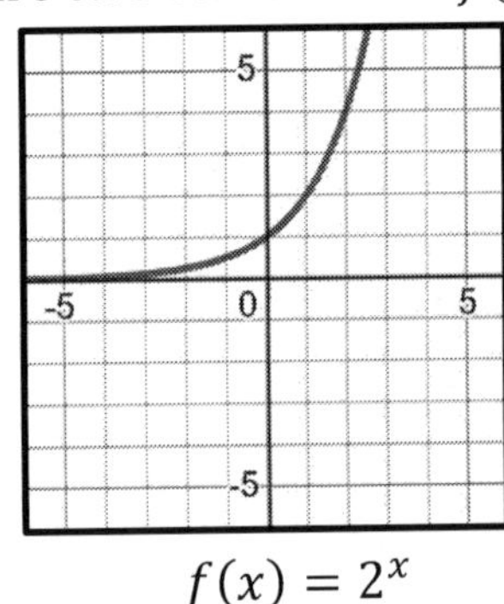

$f(x) = 2^x$

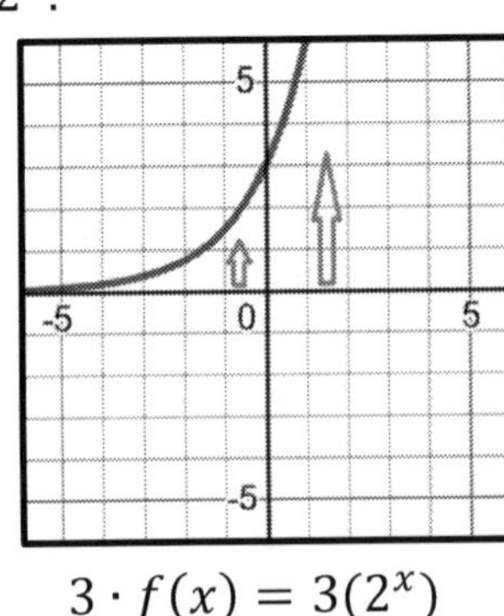

$3 \cdot f(x) = 3(2^x)$

vertically stretched by 3

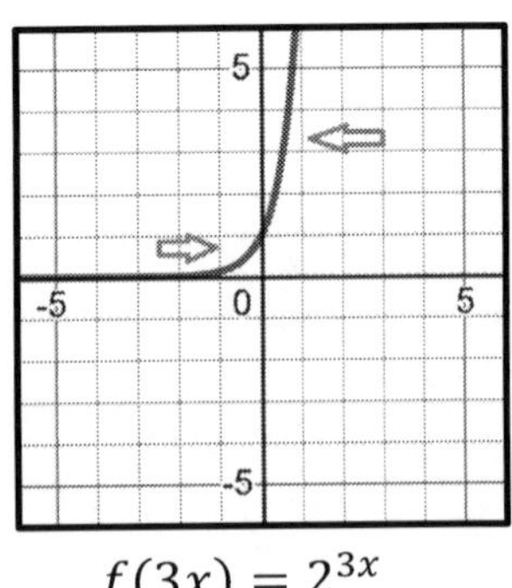

$f(3x) = 2^{3x}$

horizontally stretched by $\frac{1}{3}$

MODEL PROBLEM

Graph the function $y = \frac{3}{2} \cdot 2^x$.

Solution:

(A) x	(B) $\frac{3}{2} \cdot 2^x$	(C) y	(D) (x, y)
0	$\frac{3}{2} \cdot 2^0 = \frac{3}{2} \cdot 1$	1.5	(0,1.5)
1	$\frac{3}{2} \cdot 2^1 = \frac{3}{2} \cdot 2$	3	(1,3)
2	$\frac{3}{2} \cdot 2^2 = \frac{3}{2} \cdot 4$	6	(2,6)
3	$\frac{3}{2} \cdot 2^3 = \frac{3}{2} \cdot 8$	12	(3,12)
4	$\frac{3}{2} \cdot 2^4 = \frac{3}{2} \cdot 16$	24	(4,24)

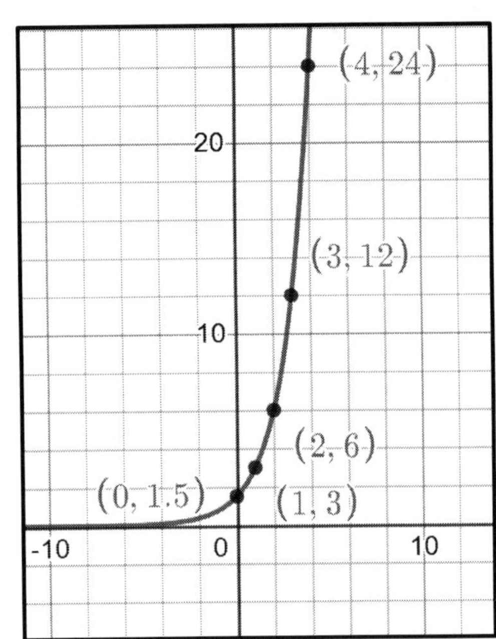

Explanation of steps:

(A) Pick values of x.

(B) Substitute the values of x into the expression.

(C) Evaluate for y.

(D) Plot the resulting points on the graph.

Practice Problems

1. On the grid below, graph $y = 2^x$ over the interval $-1 \le x \le 3$.

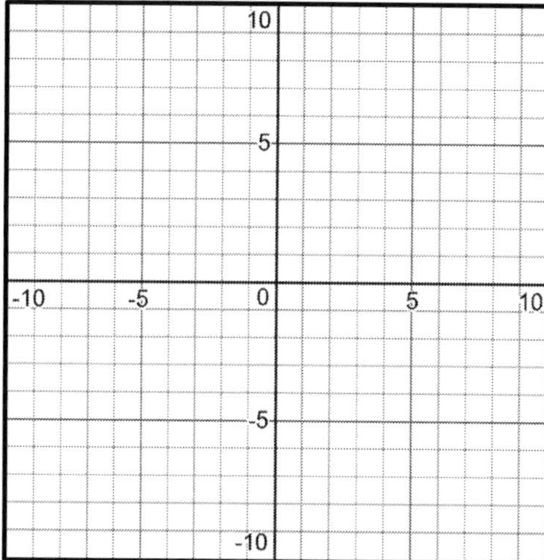

2. On the grid below, graph $y = 3^x$ over the interval $-1 \le x \le 2$.

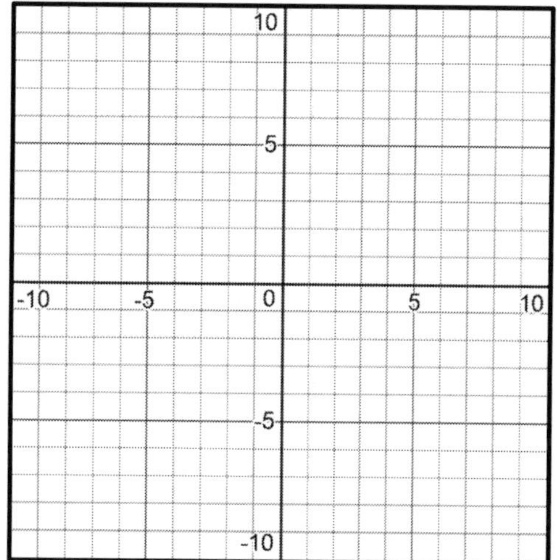

3. On the calculator, graph $y = \frac{1}{3} \cdot 2^x$. Sketch the graph below.

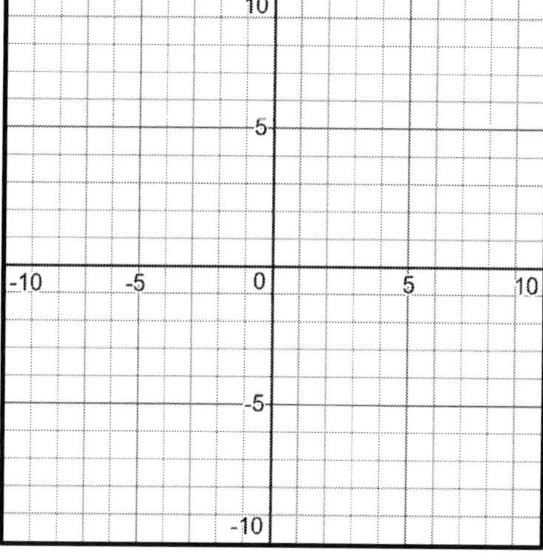

4. On the calculator, graph $y = 3 \cdot \left(\frac{1}{2}\right)^x$. Sketch the graph below.

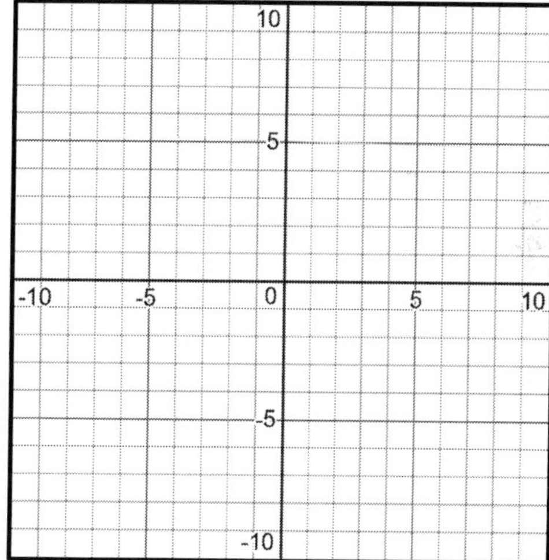

5. On the calculator, graph $y = 12(1.5)^x$. Sketch the graph below.

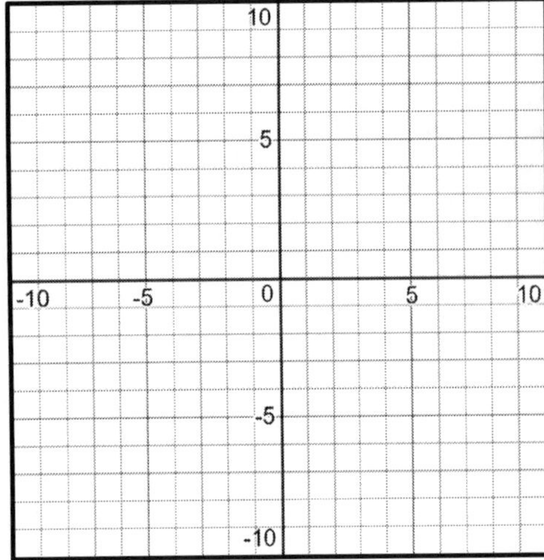

6. On the calculator, graph $y = 12(0.5)^x$. Sketch the graph below.

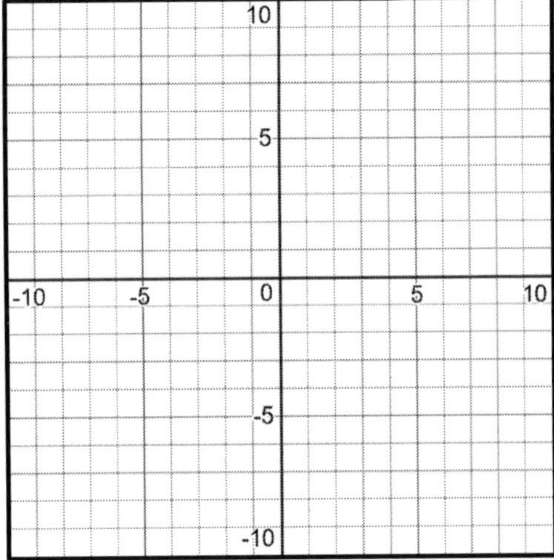

7. On the grid below, graph $y = -2^x$.

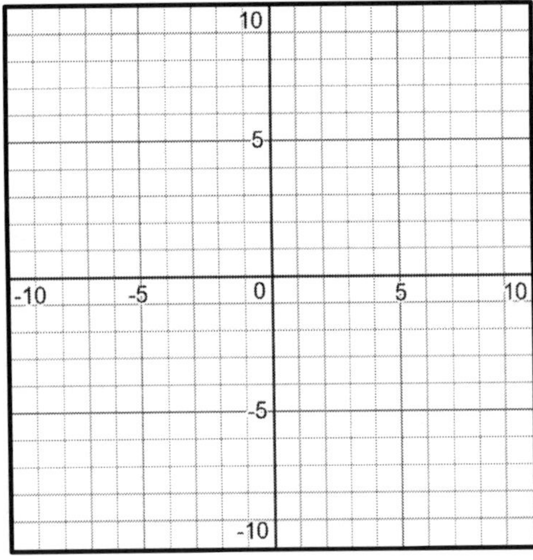

8. On the grid below, graph $y = 2^x - 5$ and explain how this graph differs from the exponential function $y = 2^x$.

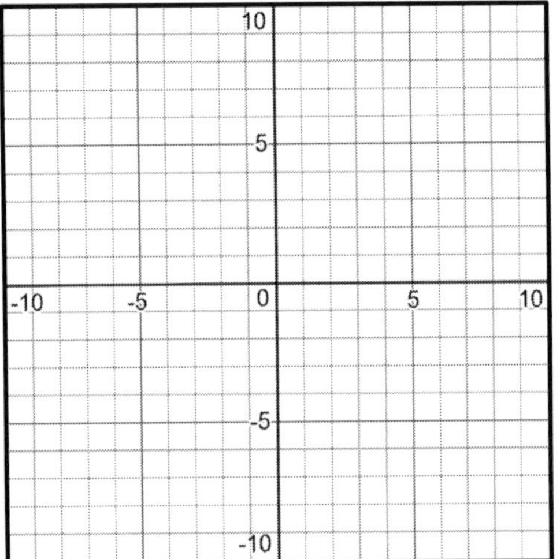

10.4 Exponential Regression

KEY TERMS AND CONCEPTS

If given a table or graph of an exponential function, you can use the calculator to find the equation. After entering the coordinates of at least two points, calculate the **exponential regression**. The exponential regression is just like a linear regression, except that it finds the *exponential* equation that best fits the set of points. The calculator instructions are given below.

Note that the exponential regression can only be calculated for a function that is always *positive* (that is, when $a > 0$). However, for a function with all negative y values, you can enter their additive inverses (positive values that produce a reflection of the function over the x-axis) and then negate the a that is found by the regression.

CALCULATOR TIP

Using the calculator to find the equation for an exponential function:

1. Enter the x and y coordinates of at least two points as L1 and L2 under [STAT][1].
2. Press [STAT] [CALC] [0] for ExpReg.
3. On the TI-84 models, if L1 and L2 are not set for XList and YList, you can set them by pressing [2nd][LIST]. To store the result in Y1, press [ALPHA][F4][1] for Store RegEQ. *[On the TI-83, you can store the result in Y1 by pressing [VARS] [Y-VARS] [1][1].]*
4. The screen will show the values of a and b for the equation $y = ab^x$.

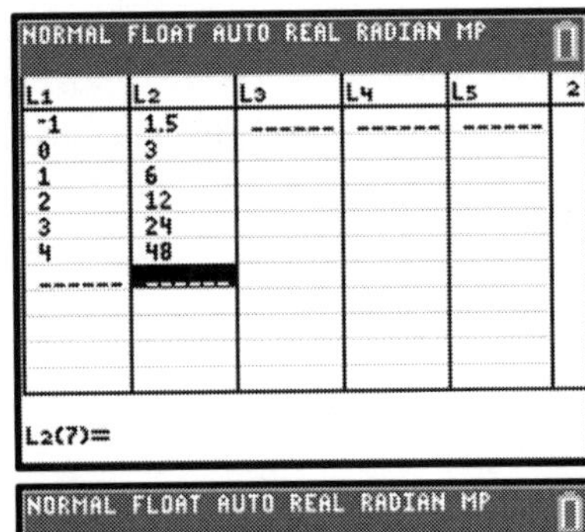

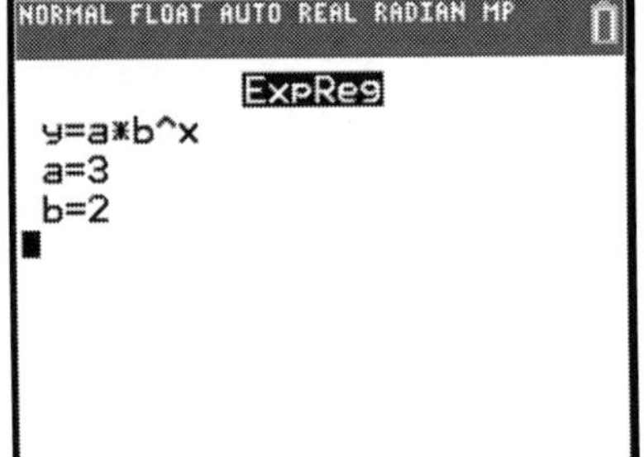

Model Problem

Write an equation for the exponential function that passes through the points $(3, 6.75)$ and $(5, 60.75)$.

Solution:

$$y = 0.25(3)^x$$

Explanation of steps:

Use the calculator's [STAT][1] to enter the coordinates of the points as L1 and L2, then use [STAT] [CALC] [0] to determine the exponential equation. *[$a = 0.25$ and $b = 3$]*

Practice Problems

1. Write an equation for the exponential function graphed below.

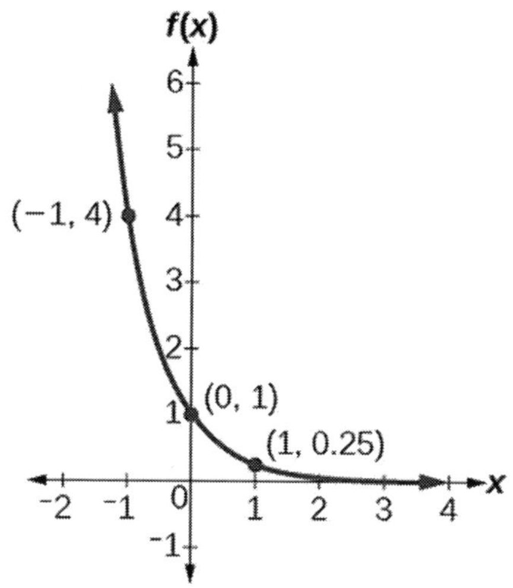

2. Write an equation for the exponential function that passes through the points $(1, 0.4)$ and $(3, 6.4)$.

3. Given exponential function $f(x)$ where $f(10) \approx 4.072$ and $f(15) \approx 5.197$, rounded to the *nearest thousandth*, find an equation for $f(x)$.

4. Given the following points, write an exponential regression equation, rounded to the *nearest thousandth*, to model the data.

x	1	3	5	7	9	11
y	5	3	2	1.25	0.8	0.5

5. The table shows the number of bacteria in a culture over time. Write an exponential regression equation, rounded to the *nearest thousandth*, to model this data.

Hours	1	2	3	4	5
Bacteria	1,746	1,960	2,215	2,490	2,805

Based on the equation, approximately how many bacteria, to the *nearest whole*, should there be after 10 hours?

6. The table shows the temperature of a cup of coffee over time. As time passes, the temperature of the coffee approaches room temperature, which is 70°F. Write an exponential function to model this data.
Note: The calculator assumes that the horizontal asymptote is $y = 0$, *so you will need to shift the function down by 70°, find the equation, and then add 70°.*

Minutes	0	5	10	15	20
Temp (°F)	210	178	154	135	120

Based on the equation, what is the temperature of the coffee, to the *nearest whole degree*, after 30 minutes?

10.5 Exponential Growth or Decay

Key Terms and Concepts

Exponential functions may be used to represent **exponential growth** and **exponential decay**. In these special cases, the equation $f(t) = ab^t$ is limited to $\boldsymbol{a > 0}$. The exponent t represents an amount of time, for $t \geq 0$.

- When $b > 1$, the function shows **exponential growth**.
- When $0 < b < 1$, the function shows **exponential decay**.

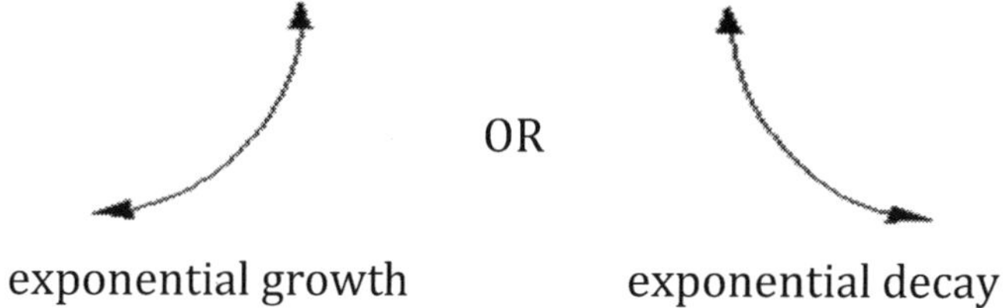

exponential growth exponential decay

The formula for exponential growth or decay is often written as $f(t) = a(1 + r)^t$. The y-intercept, the constant a, represents the starting value at time $t = 0$. The constant r represents the rate of change at each of the t intervals of time. $f(t) = a(1 + r)^t$ is an exponential growth function if $r > 0$ or an exponential decay function if $r < 0$.

Example: For the function $f(t) = 1000(1.05)^t$, where t represents a number of hours, the y-intercept (or starting value, a) is 1000, while the value 1.05 represents an exponential growth of 5% per hour.

If we change the exponent in this formula, we can represent different intervals at which the growth occurs.

Example: If t represents the number of weeks, then

(a) $f(t) = a(1 + r)^t$ represents a growth of r (as a percent) every week,

(b) $f(t) = a(1 + r)^{2t}$ represents a growth of r twice a week, and

(c) $f(t) = a(1 + r)^{\frac{t}{2}}$ represents a growth of r every two weeks.

We can use this formula to model various examples of exponential growth or decay.

Example: A certain bacteria will grow by 50% every 10 hours. If we start with a culture of 1,000 of the bacteria, a formula for the number of bacteria, B, in terms of the number of hours, t, can be written as $B(t) = 1000(1.5)^{\frac{t}{10}}$.

A common application of exponential decay is in the half-life of radioactive elements. The **half-life** of a substance is the amount of time, on average, that it takes for half of its atoms to decay. For problems involving a half-life, the rate of change, *r*, is -0.5, representing a 50% decay. So, the function for the amount of substance remaining is $f(t) = a(0.5)^{\frac{t}{h}}$, where t represents time and h represents the half-life (using the same unit of time).

Example: The half-life of carbon-14 is 5,700 years. Given 800g of carbon-14, the amount of this substance that will remain after 28,500 years can be calculated as $800(0.5)^{\frac{28,500}{5,700}} = 25\text{g}$.

Given a formula in which r represents an *annual* rate of growth or decay, $f(t) = a(1 + r)^t$, we can convert this to an equivalent formula expressed in terms of the rate of growth or decay over a shorter interval. If the shorter interval is $\frac{1}{n}$ of a year, then we can take the nth root of $(1 + r)$ and raise it to the nt power instead. This works because $ab^t = a\left(b^{\frac{1}{n}}\right)^{nt}$.

Remember, the nth root $\sqrt[n]{b}$ can be expressed as $b^{\frac{1}{n}}$.

Examples: a) In the equation $y = 250(1.06)^t$, the variable t represents a number of years. To write an equivalent equation expressed in terms of a *monthly* rate of growth, we calculate $\sqrt[12]{1.06} \approx 1.00487$, giving us $y = 250(1.00487)^{12t}$. This tells us that the approximate monthly rate of growth is 0.487%. If we let m represent the number of months, we can rewrite the equation as $y = 250(1.00487)^m$.

b) For a 20% annual rate of decay, where $(1 - 0.20) = 0.80$, we can find the monthly rate of decay as $\sqrt[12]{0.80} \approx 0.9816$, or an approximate 1.84% monthly rate of decay.

c) If given a 40% annual rate of growth, we can find the *daily* rate of growth by calculating $\sqrt[365]{1.40} \approx 1.0009$, or an approximate daily rate of 0.09%.

MODEL PROBLEM

\$5000 is deposited in a bank account that pays 4% annual interest.

a) Write a formula for $A(t)$, the amount of money in the account after t years and find the balance in the account after 10 years.

b) Write an equivalent function definition that represents the amount in the account in terms of the monthly growth rate, and state the monthly growth rate to the *nearest thousandth of a percent.*

Solution:

(A) $A(t) = 5000(1.04)^t$

$A(10) = 5000(1.04)^{10} \approx \$7401.22$

(B) $A(t) = 5000(1.00327374)^{12t}$, so the monthly rate is approximately 0.327%.

Explanation of steps:

(A) Write the function in the form $a(1+r)^t$, where a is the starting amount and r is the annual interest rate. Evaluate the function for the given value of t.

(B) For monthly growth, use the equivalent expression $a\left(\sqrt[12]{1+r}\right)^{12t}$ *[calculate* $\sqrt[12]{1.04} = 1.00327374$*].* Convert the decimal representation of the monthly growth rate into a percent.

PRACTICE PROBLEMS

1. Which function represents exponential decay? (1) $f(t) = 2\left(\frac{5}{4}\right)^t$ (2) $g(t) = \frac{1}{2}\left(\frac{5}{4}\right)^t$ (3) $h(t) = 2\left(\frac{4}{5}\right)^t$ (4) $j(t) = 2\left(\frac{5}{4}\right)^{\frac{t}{2}}$	2. An item purchased for \$500 depreciates in value by a rate of 25% every year. Find the value of the item after 5 years.

3. A tree grows 8% taller every 10 days. Given an initial height of one foot, calculate the height of the tree, to the *nearest foot*, after one year (365 days).	4. The level of salt in a compound is reduced by exposure to sulfur dioxide. If an initial amount of 250mg of salt is reduced by 15% every 3 hours, how much salt, to the *nearest milligram*, remains in the compound after 24 hours of exposure?
5. A device records the growth of a bacteria in a culture, from an initial amount of 5,000 cells. For the first few minutes, the number of bacteria has grown by 10% every 30 seconds. If this rate of growth continues, how many cells of bacteria will there be at the end of one hour?	6. An investment grows by 0.5% every two weeks. At the end of the first year (52 weeks), the value of the investment is $28,461.49. Find the original amount of the investment.

7. The radioactive isotope strontium-90 has a half-life of approximately 29 years. Write a function to model the amount of strontium-90 present after t years given an initial mass of 10 grams.

8. Sodium iodide-131 is a radioactive compound used to treat certain medical conditions. It has a half-life of 1.8 hours. If a person is given a dose of 278 MBq of sodium iodide-131, how much of the substance will remain in the body after 18 hours, to the nearest hundredth of an MBq?
   (Note: MBq stands for megabecquerels, a unit of measure of radioactivity.)

9. $V = 300(1.02)^t$ models exponential growth, where t represents the number of years.

   a) What is the annual rate of growth?

   b) Calculate the monthly rate of growth to the *nearest hundredth of a percent.*

10. The estimated population for a town is modeled by $P(t) = 2000(1.15)^t$, where 2000 is the current population and t is the number of years into the future.

    a) What is the annual rate of growth?

    b) Calculate the weekly rate of growth, to the *nearest hundredth of a percent.*

11. A bank account has a 3.3% annual interest rate, such that its value is $P(1.033)^t$, where P is the starting principal and t is the number of years. Write an equivalent expression in terms of m, the number of months. Round to the *nearest hundred-thousandth.*	12. $f(t) = 750(0.75)^t$ models an exponential decay of 25% per year, given an initial value of 750. Calculate the *daily* rate of decay, to the *nearest hundredth of a percent.*

10.6 Periodic Compound Interest

Key Terms and Concepts

Compound interest is a form of exponential growth used in finance. A principal amount is deposited into an account. Then, at regular intervals called *periods*, the account earns interest. In compound interest, the interest rate is applied not only to the principal amount, but also to any interest already added to the account.

If the period is one year, then the formula for annual compound interest is very familiar: $A = P(1+r)^t$, where A is the final amount, P is the starting principal, r is the annual interest rate, and t is the number of years.

Example: An account with an initial deposit of \$250 is compounded annually at a rate of 6%. A function for the amount in the account after t years is $f(t) = 250(1.06)^t$. For example, the value of the account after 5 years is $250(1.06)^t \approx \$334.56$.

However, instead of compounding once a year, suppose we want to compound every month, or every week, or even every day? In cases like these, we would use a more general formula for periodic compound interest.

The formula for **Periodic Compound Interest** is $A = P\left(1+\frac{r}{n}\right)^{nt}$, where A is the final amount, P is the starting principal, r is the *annual* interest rate, n is the number of compounding periods per year, and t is the number of years.

Example: Suppose a person deposits \$1,000 into an account with interest compounded monthly at an annual interest rate of 1.2% for 5 years. The final balance at the end of 5 years would be $\$1{,}000\left(1+\frac{0.012}{12}\right)^{(12)(5)} \approx \$1{,}061.80$..

Note that in finance, we are usually given a **nominal annual interest rate**. The word "nominal" means in name only, to differentiate it from the effective annual interest rate.

Example: In the above example, we are given a nominal annual interest rate of 1.2%. Since the interest is compounded monthly and there are 12 months in a year, we can calculate the *periodic* (monthly) rate by dividing 1.2% by 12, which is a 0.1% monthly interest rate. This periodic rate is expressed as $\frac{r}{n}$ in the Periodic Compound Interest formula.

The **effective annual interest** rate is the *actual* rate of growth earned by compounding periodically for one year. When the compounding period is less than one year, the effective annual interest rate will always be greater than the nominal annual rate and is calculated as $\left(1+\frac{r}{n}\right)^n - 1$, where r is the nominal annual rate and n is the number of periods.

Example: In the above example, for a nominal rate of 1.2% compounded monthly, the effective annual rate is $\left(1+\frac{0.012}{12}\right)^{12} - 1 = 0.01206622$, or only very slightly more than the nominal rate.

In the previous example, the difference between the nominal and effective annual rates was very small. However, for higher interest rates, the difference can be more significant.

Example: A bank loan has an annual interest rate of 26%, compounded weekly. The effective annual interest rate for the loan is $\left(1+\frac{0.26}{52}\right)^{52} - 1 \approx 0.29609$, or approximately 29.6%. Therefore, a bank loan of \$1000 at this rate would require a repayment of about \$1,296 after one year (*not* \$1260, as one might expect given the nominal rate of 26%).

Note that the formula for calculating an amount using monthly compound interest is *not* the same formula as estimating the monthly rate of growth based on an annual rate.

Estimating monthly growth: $f(t) = P\left(1.05^{\frac{1}{12}}\right)^{12t}$

Monthly compound interest: $f(t) = P\left(1+\frac{0.05}{12}\right)^{12t}$

More generally, the formula $f(t) = P\left(1+\frac{r}{n}\right)^{nt}$ can be used for any function involving **periodic growth or decay**, where r is the annual rate of growth ($r > 0$) or decay ($r < 0$).

Example: Candace has a collection of 1000 jewelry beads. Every month, she plans to purchase beads to increase the size of her collection by 20%. How many beads would she have after two years?

$$f(t) = 1000\left(1+\frac{0.20}{12}\right)^{12\cdot 2} \approx 1487$$

The above compound interest formulas assumed only one principal deposit at the start. Suppose we also want to add a series of equal deposits at the end of each period? These are called **regular contributions**. The value, V, on *just these contributions* (including interest) can be calculated using the formula:

$$V = \frac{c - c\left(1 + \frac{r}{n}\right)^{nt}}{1 - \left(1 + \frac{r}{n}\right)}, \text{ where } c \text{ is the periodic contribution.}$$

If an account includes *both* an initial principal deposit and regular contributions, calculate both A and V above and add the results, $FV = A + V$. In this formula, $A = P\left(1 + \frac{r}{n}\right)^{nt}$ calculates the amount in the account based on compounding the *principal only*, and V is the additional amount added to the account based on *regular contributions* (including interest).

MODEL PROBLEM

Suppose a person deposits \$1,000 into an account with interest compounded monthly at an annual interest rate of 1.2% for 5 years. Find the balance at the end of 5 years. How does this compare to compounding annually?

Solution:

(A) $\$1{,}000\left(1 + \frac{0.012}{12}\right)^{(12)(5)} \approx \$1{,}061.80$

(B) $\$1{,}000(1.012)^5 \approx \$1{,}061.46$

(C) Compounding monthly earns 34 cents more over 5 years.

Explanation of steps:

(A) Use the periodic compound interest formula $A = P\left(1 + \frac{r}{n}\right)^{nt}$.

[$P = 1{,}000$, $r = 1.2\% = 0.012$, $n = 12$ (12 months per year), and $t = 5$ (5 years)]

(B) Use the annual interest formula $A = P(1 + r)^t$.

(C) State the difference.

PRACTICE PROBLEMS

1. A deposit of \$500 earns 4% annual interest, compounded *monthly.* Calculate the amount in the account after 3 years.	2. A deposit of \$500 earns 4% annual interest, compounded *daily.* Calculate the amount in the account after 3 years. (Assume 365 days per year.)
3. \$2,000 is placed into a savings account with a 3% annual interest rate, compounded quarterly. a) What is the quarterly interest rate? b) Find the balance in the account after 5 years.	4. \$850 is placed into a savings account with a 5% annual interest rate, compounded monthly. a) What is the monthly interest rate, to the *nearest hundredth of a percent?* b) Find the balance in the account after 21 months.

5. An account with an initial deposit of \$2,000 earns 2% *quarterly* interest (every three months). a) Write a function $f(t)$, where t represents years, to model this growth. b) Find the value of the account, to the *nearest dollar*, after 18 months. c) Approximate the effective annual interest rate to the *nearest hundredth of a percent.*	6. A loan of \$100 is charged 0.1% *daily* interest compounded daily (assuming 365 days in a year). a) What is the nominal annual interest rate for the loan? b) If no payments are made throughout the year, what is the amount due after one year, to the *nearest dollar*? c) Approximate the effective annual interest rate to the *nearest percent.*
7. A savings account has an annual interest rate of 6.8% compounded monthly. If there is \$1403.60 in the account at the end of 5 years, how much was originally deposited?	8. How much would need to be deposited into an account earning 3.65% annual interest compounded daily so that the balance will be \$1,000,000 in 20 years?

9. Ann opens an account with a $1,000 initial deposit and then adds $200 per year for 5 years. The account has a 4% annual interest rate, compounded annually.

   a) Calculate the future value of the initial deposit only.

   b) Calculate the future value of the additional contributions only.

   c) Calculate the total future value.

10.7 Continuous Growth or Decay

Key Terms and Concepts

Instead of compounding interest at discrete periods, such as every year or every month or every day, suppose we wanted to compound continuously.

For this discussion, suppose we use a 100% annual interest rate for one year. Let's take a look at the expression $\left(1+\frac{r}{n}\right)^{nt}$, which for $r = 1.00$ and $t = 1$, becomes $\left(1+\frac{1}{n}\right)^{n}$.

The table below shows the value of this expression as the number of periods increase.

compounded	n	computation
annually	1	$\left(1+\frac{1}{1}\right)^{1} = 2$
semi-annually	2	$\left(1+\frac{1}{2}\right)^{2} = 2.25$
quarterly	4	$\left(1+\frac{1}{4}\right)^{4} = 2.44140625$
monthly	12	$\left(1+\frac{1}{12}\right)^{12} \approx 2.61303529022\ldots$
weekly	52	$\left(1+\frac{1}{52}\right)^{52} \approx 2.69259695444\ldots$
daily	365	$\left(1+\frac{1}{365}\right)^{365} \approx 2.71456748202\ldots$
hourly	8,760	$\left(1+\frac{1}{8760}\right)^{8760} \approx 2.71812669063\ldots$
every minute	525,600	$\left(1+\frac{1}{525600}\right)^{525600} \approx 2.7182792154\ldots$
every second	31,536,000	$\left(1+\frac{1}{31536000}\right)^{31536000} \approx 2.71828247254\ldots$

As n gets larger, we are getting closer and closer to compounding continuously. (Compounding every second isn't quite "continuously" – not even every nanosecond!)

As we get closer to continuous growth, the value of this expression approaches a constant which we call e. The constant e is an irrational number (like π) that is approximately:

2.71828182845904523536028747135266249775724709369995...

The constant is more formally called Euler's number, after Swiss mathematician Leonhard Euler, which is why the letter e is used.

CALCULATOR TIP

We can view an approximation of e on the calculator by pressing [2nd] [e] (the [÷] key).

The constant e is often called the **natural base**, and the function $f(x) = e^x$ is called the **natural exponential function**.

The graph of $y = \left(1 + \frac{1}{x}\right)^x$ for $x > 0$ has an asymptote at $y = e$, as shown below.

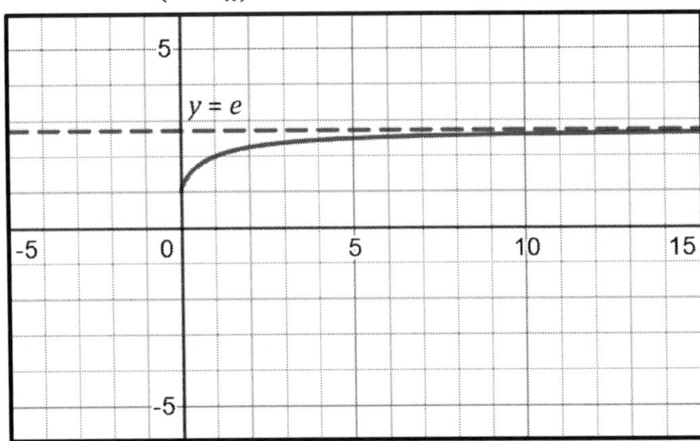

We can create a formula for **continuously compounded interest** using the natural base:

$A = Pe^{rt}$ where A is the final amount, P is the starting principal, r is the annual interest rate, and t is the number of years

More generally, this formula can be used for any function involving **continuous growth or decay**, such as population growth. Just replace A with $f(t)$, P with the initial positive value, and r with the rate of growth $(r > 0)$ or decay $(r < 0)$.

MODEL PROBLEM

How much money do you need to invest in an account that pays 2.75% interest compounded continuously in order to have \$50,000 at the end of 10 years?

Solution:

(A) $A = Pe^{rt}$

(B) $50{,}000 = Pe^{(0.0275)(10)}$

(C) $P = \dfrac{50{,}000}{e^{0.275}} \approx 37{,}979$

Explanation of steps:

(A) Start with the formula for continuously compounded interest.

(B) Substitute for the known variables.

(C) Solve for P.

PRACTICE PROBLEMS

1. Find the value of an account where $5,000 is invested for 4 years at an annual rate of 3% compounded continuously.

2. Find the value of an account where $550 is invested for 10 years at an annual rate of 6.6% compounded continuously.

3. You have $1,000 to deposit into a bank account that pays an annual interest of 2.5% for 3 years. Calculate the future value when compounding

   a) annually

   b) monthly

   c) continuously

4. How much money, to the *nearest dollar*, do you need to invest in an account that pays 5% interest compounded continuously in order to have $30,000 at the end of 6 years?

Chapter 11. Logarithms

11.1 General and Common Logarithms

KEY TERMS AND CONCEPTS

A **logarithm** is the power that a base must be raised to in order to get a value.

For a base of b, $\log_b x = y$ means $b^y = x$, where b and x are positive and $b \neq 1$.

Example: Since $2^3 = 8$, we can write $\log_2 8 = 3$. In other words, $\log_2 8$ means the power that we need to raise 2 to in order to get the value 8, which is 3.

It is often helpful to convert equations between logarithmic and exponential forms. To do so, remember to work in a counter-clockwise spiral order as shown below.

$\log_b x = y$ means $b^y = x$

and $b^y = x$ means $\log_b x = y$

Examples: $\log_3 81 = 4$ means $3^4 = 81$, and $5^x = 3125$ means $\log_5 3125 = x$.

From the definition of logarithm, the following rules apply:

a) $\log_b 1 = 0$ because $b^0 = 1$
b) $\log_b b = 1$ because $b^1 = b$
c) $\log_b b^x = x$ because $b^x = b^x$
d) $b^{\log_b x} = x$ because $b^y = x$ means $\log_b x = y$

It is not possible to evaluate the logarithm of zero or a negative number.

If the base of a logarithm is not written, then a base of 10 is assumed: $\log x$ means $\log_{10} x$. This is called the **common logarithm**.

CALCULATOR TIP

We can evaluate a common logarithm by using the calculator's [LOG] key.

Example: To calculate $\log 40$, press [LOG][4][0][)][ENTER]. $\log 40 \approx 1.602$.
This means $10^{1.602} \approx 40$.

Change of Base Formula:

We can evaluate a logarithm of any base b by changing it into an equivalent expression involving common logarithms using the conversion formula, $\log_b x = \frac{\log x}{\log b}$.

Examples: To find $\log_2 5$ to the *nearest ten-thousandth*, we can write
$\log_2 5 = \frac{\log 5}{\log 2} \approx 2.3219$.

CALCULATOR TIP

On the TI-84 calculator, you can calculate $\mathbf{log}_b\, \boldsymbol{x}$ for any base b using the logBASE function.

1. Press [ALPHA][F2][5] (or press [MATH][ALPHA][A]) to select the logBASE function.
2. Type the base (b), then press [▶] and type the value (x), then [▶] and [ENTER].

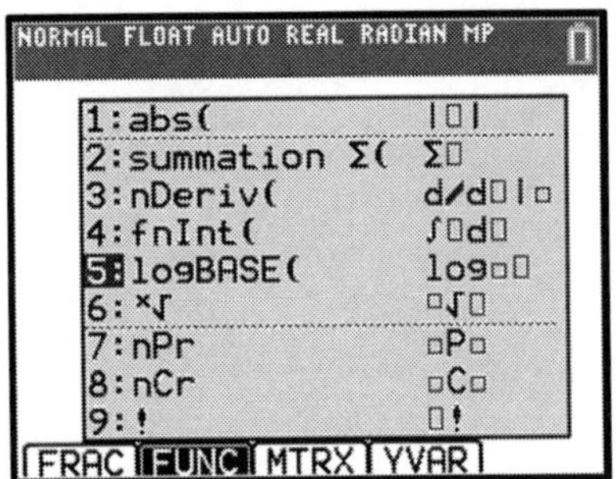

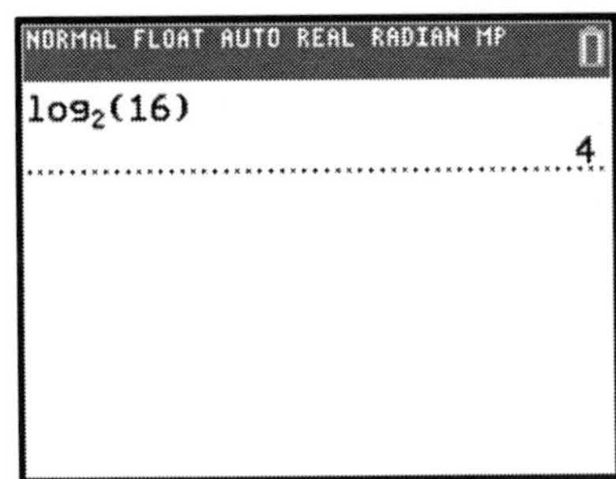

[Note to teachers: If you set calculators to Test Mode by holding down [◀][▶][ON] keys simultaneously, make sure that logBASE and summation Σ are not disabled.]

Unfortunately, the TI-83 calculator does not have the logBASE function, but we can still calculate $\mathbf{log}_b\, \boldsymbol{x}$ by using the change of base formula, $\log_b x = \dfrac{\log x}{\log b}$.

Example: To calculate $\log_2 16 = \dfrac{\log 16}{\log 2}$, type [LOG] [1][6][)] [÷] [LOG] [2][)] [ENTER].

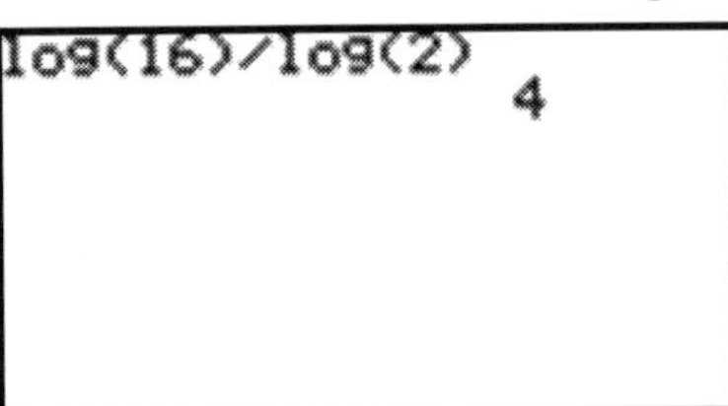

Some natural phenomena are measured using logarithmic scales. In a **logarithmic scale**, each magnitude represents the exponent of the value rather than the actual value. A logarithmic scale allows us to categorize a very large range of values in a compact way. Two well-known examples are the Richter Scale, used to measure the intensity of an earthquake, and the Decibel Scale, used to measure the intensity of a sound.

magnitude	0	1	2	3	4	5	6	7	8	9
intensity	1	10^1	10^2	10^3	10^4	10^5	10^6	10^7	10^8	10^9

For example, on a Richter Scale, a reading of a magnitude of 5 indicates a disturbance with ground motion that is 10 times larger than a reading of 4. We can calculate how many times more intense one earthquake is than another by the formula $\log\frac{I_1}{I_2} = M_1 - M_2$, where I represents the intensity and M represents the magnitude on the Richter Scale.

Example: The 1976 earthquake in China had a 7.5 magnitude and the 2004 earthquake in Indonesia had a 9.0 magnitude. Approximately how many times more intense was the Indonesian earthquake, to the *nearest whole*?

$\log\frac{I_1}{I_2} = M_1 - M_2$

$\log\frac{I_1}{I_2} = 9.0 - 7.5$

$\log\frac{I_1}{I_2} = 1.5$

$\frac{I_1}{I_2} = 10^{1.5} \approx 32$ times as strong *[by definition, $\log x = y$ means $x = 10^y$]*

For the Decibel Scale, an increase of a magnitude of 10 represents 10 times the intensity, so 50 dB is 10 times as intense as 40 dB. Therefore, the formula for calculating the ratio of intensity is $\log\frac{I_1}{I_2} = \frac{M_1 - M_2}{10}$.

Example: Normal conversation is measured as approximately 60 dB. The sound produced by Niagara Falls is about 90 dB. Approximately how many times more intense is the sound produced by Niagara Falls compared to normal conversation, to the *nearest whole*?

$\log\frac{I_1}{I_2} = \frac{M_1 - M_2}{10}$

$\log\frac{I_1}{I_2} = \frac{90 - 60}{10}$

$\log\frac{I_1}{I_2} = 3$

$\frac{I_1}{I_2} = 10^3 = 1{,}000$ times as loud

Model Problem

Solve for x: $\log_5 x = 4$

Solution:

By definition, $\log_5 x = 4$ means $x = 5^4$, so $x = 625$.

Practice Problems

1. Find the value of $\log_4 64$.	2. Find the value of $\log_5 \frac{1}{125}$.
3. Find the value of $\log_6 1$.	4. For what value of k will the graph of $y = \log_2 x$ contain the point $(1, k)$?
5. Solve for x: $\log_3 x = 4$.	6. If $\log_5 x = 2$, what is the value of $\sqrt{x}$?
7. Solve for x: $\log_2(x+1) = 3$	8. Solve for x: $\log_2(5x - 7) = 3$

9. Solve for x: $\frac{1}{2}\log_{10}(x+2) = 2$

10. Complete the table below for the values of y for the equation $y = \log_2 x$.

x	$\frac{1}{4}$	$\frac{1}{2}$	1	2	4
y					

11. How many times as intense is a magnitude 8 earthquake than a magnitude 5.5 earthquake, to the *nearest whole*? Use the formula $\log\frac{I_1}{I_2} = M_1 - M_2$, where I represents the intensity and M represents the magnitude on the Richter Scale.

12. Two sounds measure 85 dB and 47 dB. How much more intense is the louder sound, to the nearest whole? Use the formula $\log\frac{I_1}{I_2} = \frac{M_1 - M_2}{10}$, where I represents the intensity and M represents the magnitude on the Decibel Scale

11.2 Graphs of Log Functions

Key Terms and Concepts

The graph of $y = \log_2 x$ is shown below.

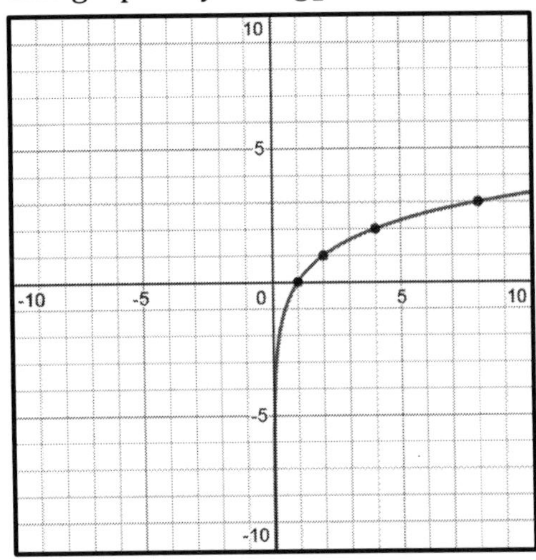

Some points on the graph of $y = \log_2 x$ include:

(1,0)	$\log_2 1 = 0$ because $2^0 = 1$
(2,1)	$\log_2 2 = 1$ because $2^1 = 2$
(4,2)	$\log_2 4 = 2$ because $2^2 = 4$
(8,3)	$\log_2 8 = 3$ because $2^3 = 8$

Note that the function $y = \log_b x$ for any positive b has a domain of $(0, \infty)$ and an asymptote at $x = 0$. It has an x-intercept at (1,0) because $\log_b 1 = 0$ for all b (that is, $b^0 = 1$ for all b). The range of the function includes all real numbers.

To the right is the graph of $y = \log_{\frac{1}{2}} x$.

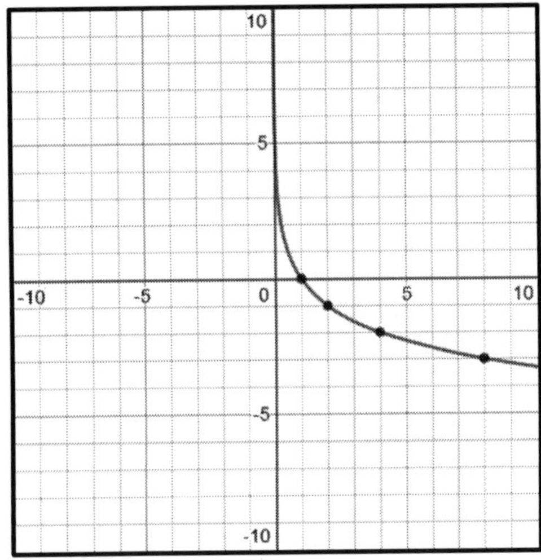

Note that for $0 < b < 1$, the graph will decrease across its entire domain, $x > 0$. Some points include $(1,0)$, $(2,-1)$, $(4,-2)$, $(8,-3)$, etc. We know, for example, that $(8,-3)$ is on the graph because $\log_{\frac{1}{2}} 8 = -3$; that is, $\left(\frac{1}{2}\right)^{-3} = 2^3 = 8$.

Here are the graphs of some other parent log functions:

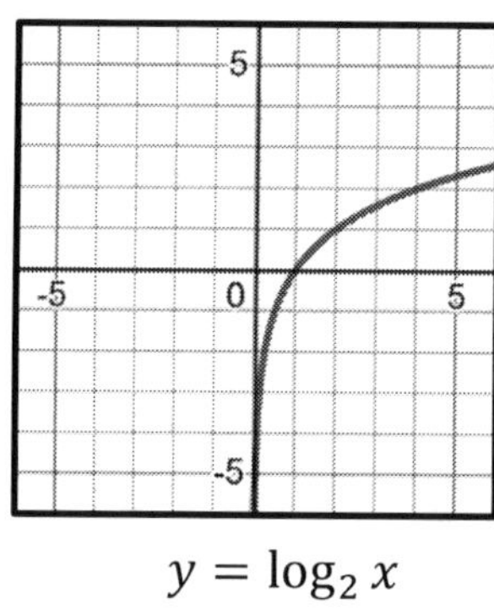

$y = \log_2 x$

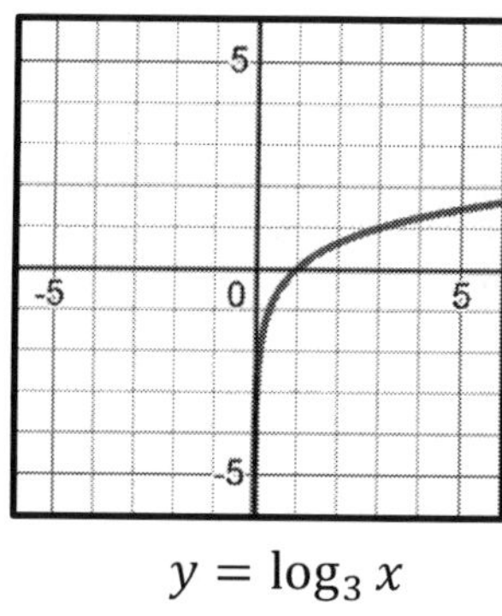

$y = \log_3 x$

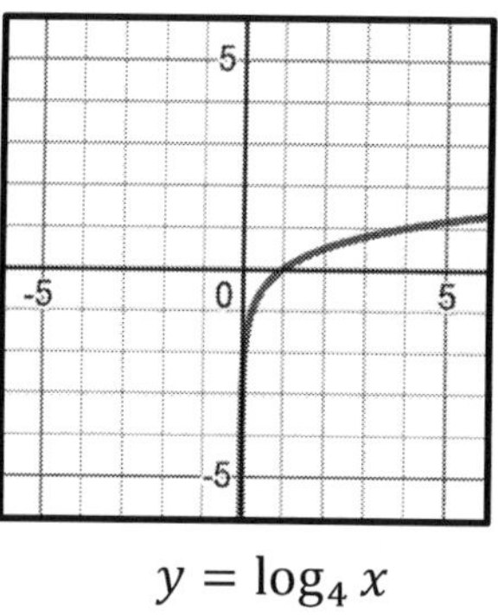

$y = \log_4 x$

Note that all three graphs include the point (1,0). However, one way we can tell the difference is that $y = \log_2 x$ includes (2,1), $y = \log_3 x$ includes (3,1), and $y = \log_4 x$ includes (4,1).

A logarithmic function $f(x)$ may be transformed in any of the following ways:

$f(x) + k$	$(x, y) \to (x, y + k)$	vertically shifts the graph up ($k > 0$) or down ($k < 0$)
$f(x + k)$	$(x, y) \to (x - k, y)$	horizontally shifts the graph left ($k > 0$) or right ($k < 0$)
$-f(x)$	$(x, y) \to (x, -y)$	reflects the graph over the x-axis
$f(-x)$	$(x, y) \to (-x, y)$	reflects the graph over the y-axis
$k \cdot f(x)$	$(x, y) \to (x, ky)$	vertically stretches by a factor of k
$f(kx)$	$(x, y) \to (kx, y)$	horizontally stretches by a factor of $\frac{1}{k}$

For example, here are two **translations** of $f(x) = \log_2 x$.

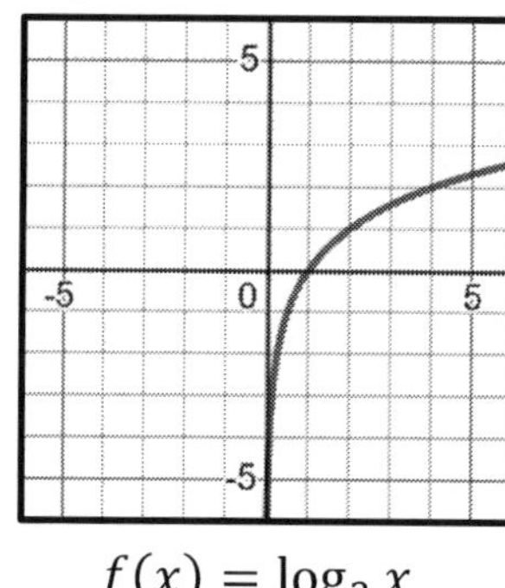

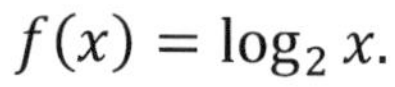
$f(x) = \log_2 x$.

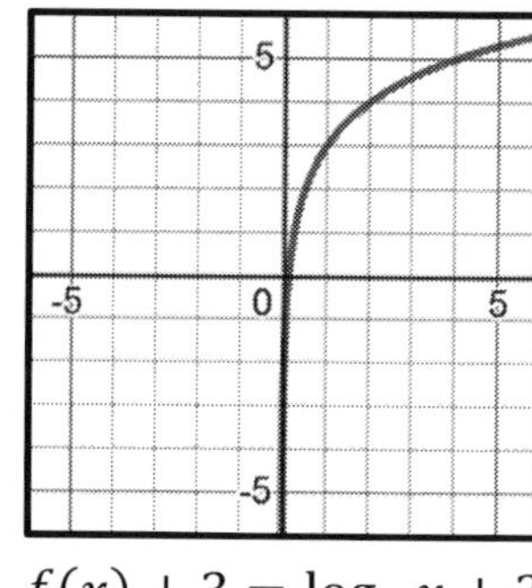

$f(x) + 3 = \log_2 x + 3$
shifted up 3

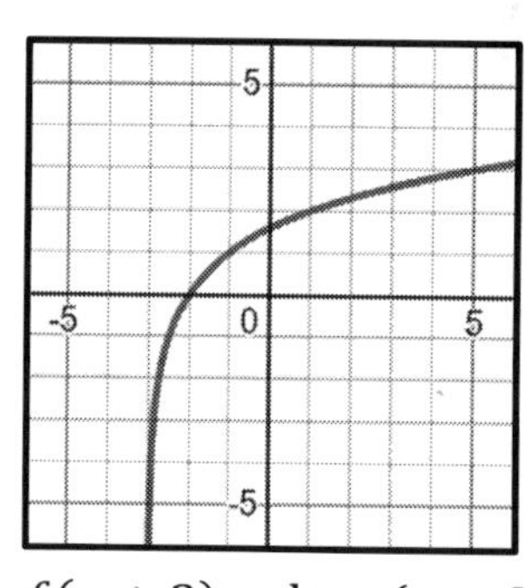

$f(x + 3) = \log_2(x + 3)$
shifted left 3

Here are two **reflections** of $f(x) = \log_2 x$.

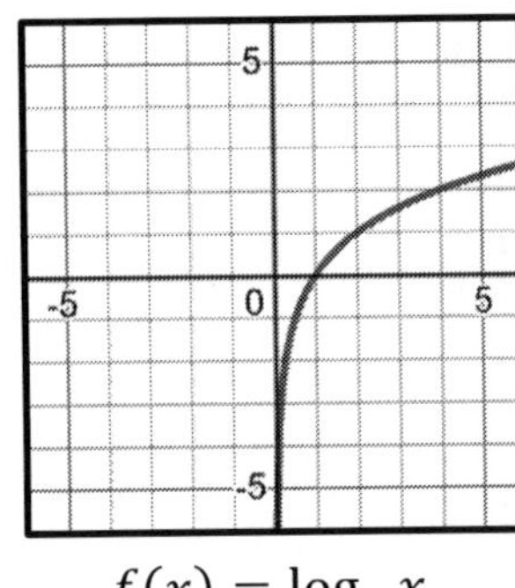

$f(x) = \log_2 x.$

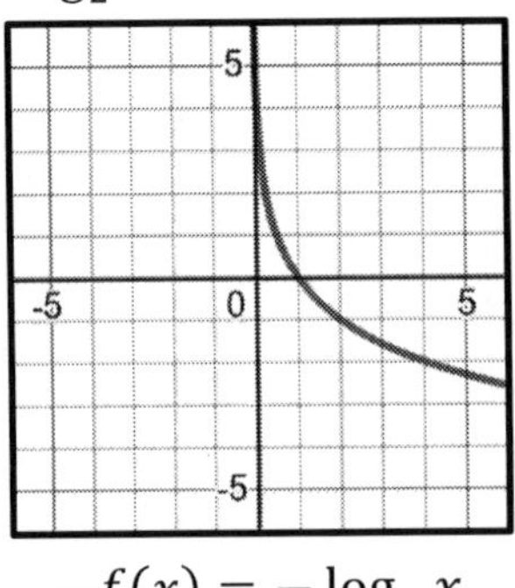

$-f(x) = -\log_2 x$

reflected over x-axis

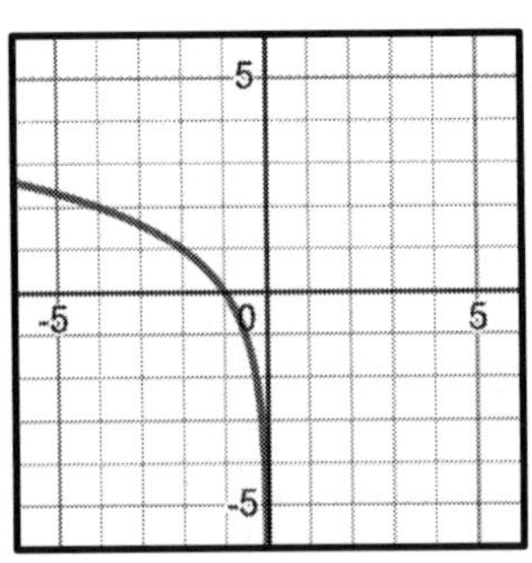

$f(-x) = \log_2(-x)$

reflected over y-axis

Here are two **stretches** of $f(x) = \log_2 x$.

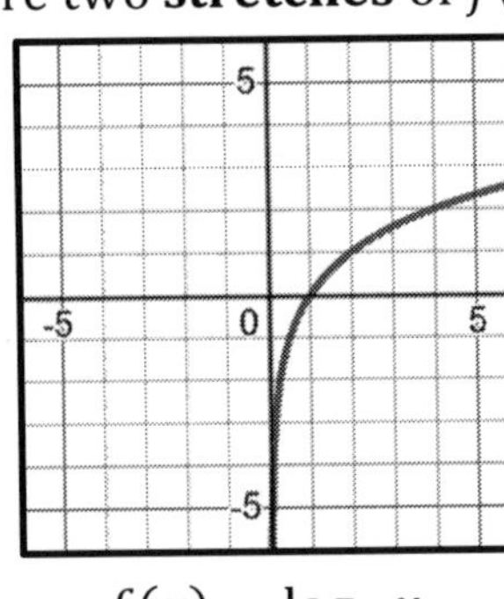

$f(x) = \log_2 x.$

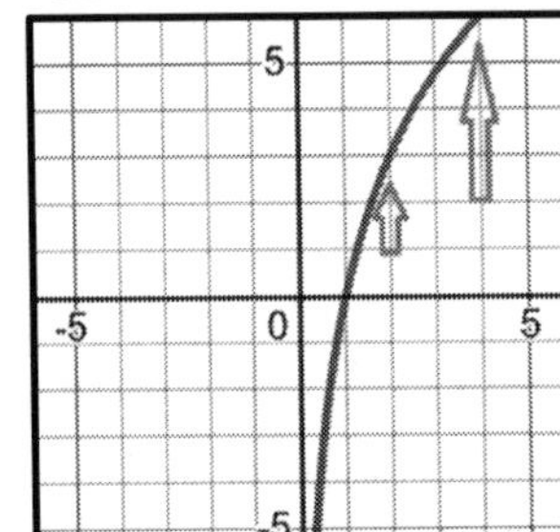

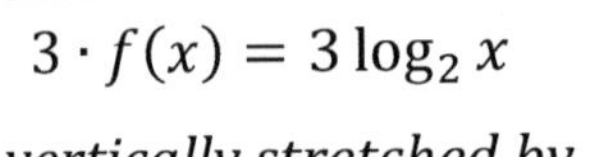

$3 \cdot f(x) = 3 \log_2 x$

vertically stretched by 3

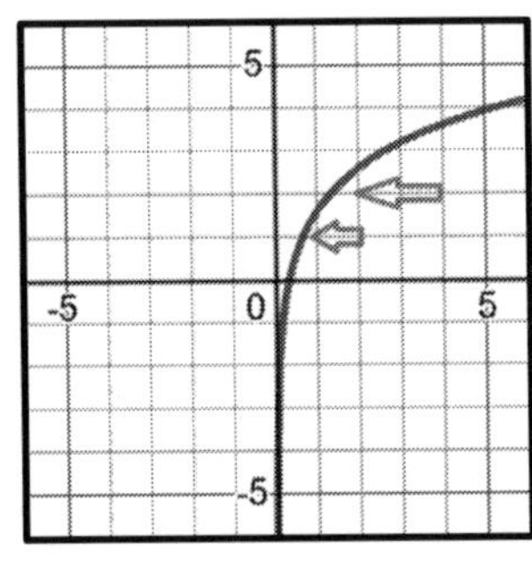

$f(3x) = \log_2(3x)$

horizontally stretched by $\frac{1}{3}$

MODEL PROBLEM 1: ***SEQUENCE OF TRANSFORMATIONS***

Given $f(x) = \log_3 x$, write the definition of $g(x)$, the image of $f(x)$ after a vertical stretch by a factor of 2 and then a translation 4 units to the right. Sketch a graph of $g(x)$.

Solution:

(A) (B)

$g(x) = 2\log_3(x - 4)$

(C)

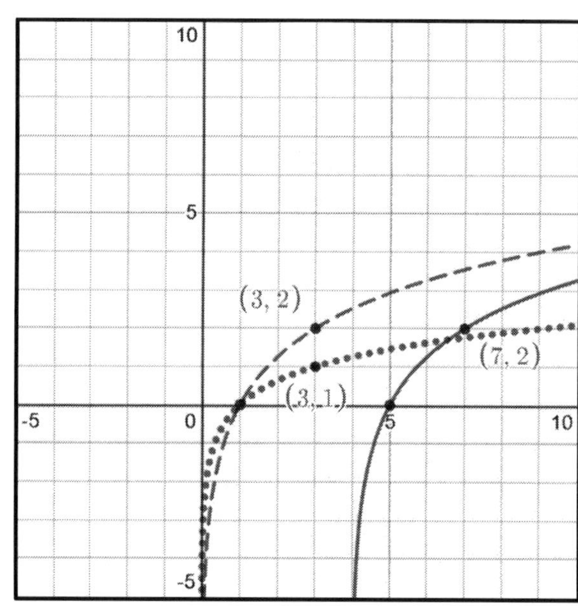

$f(x)$ is the dotted graph. $g(x)$ is the solid graph.
The dashed graph is the image of $f(x)$ after the vertical stretch only.

Explanation of steps:

(A) To vertically stretch $f(x)$ by a factor of k, multiply the function by k.
[Multiply $log_3\, x$ by 2.]

(B) To translate a function horizontally by h units, add h (to go left) or subtract h (to go right) inside parentheses. *[Move 4 units right by subtracting 4 in the parentheses.]*

(C) To graph the image of a function, it is helpful to graph the images of a few points.
[Let's look at two points, $(1,0)$ *and* $(3,1)$ *on the dotted graph of* $f(x)$*:*
For $(1,0)$*, a vertical stretch of 2 maps* $(x, y) \rightarrow (x, 2y)$*, which maps* $(1,0)$ *to itself, and then a translation 4 units right maps* $(x, y) \rightarrow (x + 4, y)$*, yielding an image of* $(5,0)$*.*
For $(3,1)$*, a vertical stretch of 2 maps* $(x, y) \rightarrow (x, 2y)$*, which maps* $(3,1)$ *to* $(3,2)$*, shown plotted on the dashed graph, and then a translation 4 units right maps* $(x, y) \rightarrow (x + 4, y)$*, yielding an image of* $(7,2)$*.*
Sketch the resulting curve through $(5,0)$ *and* $(7,2)$*.]*

Practice Problems

1. Match all four equations with their graphs.

   a) $y = \log_3 x$ ____

   b) $y = 3^x$ ____

   c) $y = \log_{0.25} x$ ____

   d) $y = 0.25^x$ ____

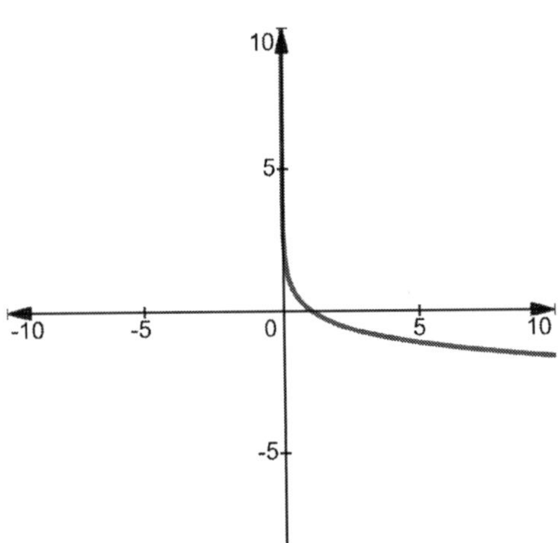

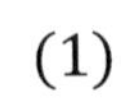

(1)

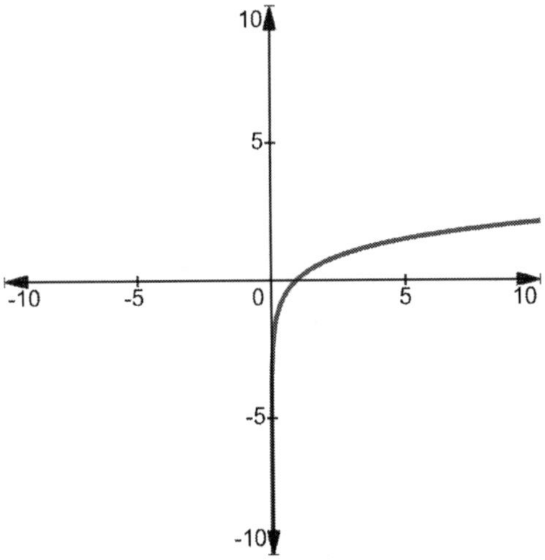

(3)

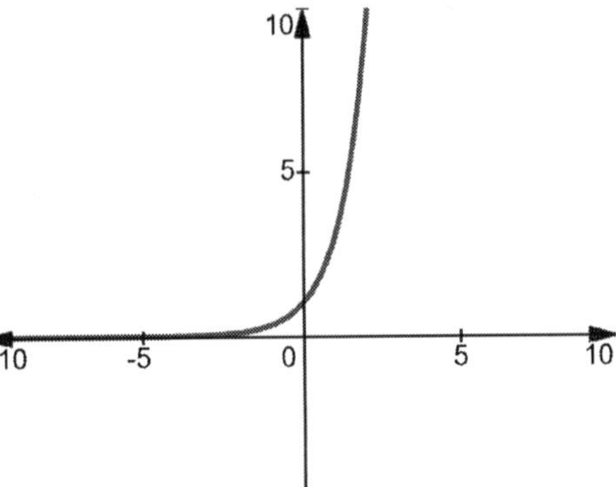

(2)

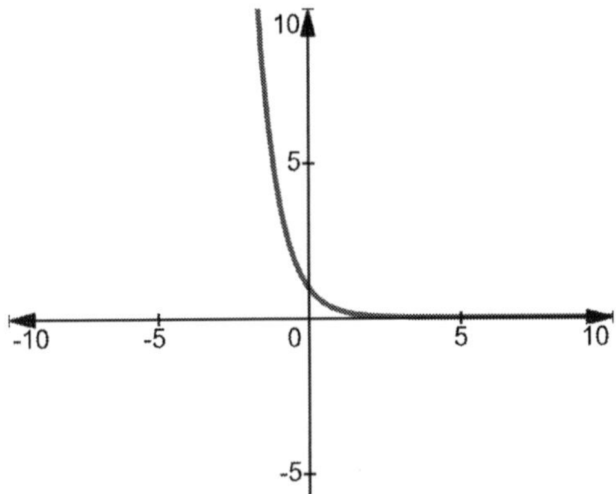

(4)

2. Given $f(x) = \log x$, write the definition of $g(x)$, the image of $f(x)$ after a translation 5 units to the left and 2 units down.	3. Given $f(x) = \log_2 x$, write the definition of $g(x)$, the image of $f(x)$ after a vertical stretch by a factor of $\frac{1}{2}$ and a translation 3 units up.
4. Sketch the graph of $y = \log_2(x - 1) + 2$. 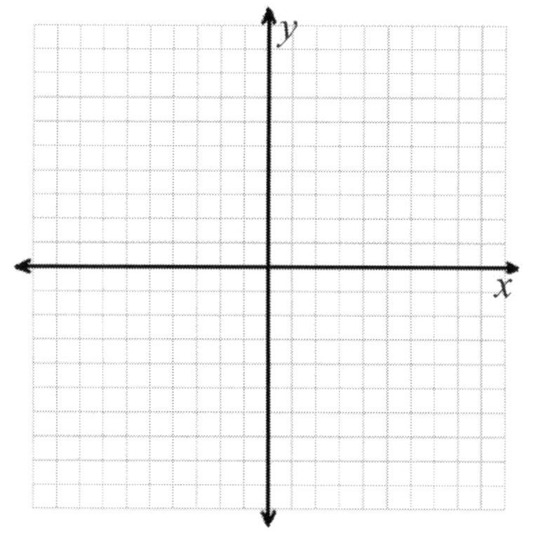	5. Sketch the graph of $y = 2\log_2(x + 2)$. 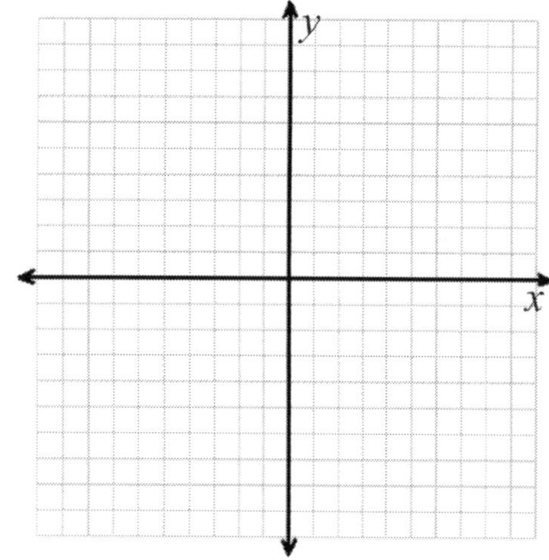

MODEL PROBLEM 2: *FIND INTERCEPTS*

To the right is a graph of $f(x) = \log_2 x$.

a) On the same set of axes, graph $g(x) = \log_2(x+5) - 2$.

b) Find the y-intercept, if one exists, for $g(x)$.

c) Find the x-intercept for $g(x)$.

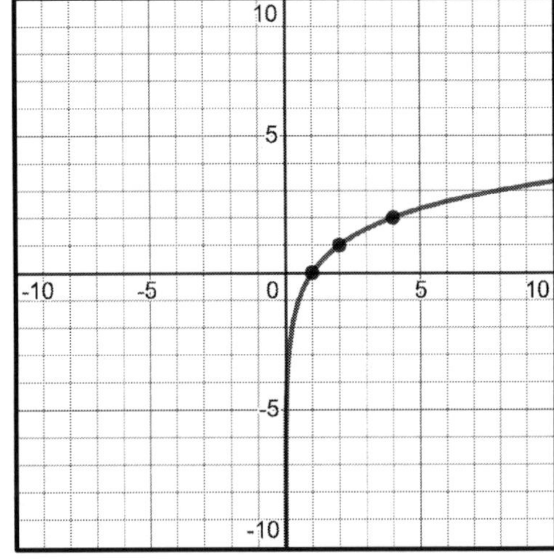

Solution:

(A)

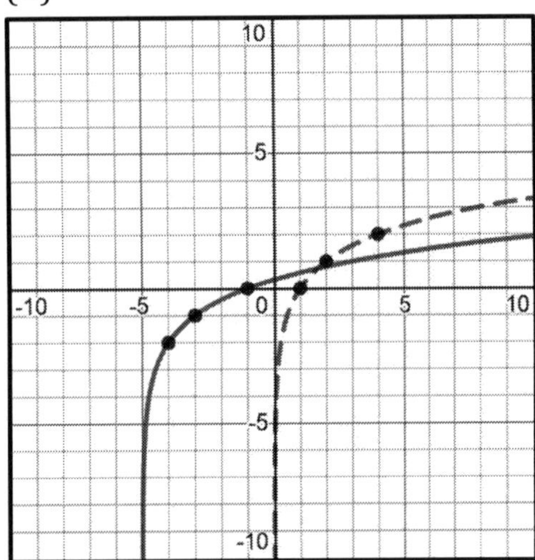

(B) y-intercept at $(0, 0.32)$

(C) x-intercept at $(-1, 0)$

Explanation of steps:

(A) Map the images of at least three selected points of the pre-image. *[The image is shifted 5 units left and 2 units down. So,* $(1,0) \to (-4,-2)$, $(2,1) \to (-3,-1)$, *and* $(4,2) \to (-1,0)$.*]* Copy the shifted graph through these points. *[Note that the vertical asymptote of the image has also shifted 5 units left to* $x = -5$.*]*

(B) We can find the y-intercept by evaluating the function for $x = 0$.
[$y = log_2(0+5) - 2 = log_2 5 - 2 \approx 2.32 - 2 \approx 0.32$*]*

(C) We can find the x-intercept by setting $y = 0$ and solving for x.
[$0 = \log_2(x+5) - 2$

$2 = \log_2(x+5)$

$2^2 = x + 5$

$4 = x + 5$

$x = -1$*]*

Practice Problems

6. For the graph of $y = \log(x+1) + 2$, find the y-intercept, if it exists, and the x-intercept.	7. For the graph of $y = 2\log_2(x+5)$, find the y-intercept, if it exists, and the x-intercept.
8. For the graph of $y = \log x - 2$, find the y-intercept, if it exists, and the x-intercept.	9. For the graph of $y = \log_3(x-5) - 3$, find the y-intercept, if it exists, and the x-intercept.

11.3 Properties of Logarithms

Key Terms and Concepts

When working with expressions involving logarithms, we can use the following rules, known as the **Properties of Logarithms**.

	General Logarithms	Common Logarithms
Product Rule:	$\log_b(xy) = \log_b x + \log_b y$	$\log(xy) = \log x + \log y$
Quotient Rule:	$\log_b\left(\frac{x}{y}\right) = \log_b x - \log_b y$	$\log\left(\frac{x}{y}\right) = \log x - \log y$
Power Rule:	$\log_b x^y = y\log_b x$	$\log x^y = y\log x$

To **expand** a logarithmic expression, we would use the above rules from left to right. Conversely, to **condense** a logarithmic expression, we would use the above rules from right to left. Note that the bases of the logs must match.

Examples: (a) The expression $\log\frac{xy}{z}$ can be expanded to $\log x + \log y - \log z$.

(b) To condense the expression $\log 4 + \log 5$ into a single logarithm, we can write $\log 4 + \log 5 = \log(4 \cdot 5) = \log 20$.

The Properties of Logarithms are derived from the laws of exponents, as follows.

Product Rule:

1. Let $x = b^m$ and $y = b^n$.
2. So, by the law of exponents, $xy = b^{m+n}$.
3. By definition from line 2, $\log_b(xy) = m + n$.
4. By definition from line 1, $m = \log_b x$ and $n = \log_b y$.
5. Therefore, by substitution, $\log_b(xy) = \log_b x + \log_b y$.

Quotient Rule:

1. Let $x = b^m$ and $y = b^n$.
2. So, by the law of exponents, $\frac{x}{y} = b^{m-n}$.
3. By definition from line 2, $\log_b\left(\frac{x}{y}\right) = m - n$.
4. By definition from line 1, $m = \log_b x$ and $n = \log_b y$.
5. Therefore, by substitution, $\log_b\left(\frac{x}{y}\right) = \log_b x - \log_b y$.

Power Rule:

1. Let $x = b^m$.
2. So, by the law of exponents, $x^y = (b^m)^y = b^{ym}$.
3. By definition from line 2, $\log_b x^y = ym$.
4. By definition from line 1, $m = \log_b x$.
5. Therefore, by substitution, $\log_b x^y = y \log_b x$.

We can use properties of logarithms to derive the change of base formula: $\log_b m = \frac{\log m}{\log b}$.

$\log_b m \left(\frac{\log b}{\log b}\right)$	multiply $\log_b m$ by $\frac{\log b}{\log b}$, a form of 1
$= \frac{(\log_b m)(\log b)}{\log b}$	write as a single fraction
$= \frac{\log b^{\log_b m}}{\log b}$	by the Power Rule, $y \log b = \log b^y$
$= \frac{\log m}{\log b}$	by the rule, $b^{\log_b x} = x$

MODEL PROBLEM

Expand the logarithmic expression $\log 5(3)^x$ completely.

Solution:

$\log 5(3)^x =$

(A) $\log 5 + \log 3^x =$

(B) $\log 5 + x \log 3$

Explanation of steps:

(A) To expand a logarithmic expression completely, first rewrite logs of products as sums of logs and logs of quotients as differences of logs.

(B) Then, rewrite each log of a power as the product of the power times the log of the base.

PRACTICE PROBLEMS

1. Expand the expression $\log x^2y$ completely.	2. Expand $\log\frac{a^3}{b}$ completely.
3. Expand $\log 2xy^3$ completely.	4. Expand $\log\sqrt{xy}$ completely.
5. Expand $\log\left(4^{2x}\sqrt{y}\right)$ completely.	6. Expand $\log\frac{5(2x)^2}{(x+1)^3}$ completely.

7. Condense the expression $2\log_3 10 - \log_3 20$ into a single logarithm.	8. Condense the expression $3\log_b x + \log_b y - 2\log_b z$ into a single logarithm.
9. Condense the expression $\frac{1}{2}\log x - 2\log y$ into a single logarithm.	10. Condense the expression $\frac{2\log x}{3} + \frac{3\log y}{4}$ into a single logarithm.
11. Find the value of $\log 3{,}000 - \log 3$.	12. Find the value of $\log_2 8 + \log_2 2$.

13. The magnitude, R, of an earthquake can be represented by the formula $R = \log\left(\frac{I}{T}\right)$, where I is the intensity and T is the lower threshold. If the intensity is doubled, its magnitude can be represented by

(1) $\log I - \log T$

(2) $2(\log I - \log T)$

(3) $2\log I - \log T$

(4) $\log 2 + \log I - \log T$

14. The formula for annual compound interest is $A = P(1+r)^t$. Which equation represents $\log A$?

(1) $t\log P + \log(1+r)$

(2) $\log P + t\log(1+r)$

(3) $\log P + t\log 1 + r$

(4) $\log P + \log t + \log(1+r)$

11.4 Use Logarithms to Solve Equations

Key Terms and Concepts

One of the properties of equality that will help us to solve exponential equations is the fact that we can take the logarithms of both sides of an equation and maintain equality. In other words, if $x = y$, then $\log_b x = \log_b y$.

The converse of this is also true: if $\log_b x = \log_b y$, then $x = y$.

This allows us to use common logarithms to **solve exponential equations.**

Example: To solve $6^x = 40$, rounded to the *nearest ten-thousandth*, we can write

$\log 6^x = \log 40$ (take the common log of both sides)

$x \log 6 = \log 40$ (power-product rule for logs)

$x = \dfrac{\log 40}{\log 6}$ (isolate x)

$x \approx 2.0588$ (evaluate using the calculator)

You can check your answer by evaluating $6^{2.0588}$ on the calculator and finding that it is approximately equal to 40.

Note that, in the example above, $6^x = 40$ could have been written as $x = \log_6 40$. We were able to solve it by calculating $x = \frac{\log 40}{\log 6}$. This shouldn't come as a surprise; it follows from the change of base formula that we saw earlier: $\log_6 40 = \frac{\log 40}{\log 6}$.

The steps for solving an exponential equation using logarithms are:

1. Isolate the power.
2. Take the log of both sides.
3. Use the Power Rule to bring down the power.
4. Isolate the variable.
5. Use the calculator.

Example: To solve $3{,}000(4)^{3x} = 18{,}000$

$4^{3x} = 6{,}000$ *[isolate the power]*

$\log 4^{3x} = \log 6000$ *[take the log of both sides]*

$3x \log 4 = \log 6000$ *[use the Power Rule]*

$x = \dfrac{\log 6000}{3 \log 4} \approx 2.09$ *[isolate the variable and use the calculator]*

If the variable appears in two exponents, it may help to use the rules of exponents to isolate the power.

Example: To solve $5(1.75)^x = (1.75)^{5x}$,

$5 = \frac{(1.75)^{5x}}{(1.75)^x}$ *[move the powers to the same side]*

$5 = (1.75)^{4x}$ *[use the rule for exponents, $\frac{b^x}{b^y} = b^{x-y}$]*

$\log 5 = \log(1.75)^{4x}$ *[take the log of both sides]*

$\log 5 = 4x \log 1.75$ *[use the Power Rule]*

$x = \frac{\log 5}{4 \log 1.75} \approx 0.72$ *[isolate the variable and use the calculator]*

In some cases, it may not be easy to isolate the power, but we may still be able to isolate the variable after taking the log of both sides.

Example: To solve $1.602(8)^x = (3)^{2x}$,

$\log[1.602(8)^x] = \log[(3)^{2x}]$ *[take the log of both sides]*

$\log 1.602 + \log 8^x = \log 3^{2x}$ *[use the Product Rule]*

$\log 1.602 + x \log 8 = 2x \log 3$ *[use the Power Rule]*

$\log 1.602 = 2x \log 3 - x \log 8$ *[move the variables to the same side]*

$\log 1.602 = x(2 \log 3 - \log 8)$ *[pull out the variable as a common factor]*

$x = \frac{\log 1.602}{2 \log 3 - \log 8} \approx 4.00$ *[isolate the variable and use calculator]*

MODEL PROBLEM 1: ***VARIABLE IN ONE EXPONENT***

Solve for x to the *nearest ten-thousandth*: $3^{x+1} = 25$.

Solution:

(A) $\log 3^{x+1} = \log 25$

(B) $(x + 1) \log 3 = \log 25$

(C) $x + 1 = \frac{\log 25}{\log 3}$

(D) $x = \frac{\log 25}{\log 3} - 1 \approx 1.9299$

Explanation of steps:

(A) Take the log of both sides.

(B) Use the Power Rule, $\log b^x = x \log b$.

(C) Solve for x by isolating the variable.

(D) Use the calculator to find the solution.

PRACTICE PROBLEMS

1. Solve for *x* to the *nearest hundredth*: $2^x = 5$	2. Solve for *x* to the *nearest hundredth*: $16^x = 88$
3. Solve for *x* to the *nearest hundredth*: $13^x = 76$	4. Solve for *x* to the *nearest hundredth*: $2^x = \frac{3}{16}$
5. Solve for *x* to the *nearest hundredth*: $3(20)^x = 27$	6. Solve for *x* to the *nearest hundredth*: $3^x - 4 = 9$
7. Solve for *x* to the *nearest hundredth*: $2^{3x} = 7$	8. Solve for *x* to the *nearest thousandth*: $3^{2x-1} = 20$

9. Solve for x to the *nearest thousandth*: $3 \cdot (5^{x+1}) - 25 = 100$	10. The deer population at a land reserve is 1,200. If the population grows at 2.4% per week, how long will it take for the deer population to double?
11. Bruce deposits money into an account that provides a 4% annual interest rate compounded quarterly. How long does it take for the value of the account to double, to the *nearest tenth of a year*?	12. Clark drinks a cup of coffee containing 95mg of caffeine. Each hour, the caffeine in his system decreases by 10%. How long will it take until he has 5mg of caffeine?

13. Soria wants to invest \$8,000 into an account that pays a 6% annual interest rate compounded monthly. How long will it take before the account reaches a value of \$10,000, to the *nearest tenth of a year*?	14. Suppose Soria invests \$8,000 into an account that pays 7% annual interest compounded quarterly. How long will it take before the account reaches a value of \$10,000, to the *nearest tenth of a year*?

MODEL PROBLEM 2: *VARIABLE APPEARS IN TWO EXPONENTS*

Solve for *t*: $900(1.05)^t = 600(1.25)^t$

Solution:

(A) $1.5 = \dfrac{(1.25)^t}{(1.05)^t}$

(B) $1.5 = \left(\dfrac{1.25}{1.05}\right)^t$

(C) $\log(1.5) = \log\left(\frac{1.25}{1.05}\right)^t$

(D) $\log(1.5) = t\log\left(\frac{1.25}{1.05}\right)$

(E) $t = \dfrac{\log(1.5)}{\log\left(\frac{1.25}{1.05}\right)} \approx 2.33$

Explanation of steps:

(A) Move the powers to the same side. *[Also, divide both sides by 600.]*

(B) Use a rule of exponents to isolate the power. $\left[\frac{a^x}{b^x} = \left(\frac{a}{b}\right)^x\right]$

(C) Take the log of both sides.

(D) Use the Power Rule.

(E) Isolate the variable and use the calculator.

Practice Problems

15. Solve for *x* to the *nearest tenth*: $97.656(2)^x = (5)^x$	16. Solve for *x* to the *nearest tenth*: $551(5)^x = 5(9)^x$
17. Solve for *x*: $\frac{40{,}353{,}607}{7^x} = 7^{\frac{x}{2}}$	18. Solve for *x* to the *nearest tenth*: $3.72(18)^x = 5^{2x}$

11.5 Natural Logarithms

Key Terms and Concepts

The **natural base**, e, is an irrational number whose value is 2.71828.... It is often used to model functions of continuous exponential growth ($f(x) = e^x$) or continuous exponential decay ($f(x) = e^{-x}$).

A **natural logarithm** is a logarithm that uses e as its base. It is usually written as **ln**.
Example: $\ln 5$ means the same as $\log_e 5$. So, $\ln 5 = x$ means $\log_e 5 = x$, or $e^x = 5$.

The **natural logarithmic function** may be written as $f(x) = \log_e x$ or as $f(x) = \ln x$. Below is a graph of this function.

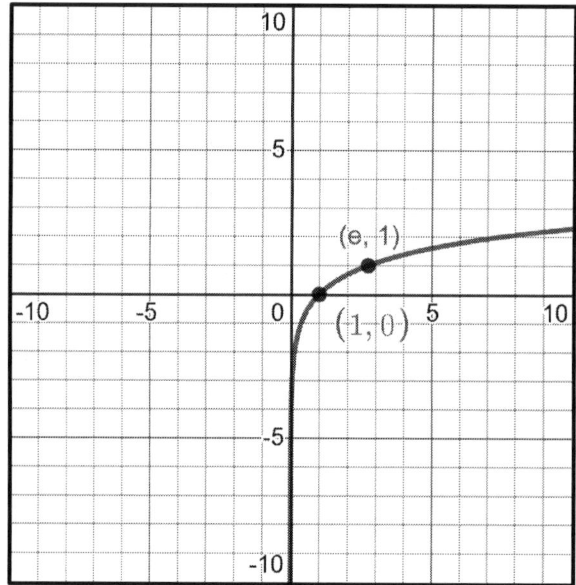

Note that the following rules apply to natural logs:

a) $\ln 1 = 0$ because $e^0 = 1$
b) $\ln e = 1$ because $e^1 = e$
c) $\ln e^x = x$ because $e^x = e^x$

We will use this last rule to solve equations involving e.

We can solve an equation of the form $e^x = y$ for x by taking the ln of both sides: $x = \ln y$.
Example: To solve $e^x = 10$

$\ln(e^x) = \ln 10$

$x = \ln 10 \approx 2.30$

CALCULATOR TIP

The natural logarithm can be evaluated on your calculator using the [LN] key.

Example: To calculate ln 10, enter [LN][1][0][)][ENTER] on the calculator. $x \approx 2.30$.

The properties of logarithms apply to natural logarithms, as well. Specifically:

Product Rule: $\ln(xy) = \ln x + \ln y$

Quotient Rule: $\ln\left(\frac{x}{y}\right) = \ln x - \ln y$

Power Rule: $\ln x^y = y \ln x$

The **Change of Base Formula** also applies to natural logarithms: $\log_b x = \frac{\ln x}{\ln b}$.

Just as we used common logarithms to solve exponential equations, we can use natural logarithms in the same way.

Example: We can solve $6^x = 40$ as follows.

$\ln 6^x = \ln 40$

$x \ln 6 = \ln 40$

$x = \frac{\ln 40}{\ln 6} \approx 2.0588$

We can use the properties of logarithms when solving equations.

Example: To solve $2 \ln x = \ln 25$, we know that $2 \ln x = \ln x^2$, so

$\ln x^2 = \ln 25$

$x^2 = 25$

$x = 5$ [reject $x = -5$ because $\ln x$ is only defined for $x > 0$.]

Natural logarithms can be used to solve exponential equations involving a base of e.

Example: To solve $4e^{3x} = 16$,

$e^{3x} = 4$ [isolate the power]

$\ln e^{3x} = \ln 4$ [take the natural log of both sides]

$3x \ln e = \ln 4$ [use the Power Rule]

$3x = \ln 4$ [$\ln e = 1$]

$x = \frac{\ln 4}{3} \approx 0.4621$ [isolate the variable]

Natural logs can often be used to solve problems involving continuous growth or decay. One common application of continuous exponential decay is Newton's Law of Cooling, $T(t) = (T(0) - a)e^{-rt} + a$, where $T(t)$ is the temperature at time t, a is the ambient (room) temperature, and r is the rate of decay.
[You are not expected to memorize this function.]

Model Problem

The temperature of a cup of coffee is modeled by the function $T(t) = 110e^{-0.045t} + 70$ where $T(t)$ is the temperature in degrees Fahrenheit after t minutes. The graph of $T(t)$ is shown below.

a) What is the temperature of the coffee at $T(0)$?
b) What is the temperature of the coffee after 10 minutes?
c) How long will it take for the coffee to cool down to 100°?

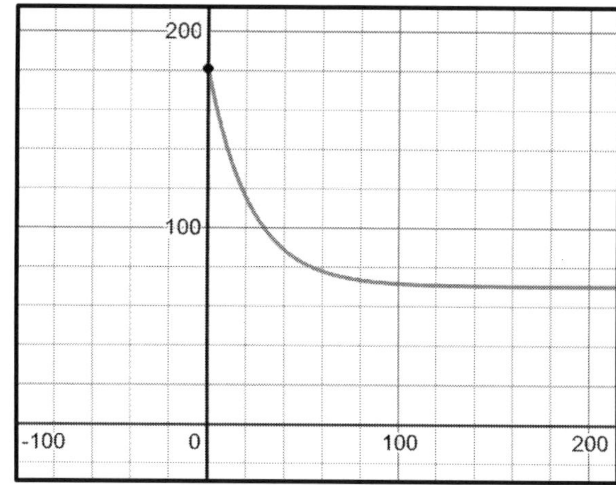

Solution:

(A) $110 + 70 = 180°$

(B) $T(10) = 110e^{-0.045(10)} + 70$
$\approx 140°$

(C) $100 = 110e^{-0.045t} + 70$
$30 = 110e^{-0.045t}$
$\frac{30}{110} = e^{-0.045t}$
$\ln\frac{30}{110} = -0.045t$
$-1.299 \approx -0.045t$
$t \approx 29$ minutes

Explanation of steps:

(A) The coefficient of e^{-rt} represents $T(0) - a$, where a is the constant ambient temperature. [We can tell from the function that the ambient temperature is 70° and the starting temperature is $110 + 70 = 180°$.]

(B) Evaluate $T(t)$ [where $t = 10$ minutes].

(C) To find how long it would take to reach 100°, substitute 100 for $T(t)$ and solve for t.

Practice Problems

1. Graph the function $h(x) = \ln e^x$ without using the graphing feature of your calculator.

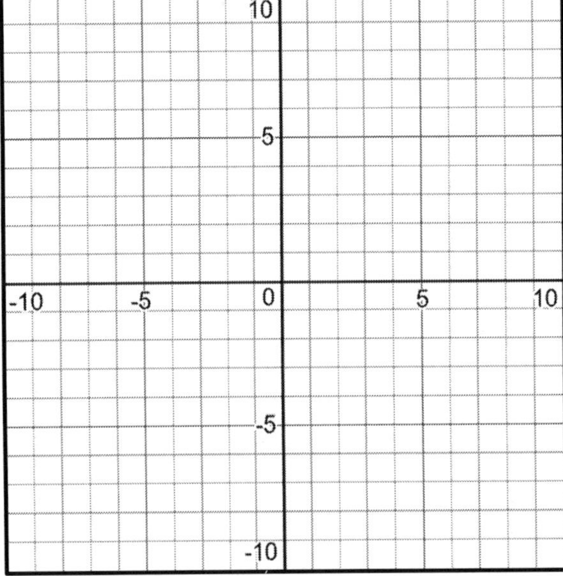

2. Solve for *x* to the *nearest hundredth*:
   $e^x = 15$

3. Solve for *x* to the *nearest hundredth*:
   $e^{2x} = 13$

4. Solve for *x* to the *nearest hundredth*:
   $4e^{2x} - 5 = 3$

5. Solve for *x* to the *nearest thousandth*:
   $2{,}050 = 1{,}500e^{4x}$

6. Solve for x: $2\ln x = \ln 16$	7. Solve for x: $\ln(2x-3) + \ln(x+2) = 2\ln x$
8. Solve for x to the *nearest hundredth*: $2\ln x = \ln 9 + 1$	9. Given $e^{kt} = 100^{2t}$, find the value of k to the *nearest hundredth*.

10. Kenny invests some money in an account that is compounded continuously. The value of his account increases by 25% over the course of 3 years. Find the annual interest rate of the account, to the *nearest tenth of a percent.*

11. Luigi invests $2500 in an account that is compounded continuously. What annual interest rate, to the *nearest tenth of a percent,* would be needed for the account to grow to $3000 in 5 years?

12. A cup of tea that is 100°F is placed in a room that is 60°F. According to Newton's Law of Cooling, the temperature of the tea decreases according to the function $f(t) = 40e^{-0.02877t} + 60$, where t is the time in minutes.

(a) Calculate the temperatures of the tea after 10 minutes and after 20 minutes.

(b) How long will it take to reach 70°F?

13. Marcus and Naomi both have retirement investments. Marcus' investment is worth $90,000 and is growing at a rate of 10% compounded continuously. Naomi's investment is worth $80,000 and is growing at a rate of 11% compounded continuously. After how many years, to the *nearest tenth of a year*, will the two investments have the same value?

11.6 Evaluate Loan Formulas

Key Terms and Concepts

Formulas involving loans can get quite complicated. For example, the following formula is used to calculate the monthly payment, M, for a mortgage. A **mortgage** is a loan for the purpose of buying real estate property, such as a home.

$$M = P \cdot \left(\frac{r(1+r)^n}{(1+r)^n - 1} \right)$$

In the formula above, M represents the fixed monthly payment, P represents the **principal** (the amount borrowed), n represents the number of months, and r represents the monthly interest rate.

If the annual interest rate is given instead, we can replace each r in the formula with $\frac{r}{12}$:

$$M = P \cdot \frac{\left(\frac{r}{12}\right)\left(1 + \frac{r}{12}\right)^n}{\left(1 + \frac{r}{12}\right)^n - 1},$$

where r is now the *annual* interest rate, but n is still the number of *monthly* payments.

For most mortgages, the buyer will make a **down payment** (an initial cash payment) and then borrow the remaining principal. For example, if a house is sold for \$400,000, the buyer may make a down payment of \$40,000 in cash. In this case, the mortgage principal will be the difference after the down payment is subtracted from the purchase price: $P = \$400{,}000 - \$40{,}000 = \$360{,}000$.

If we are given enough information, we can evaluate the formula to find an unknown value.

Example: Using the formula above, if a person borrows \$360,000 at a 0.25% monthly interest rate for 15 years, we can substitute the known values for the variables in order to find the monthly payment, rounded to the nearest dollar. In this case, $P = 360{,}000$, $r = 0.0025$, and $n = 15 \times 12 = 180$.

$$M = 360{,}000 \cdot \left(\frac{0.0025(1.0025)^{180}}{(1.0025)^{180} - 1}\right) \approx \$2{,}486$$

On the calculator, we can enter this expression as shown below.

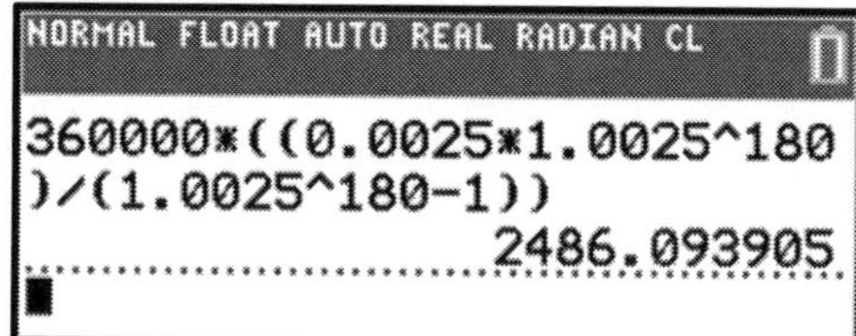

Note that we can now calculate the amount of interest that would be paid. Since we are paying \$2,486 per month for 180 months, the total payments would be \$447,480. Of this amount, \$360,000 is applied to the principal and the remaining \$87,480 is applied to the interest.

To find the number of months, n, using a formula like this, we will need the log function.

Example: Given the same formula and scenario above (\$360,000 borrowed at a 0.25% monthly interest rate), how many months would it take to pay off the mortgage if the borrower is able to pay \$3,000 per month?

$$3{,}000 = 360{,}000 \cdot \left(\frac{0.0025(1.0025)^{n}}{(1.0025)^{n} - 1}\right)$$

We need to solve for n.

$3{,}000(1.0025^n - 1) = 360{,}000(0.0025)(1.0025^n)$ Cross multiply

$1.0025^n - 1 = 120(0.0025)(1.0025^n)$ Divide by 3,000

$1.0025^n - 1 = 0.3(1.0025^n)$ Simplify

$0.7(1.0025^n) = 1$ Isolate 1.0025^n

$1.0025^n = \frac{10}{7}$

$$n = \log_{1.0025}\left(\frac{10}{7}\right) = \frac{\log\frac{10}{7}}{\log 1.0025} \approx 143 \text{ months}$$

There are other formulas that may be used to calculate loan repayments. Although the formulas may differ, the process for calculating is the same: substitute for all the known variables and solve for the unknown.

Model Problem

A homebuyer makes a \$25,000 down payment to purchases a home for a selling price of \$200,000. The buyer chooses a mortgage with a 0.35% monthly interest rate with monthly payments of \$1,750. To the nearest month, how many months will it take to pay the mortgage? Use the formula,

$$M = P \cdot \left(\frac{r(1+r)^n}{(1+r)^n - 1}\right),$$

where M is the monthly payment, P is the principal, n is the number of months, and r is the monthly interest rate.

Solution:

(A) $1{,}750 = 175{,}000 \cdot \left(\frac{0.0035(1.0035)^n}{(1.0035)^n - 1}\right)$

(B) $0.01 = \frac{0.0035(1.0035)^n}{(1.0035)^n - 1}$

(C) $0.01(1.0035^n - 1) = 0.0035(1.0035^n)$

(D) $0.01(1.0035^n) - 0.01 = 0.0035(1.0035^n)$

(E) $0.0065(1.0035^n) = 0.01$

(F) $1.0035^n = 1.53846$

(G) $\log 1.0035^n = \log 1.53846$

(H) $n \log 1.0035 = \log 1.53846$

(I) $n = \frac{\log 1.53846}{\log 1.0035}$

(J) $n \approx 123$ months

Explanation of steps:

(A) Substitute the known values for the variables.

(B) Divide both sides by the principal.

(C) Multiply both sides by the denominator.

(D) Distribute.

(E) Move terms with the variable n to the left and terms without to the right.

(F) Divide both sides by the factor.

(G) Take the log of both sides.

(H) Use the power rule to bring down the exponent ($\log x^a = a \log x$).

(I) Isolate n.

(J) Use the calculator to calculate the approximate value of n.

Practice Problems

1. Christina borrows \$300,000 to purchase a home. She signs a 30-year mortgage at a 0.25% monthly interest rate.

   a) Use the formula below to calculate her monthly payment to the *nearest dollar*.

   $$M = P \cdot \left(\frac{r(1+r)^n}{(1+r)^n - 1}\right)$$

   where M is the monthly mortgage payment, P is the principal amount of the loan, r is the monthly interest rate, and n is the number of monthly payments.

   b) By how much can she reduce her monthly payment if she makes a \$25,000 down payment?

2. Jack will borrow \$650,000 to purchase his home. He arranges for a 25-year mortgage at a 2.5% annual interest rate.

   Use the formula below to calculate his monthly payment to the *nearest dollar.*

$$M = P \cdot \frac{\left(\frac{r}{12}\right)\left(1 + \frac{r}{12}\right)^n}{\left(1 + \frac{r}{12}\right)^n - 1},$$

   where M is the monthly payment, P is the principal, r is the annual interest rate, and n is the number of monthly payments.

3. Maurice wants to purchase a home worth \$500,000. He can secure a 30-year mortgage with a 0.2% monthly interest rate. Monthly payments are calculated using the formula,

$$M = P \cdot \left(\frac{r(1 + r)^n}{(1 + r)^n - 1}\right)$$

   where M is the monthly mortgage payment, P is the principal amount of the loan, r is the monthly interest rate, and n is the number of monthly payments.

   How much of a down payment, to the *nearest dollar*, should Maurice make to keep his monthly payments below \$1,500 per month?

4. The Gibbs family borrows \$250,000 to purchase a home. They take out a mortgage with a 3% annual interest rate and will make monthly payments of \$1386.50, which is calculated using the formula,

$$M = P \cdot \frac{\left(\frac{r}{12}\right)\left(1+\frac{r}{12}\right)^n}{\left(1+\frac{r}{12}\right)^n - 1}$$

where M is the monthly payment, P is the principal, r is the annual interest rate, and n is the number of monthly payments.

How many years, to the *nearest year*, will it take pay off the mortgage?

Chapter 12. Trigonometric Functions

12.1 Trigonometric Ratios

Key Terms and Concepts

In addition to the trigonometric functions learned in Geometry – sine, cosine, and tangent – we can add three more. These functions, **cosecant**, **secant**, and **cotangent**, are the *reciprocals* of sine, cosine, and tangent, respectively.

Sine:	$\sin\theta = \dfrac{opp}{hyp}$	Cosecant:	$\csc\theta = \dfrac{hyp}{opp}$
Cosine:	$\cos\theta = \dfrac{adj}{hyp}$	Secant:	$\sec\theta = \dfrac{hyp}{adj}$
Tangent:	$\tan\theta = \dfrac{opp}{adj}$	Cotangent:	$\cot\theta = \dfrac{adj}{opp}$

A **trigonometric identity** is an equation involving trigonometric functions and angles.

Since the pairs of functions above are reciprocal functions, the following **reciprocal identities** are true, where the denominators are not zero:

$$\sin\theta = \frac{1}{\csc\theta} \qquad \csc\theta = \frac{1}{\sin\theta}$$

$$\cos\theta = \frac{1}{\sec\theta} \qquad \sec\theta = \frac{1}{\cos\theta}$$

$$\tan\theta = \frac{1}{\cot\theta} \qquad \cot\theta = \frac{1}{\tan\theta}$$

The following **quotient identities** are also true, where the denominators are not zero:

$$\tan\theta = \frac{\sin\theta}{\cos\theta} \qquad \cot\theta = \frac{\cos\theta}{\sin\theta}$$

We can show the quotient identities are true by the definitions of the functions.

Example: $$\frac{\sin\theta}{\cos\theta} = \frac{\frac{opp}{hyp}}{\frac{adj}{hyp}} = \frac{opp}{\cancel{hyp}} \cdot \frac{\cancel{hyp}}{adj} = \frac{opp}{adj} = \tan\theta.$$

CALCULATOR TIP

When working with angle measures, always make sure that you know whether your calculator is in Degree or Radian mode. For this chapter, it should be in Degree mode. To set it to Degree mode, enter [MODE] [Degree] [ENTER] [2nd] [QUIT].

Most calculators do not have buttons for the csc, sec and cot functions, so to calculate them, you will need to enter their reciprocal functions.

Example: To calculate $\cot 60°$, you would enter [1] [÷] [TAN] [6][0] [)].

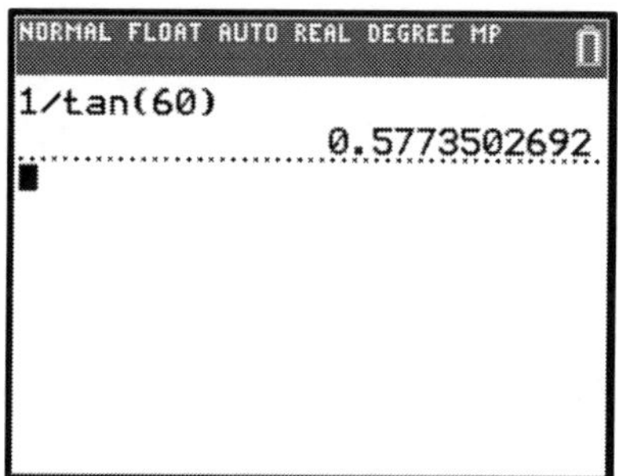

You may have noticed that the names of the functions are in pairs, with the letters "co" before one of each pair, as in sine and cosine, tangent and cotangent, and secant and cosecant. These pairs are fittingly called **cofunctions**. The "co" is short for "complementary," due to a special relationship between them: cofunctions of complementary angles are equal.

$$\sin x = \cos(90° - x)$$
$$\tan x = \cot(90° - x)$$
$$\sec x = \csc(90° - x)$$

Examples: $\sin 30° = \cos 60°$, $\tan 30° = \cot 60°$, and $\sec 30° = \csc 60°$.

For this course, it will be important to memorize the dimensions of two special triangles: the 30-60-90 right triangle and the 45-45-90 isosceles right triangle, both shown below.

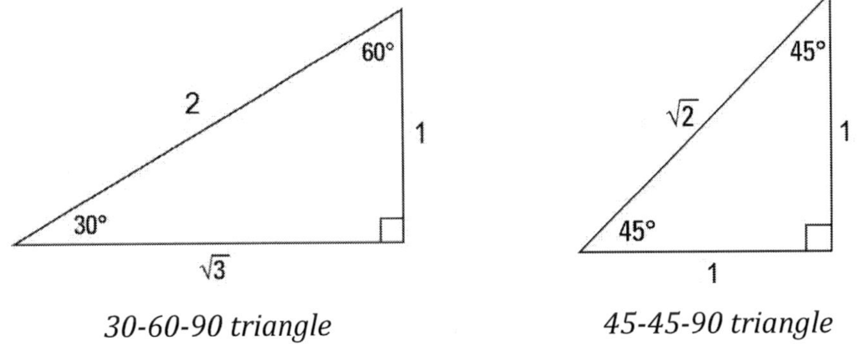

30-60-90 triangle *45-45-90 triangle*

From these diagrams, we can derive the sine, cosine, and tangent of their acute angles. These are shown in the table below left, with rationalized denominators:

	30°	45°	60°
sin	$\frac{1}{2}$	$\frac{\sqrt{2}}{2}$	$\frac{\sqrt{3}}{2}$
cos	$\frac{\sqrt{3}}{2}$	$\frac{\sqrt{2}}{2}$	$\frac{1}{2}$
tan	$\frac{\sqrt{3}}{3}$	1	$\sqrt{3}$

	30°	45°	60°
csc	2	$\sqrt{2}$	$\frac{2\sqrt{3}}{3}$
sec	$\frac{2\sqrt{3}}{3}$	$\sqrt{2}$	2
cot	$\sqrt{3}$	1	$\frac{\sqrt{3}}{3}$

We also know that the cosecant, secant, and cotangent of these angles are the reciprocals of the sine, cosine, and tangent, respectively. These are shown in the table above right.

Example: $\sin 45° = \frac{\sqrt{2}}{2}$, so $\csc 45° = \frac{2}{\sqrt{2}} = \sqrt{2}$.

Model Problem

For the diagram below, express $\sec T$ as a ratio in simplest form.

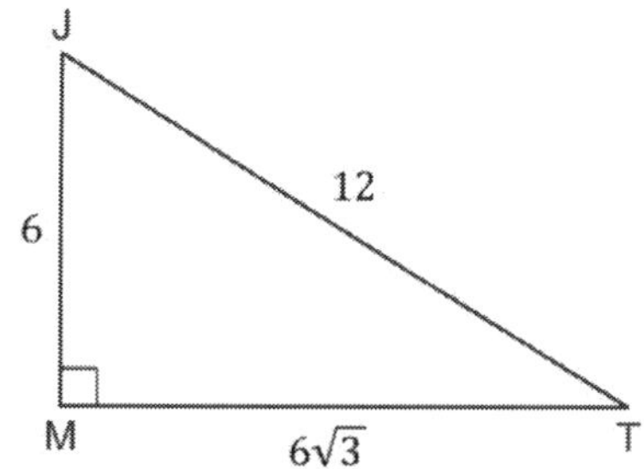

Solution:

(A) (B) (C) (D)

$$\sec T = \frac{hyp}{adj} = \frac{12}{6\sqrt{3}} = \frac{2}{\sqrt{3}} = \frac{2\sqrt{3}}{3}$$

Explanation of steps:

(A) Express the trigonometric function as a ratio of sides *[sec is the reciprocal of cos]*.

(B) Substitute values.

(C) Reduce.

(D) Rationalize the denominator.

Practice Problems

1. Find $\cot A$. 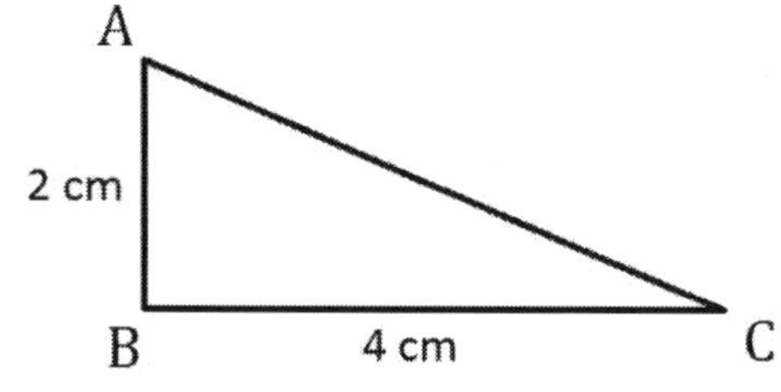 	2. Find $\csc A$. 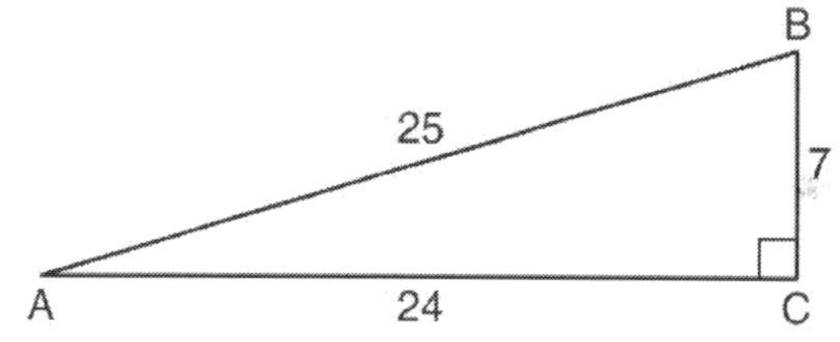
3. If $\sin 45° = \frac{\sqrt{2}}{2}$, find $\csc 45°$ in simplest radical form.	4. If $\sin\theta = \frac{1}{4}$ and $\cos\theta = \frac{\sqrt{15}}{4}$, find $\tan\theta$.

5. Find the exact value of $\cot 60°$ in simplest radical form.	6. Find the exact value of $\csc 60°$ in simplest radical form.
7. Find $\sec 35°$ to the *nearest thousandth.*	8. Find $\csc 35°$ to the *nearest thousandth.*
9. If $\sin 70° \approx 0.9397$, find $\cos 20°$ without using the calculator.	10. If $\sec 28° = \csc a$, find a.

11. A diagram of a regular hexagon, with side lengths of x, is shown below.

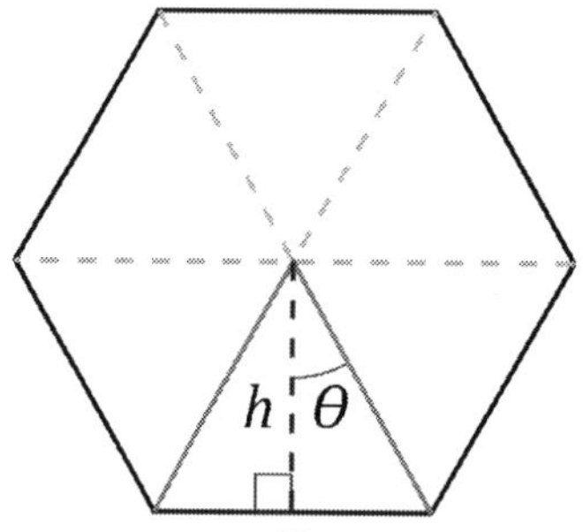

a) Find the measure of angle θ.

b) Write an expression for h in terms of x and $\cot\theta$.

c) Find $\cot\theta$ and substitute its value into the expression from part b.

d) Write an expression for the area of each of the six triangles shown.

e) Write an expression for the area of the regular hexagon.

12.2 Radians

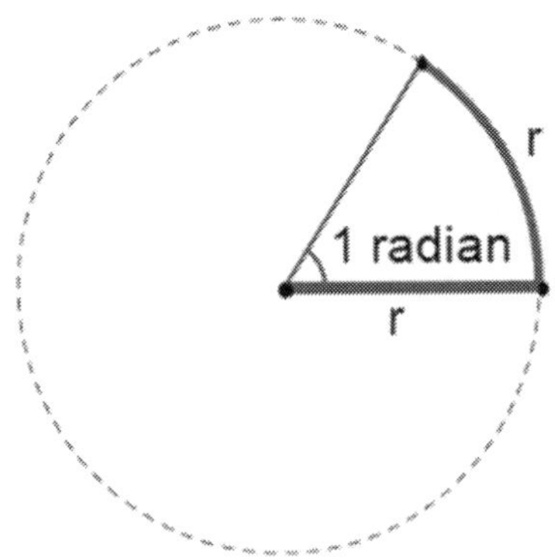

Key Terms and Concepts

Angles may be measured in degrees or in radians. A central angle of **one radian** intercepts an arc whose length is equal to the radius of the circle.

1 radian is approximately 57.2958°. The abbreviation for radians is rad.

The circumference of a circle is $2\pi r$, so there are 2π radians in a circle. So, 2π radians = 360°, and π radians = 180°. Therefore:

$$1 \text{ radian} = \frac{180}{\pi} \text{ degrees} \qquad 1 \text{ degree} = \frac{\pi}{180} \text{ radians}$$

To convert from radians to degrees, just multiply by $\frac{180°}{\pi \text{ rad}}$.

Example: To convert $\frac{\pi}{3}$ rad to degrees, $\frac{\cancel{\pi}}{3} \cancel{\text{rad}} \cdot \frac{180°}{\cancel{\pi} \cancel{\text{rad}}} = 60°$.

 CALCULATOR TIP

To convert from radians to degrees on the calculator:

While the calculator is in Degree mode, enter the radians in parentheses, then press [2nd][ANGLE][3][ENTER].

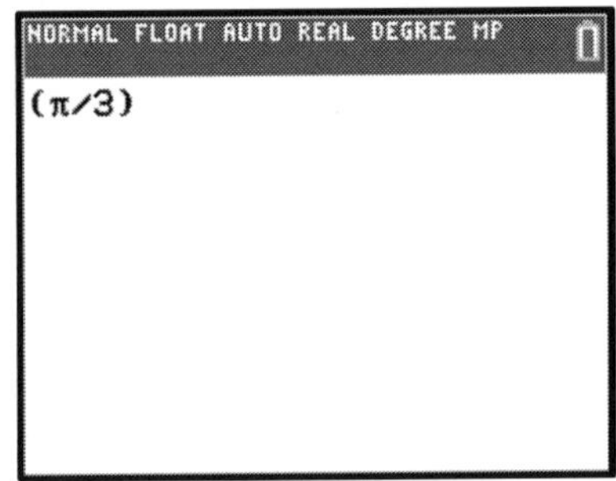

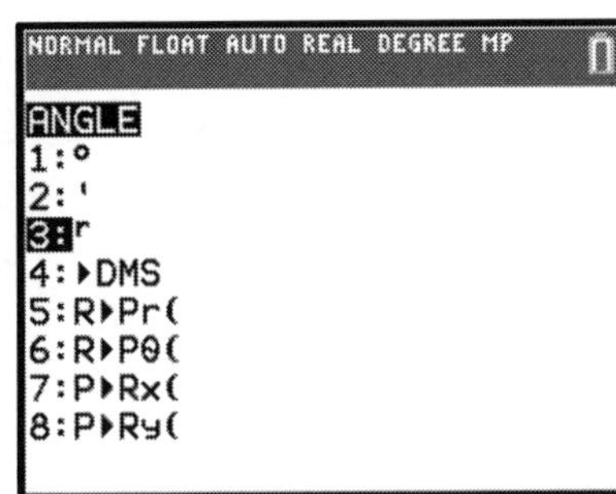

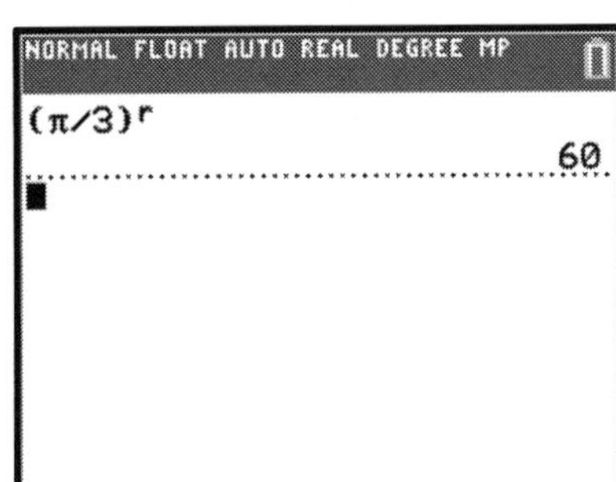

To convert from degrees to radians, just multiply by $\frac{\pi \text{ rad}}{180^\circ}$.

Example: To convert 60° to radians, $60^\circ \cdot \frac{\pi \text{ rad}}{180^\circ} = \frac{\pi}{3}$ rad.

CALCULATOR TIP

To convert from degrees to radians on the calculator:

1. Set your calculator to Radian mode temporarily. To do so, press [MODE] [Radian] [ENTER] [2nd] [QUIT].
2. Enter the number of degrees, and press [2nd][ANGLE][1][÷][2nd][π][MATH][1][ENTER]. The fractional answer will represent the factor that needs to be multiplied by π.
   [On the TI-84, you may enter the fraction using [ALPHA][F1][1] or [ALPHA][n/d], but on some models, you may still need to press [MATH][1] to show the result as a fraction.]
3. Set your calculator back to Degree mode.

Example: Entering 30, then the keystrokes above, will calculate that $30^\circ = \frac{1}{6}\pi$ radians.

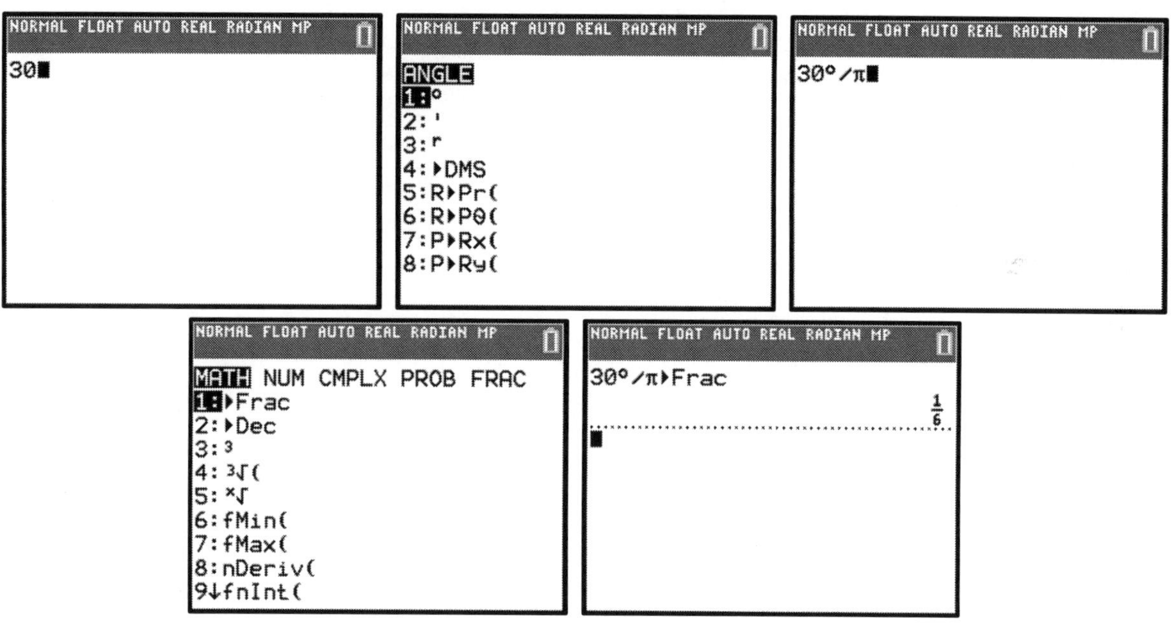

CALCULATOR TIP

No matter which mode your calculator is in, you can specify degrees or radians by using [ANGLE] on your calculator. When entering radian expressions (e.g., division) while in Degree mode, an extra set of parentheses is necessary around the radian measure.

Examples: a) In Radian mode, you can find sin 30° by entering [SIN][3][0][2nd][ANGLE][1][)][ENTER], as shown below left.

b) In Degree mode, you can find $\sin\frac{\pi}{6}$ by entering [SIN][ALPHA][F1][1][2nd][π][▼][6][▶][2nd][ANGLE][3][)][ENTER], as shown below center. *[On the TI-83, you will need to type the fraction in another set of parentheses by entering [SIN][(][2nd][π][÷][6][)][2nd][ANGLE][3][)][ENTER], as shown below right.]*

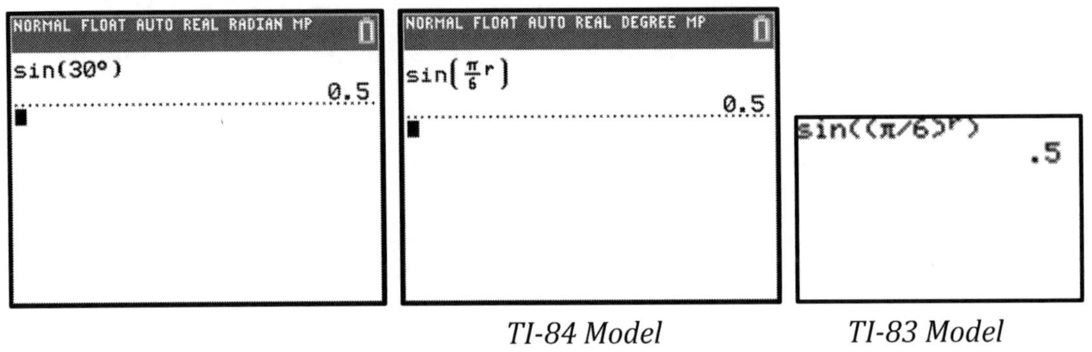

TI-84 Model *TI-83 Model*

It is helpful to begin to memorize the radian equivalents of some special angles:

$0° = 0$ rad	$90° = \frac{\pi}{2}$ rad
$30° = \frac{\pi}{6}$ rad	$180° = \pi$ rad
$45° = \frac{\pi}{4}$ rad	$360° = 2\pi$ rad
$60° = \frac{\pi}{3}$ rad	

Calculating the length of an arc:

There is a basic relationship among the central angle (measured in radians), the radius of the circle, and the length of the intercepted arc. When a central angle θ is measured in radians, we can set up the proportion, $\frac{\theta}{2\pi} = \frac{L}{2\pi r}$, where L is the length of the intercepted arc. Solving for L gives us the formula, $L = \theta r$.

Model Problem

In a circle with a 3-inch radius, what is the measure of a central angle, in radians, that intercepts an arc of:

(a) 6 inches? (b) 3π inches?

Solution:

(a) $6 = 3 \cdot \theta \quad \theta = 2$ radians

(b) $3\pi = 3 \cdot \theta \quad \theta = \pi$ radians

Explanation of steps:

Substitute the known values *[the arc length L and the radius r]* into the formula $L = \theta r$ and solve.

Practice Problems

1. Convert to radians: a) 45° b) 270° c) 150° d) −210°	2. Convert to degrees: a) $\frac{\pi}{6}$ rad b) $\frac{5\pi}{4}$ rad c) $-\frac{3\pi}{5}$ rad d) $\frac{5\pi}{9}$ rad e) $\frac{8\pi}{5}$ rad
3. Find $\sin\frac{\pi}{3}$ to the *nearest thousandth.*	4. Find $\csc\left(-\frac{5\pi}{6}\right)$.
5. If $\tan\frac{\pi}{6} = \cot x$, find x.	6. If $\sec\theta = \csc\left(\theta + \frac{\pi}{3}\right)$, find θ.

7. A circle has a radius of 12 inches. What is the length of the arc intercepted by a central angle of $\frac{\pi}{4}$ rad?	8. A circle has a radius of 4 inches. What is the length of the arc intercepted by a central angle of 2 radians?
9. In a circle with a 10-inch radius, an arc measures 8π inches in length. Find the measure of its central angle in radians.	10. Find the radius of a circle in which a central angle of 5 radians intercepts an arc of 65 feet.

12.3 Unit Circle

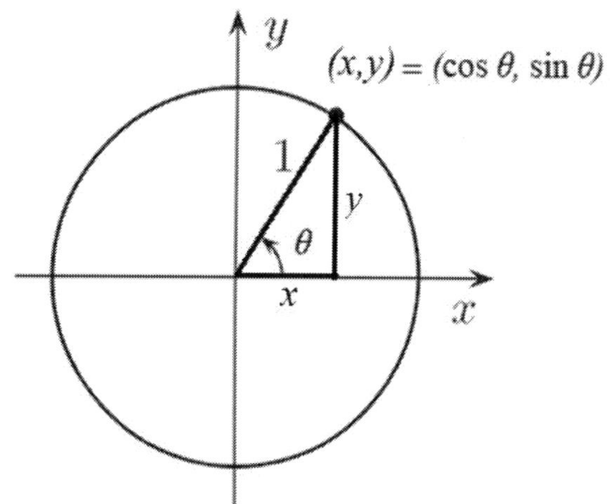

Key Terms and Concepts

A **unit circle** is a circle centered at the origin with a radius of 1. For any point (x, y) on the circle but not on an axis, we can draw a right triangle with the radius as the hypotenuse, as shown below. The angle of the triangle at the origin is labelled θ.

The point $P(x, y)$ is called the **terminal point**. The angle θ is in **standard position** if it has its vertex on the origin and one of its sides, called the **initial side**, lying on the positive x-axis. The other side of angle θ (the hypotenuse of the right triangle) is called the **terminal side**.

If we measure the angle from the initial side moving *counterclockwise* to the terminal side, the measure of the angle is *positive*. Moving in a *clockwise* direction would represent a *negative* angle measure. The terminal side may rotate more than a complete circle.

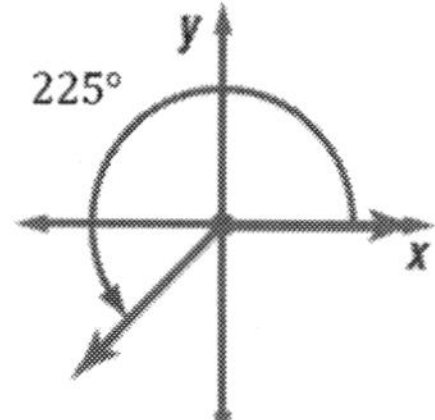

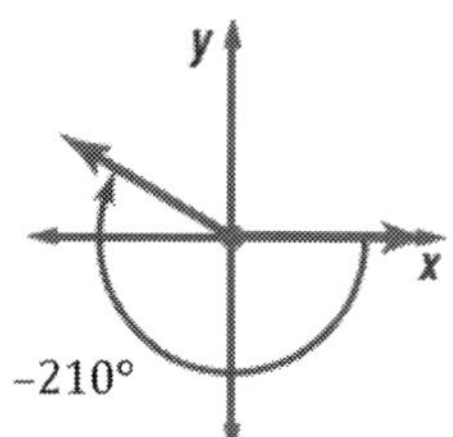

Since $\cos\theta = \dfrac{adj}{hyp} = \dfrac{x}{1} = x$ and $\sin\theta = \dfrac{opp}{hyp} = \dfrac{y}{1} = y$, the coordinates of the point can also be written as $(\cos\theta, \sin\theta)$. Remember that $\cos\theta$ comes before $\sin\theta$ in the coordinates, just as it would alphabetically.

Also, since $\tan\theta = \dfrac{\sin\theta}{\cos\theta}$, we can express $\tan\theta$ as $\dfrac{y}{x}$. Note that $\tan\theta$ is actually the **slope of the terminal side**.

The terminal point may lie in any of the four quadrants depending on the measure of θ. It will lie in Quadrant I if $0° < \theta < 90°$, in Quadrant II if $90° < \theta < 180°$, in Quadrant III if $180° < \theta < 270°$, or in Quadrant IV if $270° < \theta < 360°$.

In Quadrant I, both x and y are positive. Therefore, both $\cos\theta$ and $\sin\theta$ are positive, and since $\tan\theta = \frac{y}{x}$, it is also positive.

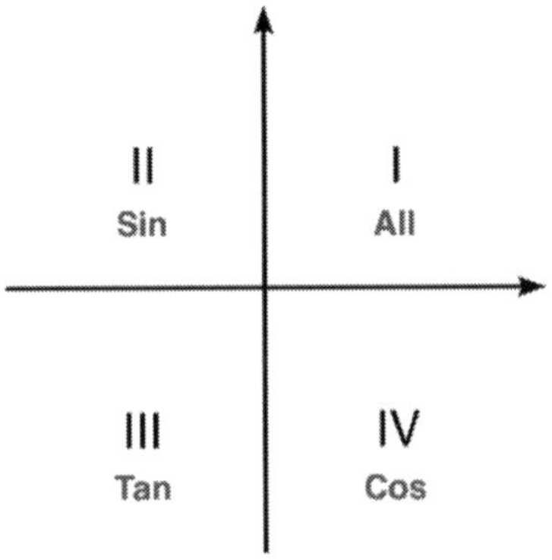

II Sin | I All | III Tan | IV Cos

In Quadrant II, x is negative and y is positive, so the sin is positive but the cos and tan are negative.

In Quadrant III, both x and y are negative, so the cos and sin are negative, but the tan is positive.

Finally, in Quadrant IV, x is positive and y is negative, so the cos is positive but the sin and tan are negative.

The letters ASTC, in counter-clockwise (quadrant number) order, are often used to express which of these functions are positive in each quadrant, as shown in the diagram above. Students will often remember these letters by memorizing the phrase,

"All Students Take Calculus."

Every angle in the unit circle has a reference angle, which is always between 0° and 90°. The **reference angle** is the acute angle between the terminal side and the x-axis.

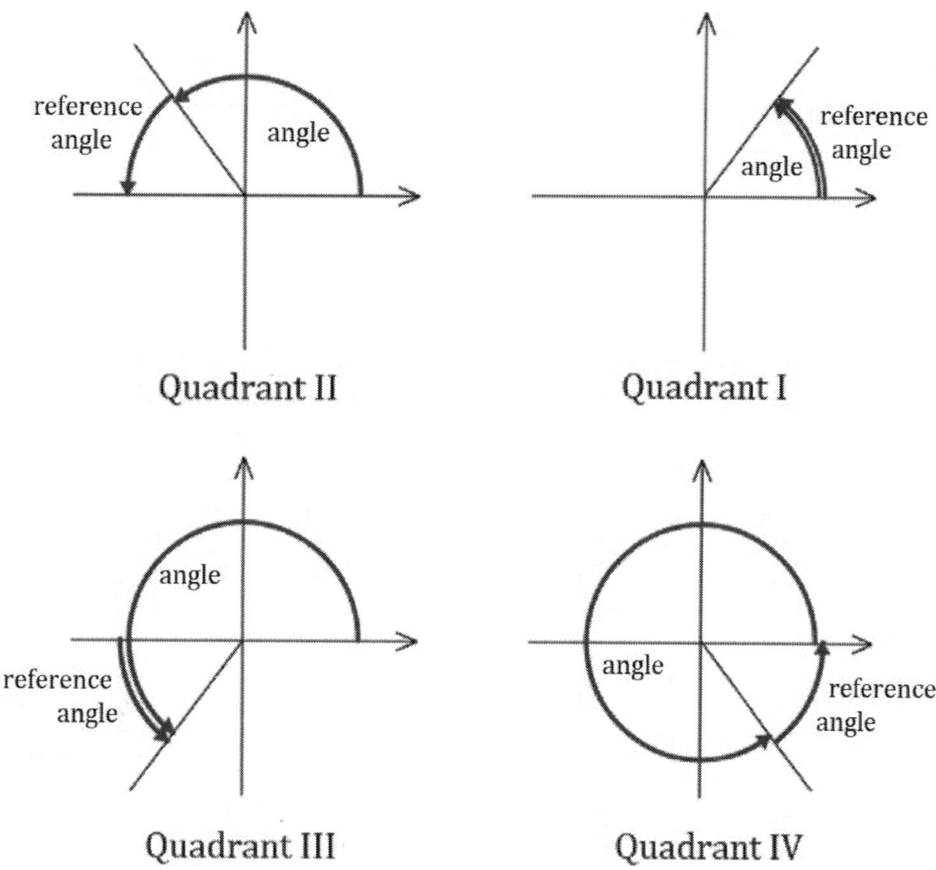

For any angle $0° \leq \theta \leq 360°$, we can calculate the reference angle for θ:

- In Quadrant I, the reference angle is θ.
- In Quadrant II, the reference angle is $(180 - \theta)°$.
- In Quadrant III, the reference angle is $(\theta - 180)°$.
- In Quadrant IV, the reference angle is $(360 - \theta)°$.

For any given angle θ, we can use its reference angle and its quadrant to express a trigonometric function of θ as an equivalent function of a positive acute angle.

Examples: (a) For $\sin 100°$, the angle lies in Quadrant II, so the reference angle is $(180 - \theta)° = 80°$. The sin is positive in Quadrant II, so $\sin 100° = \sin 80°$.

(b) For $\cos 210°$, the angle lies in Quadrant III, so the reference angle is $(210 - 180)° = 30°$. The cos is negative in Quadrant III, so $\cos 210° = -\cos 30°$.

(c) For $\tan 315°$, the angle lies in Quadrant IV, so the reference angle is $(360 - \theta)° = 45°$. The tan is negative in Quadrant IV, so $\tan 315° = -\tan 45°$.

The measure of angle θ may be expressed in degrees or radians. Remember that a full circle is $360°$ or 2π radians. So, for example, an angle of $45°$ is equivalent to $\frac{\pi}{4}$ radians.

For any angle $0 \leq \theta \leq 2\pi$ radians, we can calculate the reference angle for θ:

- In Quadrant I, the reference angle is θ.
- In Quadrant II, the reference angle is $(\pi - \theta)$ radians.
- In Quadrant III, the reference angle is $(\theta - \pi)$ radians.
- In Quadrant IV, the reference angle is $(2\pi - \theta)$ radians.

Example: For $\cos\frac{7\pi}{6}$, the reference angle is $\left(\frac{7\pi}{6} - \pi\right) = \frac{\pi}{6}$ radians.

$\frac{7\pi}{6}$ is between π and $\frac{3\pi}{2}$, so it lies in Quadrant III where cos is negative.

Therefore, $\cos\frac{7\pi}{6} = -\cos\frac{\pi}{6}$.

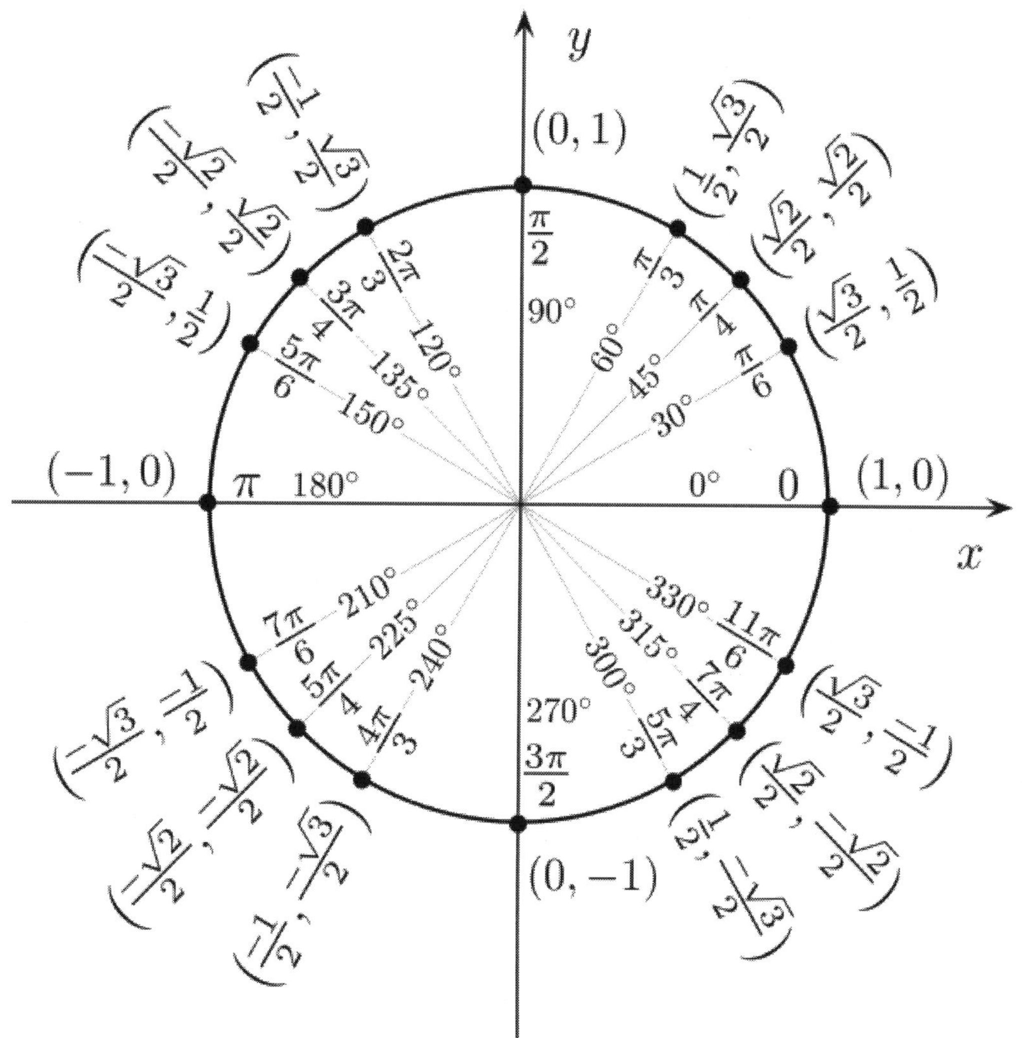

The diagram above shows a number of important points on the unit circle. On the inside of the circle are the measures of the angles in both degrees and radians. Each point gives the cosine (x) and sine (y) of the corresponding angle.

Example: Locate the terminal side for an angle of 300°. The diagram shows that 300° is equivalent to $\frac{5\pi}{3}$ radians, and that $\cos 300° = \frac{1}{2}$ and $\sin 300° = -\frac{\sqrt{3}}{2}$.

The degree measures in the first quadrant have the following radian equivalents:

$0° = 0 \text{ rad} \quad 30° = \frac{\pi}{6} \text{ rad} \quad 45° = \frac{\pi}{4} \text{ rad} \quad 60° = \frac{\pi}{3} \text{ rad} \quad 90° = \frac{\pi}{2} \text{ rad}$

The coordinates of the points in Quadrant I come from the trigonometric ratios of the special 30-60-90 and 45-45-90 right triangles we reviewed in Section 12.1. It is helpful to memorize these ratios, but there is another way to derive the coordinates of the unit circle's first quadrant if you don't recall them by heart, as shown in the table below.

Start with the values for $\theta = 0°, 30°, 45°, 60°$, and $90°$. Label these angles as $n = 0$ to 4. Each y ($\sin\theta$) value will be $\frac{\sqrt{n}}{2}$. Then, the x ($\cos\theta$) values are just the y values in reverse.

Angle θ	$0°$	$30°$	$45°$	$60°$	$90°$
n	0	1	2	3	4
$y = \sin\theta = \frac{\sqrt{n}}{2}$	$\frac{\sqrt{0}}{2} = 0$	$\frac{\sqrt{1}}{2} = \frac{1}{2}$	$\frac{\sqrt{2}}{2}$	$\frac{\sqrt{3}}{2}$	$\frac{\sqrt{4}}{2} = 1$
$x = \cos\theta$ (y values in reverse)	1	$\frac{\sqrt{3}}{2}$	$\frac{\sqrt{2}}{2}$	$\frac{1}{2}$	0

To fill in the other quadrants:

- The points in Quadrant II are a reflection over the *y-axis* of the points in Quadrant I, so use $(x, y) \to (-x, y)$.
- The points in Quadrant III are a reflection over the *origin* of the points in Quadrant I, so use $(x, y) \to (-x, -y)$.
- The points in Quadrant IV are a reflection over the *x-axis* of the points in Quadrant I, so use $(x, y) \to (x, -y)$.

Note that the signs of the functions correspond to ASTC. In Quadrant II, sin is positive and cos is negative. In Quadrant III, sin and cos are both negative. In Quadrant IV, sin is negative and cosine is positive.

We can use the unit circle diagram to find the values of other trigonometric functions.

Examples: (1) $\tan\theta = \frac{\sin\theta}{\cos\theta}$, so $\tan 60° = \frac{\sin 60°}{\cos 60°} = \frac{\frac{\sqrt{3}}{2}}{\frac{1}{2}} = \sqrt{3}$

(2) $\sec\theta = \frac{1}{\cos\theta}$, so $\sec 150° = \frac{1}{\cos 150°} = \frac{1}{-\frac{\sqrt{3}}{2}} = -\frac{2}{\sqrt{3}} = -\frac{2\sqrt{3}}{3}$

All of the six trigonometric functions of angle θ represent various segment lengths within the unit circle, as shown below.

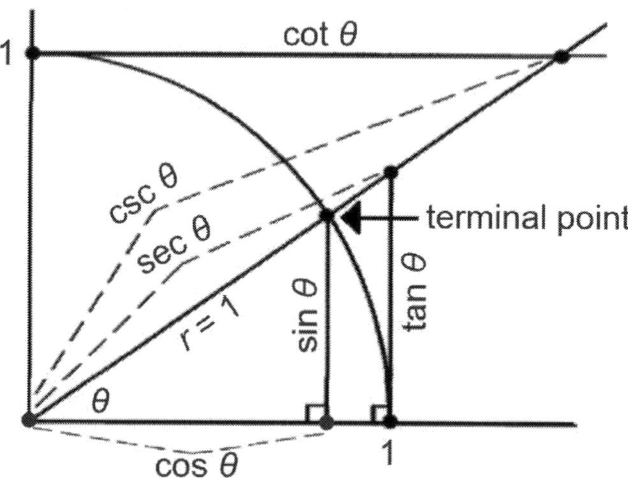

For the sine, cosine, or tangent of **negative angles**, we can apply the following rules:

$\sin(-x) = -\sin x$ $\quad\quad$ $\cos(-\theta) = \cos\theta$ $\quad\quad$ $\tan(-x) = -\tan x$

We can see that these are true by looking at the unit circle, below, with terminal points marked for angles of θ and $-\theta$. Note that $\cos(-\theta) = \cos\theta$ because they both represent the same x-coordinates and $\sin(-x) = -\sin x$ because they represent y-coordinates that are opposites of each other. Finally, remembering that tangent represents the slope of the terminal sides, we see that $\tan(-x) = -\tan x$ because they have opposite slopes. These relationships would be true no matter which quadrant we selected for angle θ.

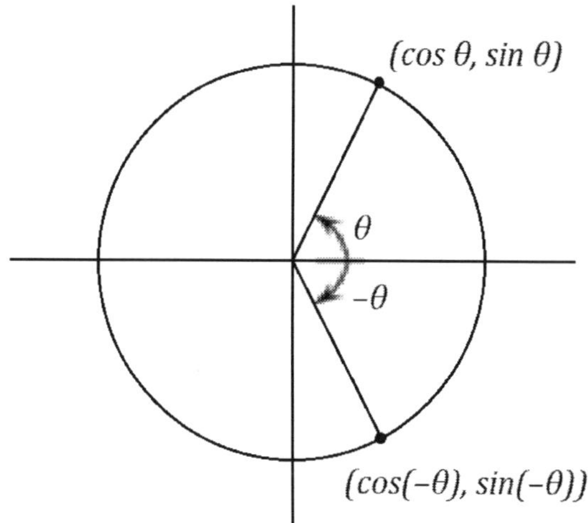

MODEL PROBLEM

In the diagram of the unit circle to the right, $m\angle\theta = 50°$. State the coordinates of the terminal point (x, y), to the *nearest thousandth.*

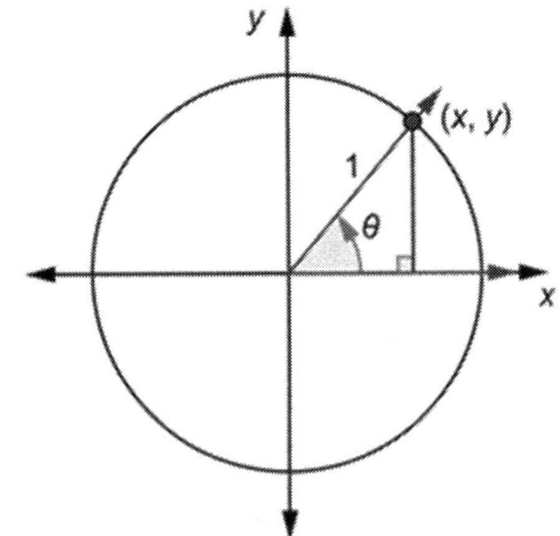

Solution:

$\cos 50° \approx 0.643$ and $\sin 50° \approx 0.766$,
so the coordinates are $(0.643, 0.766)$.

Explanation of steps:

Use $(x, y) = (\cos\theta, \sin\theta)$ to find the coordinates. Be sure to use the correct sign for each coordinate *[in Quadrant I, x and y are both positive]*.

PRACTICE PROBLEMS

1. The accompanying diagram shows unit circle O, with radius $OB = 1$.

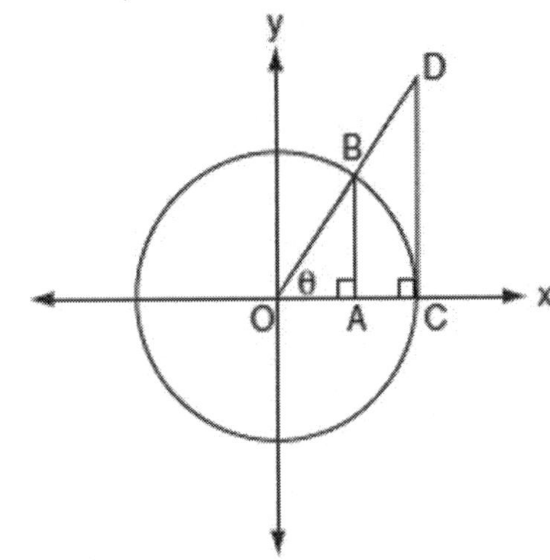

Which line segment has a length equivalent to $\cos\theta$?

(1) $\overline{AB}$ (3) $\overline{OC}$

(2) $\overline{CD}$ (4) $\overline{OA}$

2. The accompanying diagram shows unit circle O, with radius $OQ = 1$.

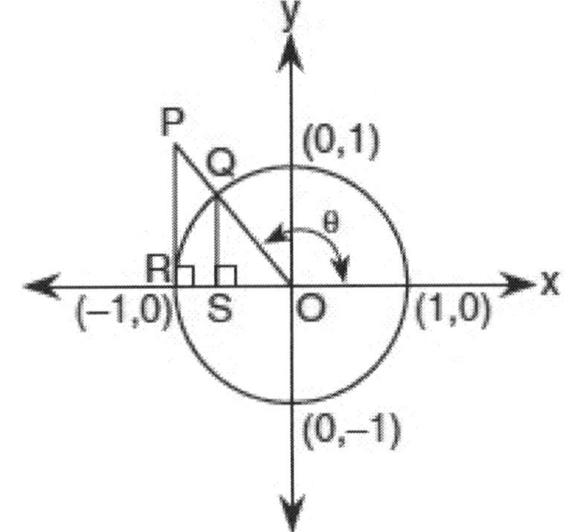

Which line segment has a length equivalent to $\sin\theta$?

(1) $\overline{PR}$ (3) $\overline{SO}$

(2) $\overline{QS}$ (4) $\overline{RO}$

3. In the accompanying diagram of a unit circle, the ordered pair (x, y) represents the locus of points forming the circle. Which ordered pair is equivalent to (x, y)?

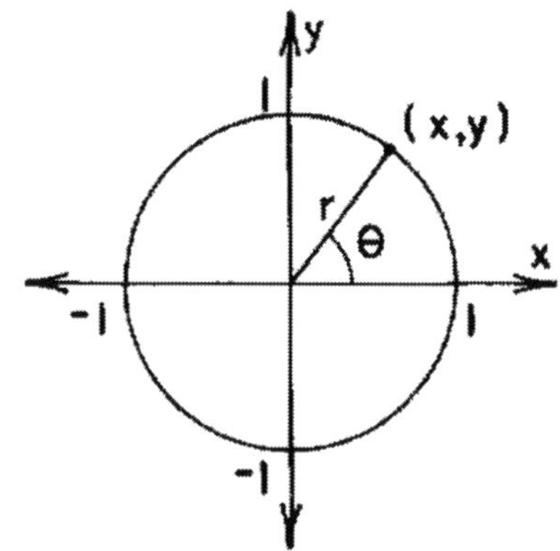

(1) $(\sin\theta, \cos\theta)$ (3) $(\tan\theta, \cot\theta)$

(2) $(\cot\theta, \tan\theta)$ (4) $(\cos\theta, \sin\theta)$

4. In the accompanying diagram of circle O, $\overline{COA}$ is a diameter, O is the origin, $OA = 1$, and $m\angle BOA = 30°$. What are the coordinates of B?

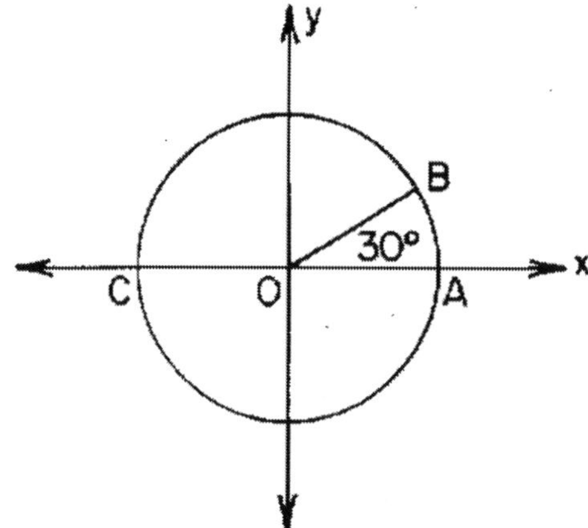

(1) $\left(\frac{1}{2}, \frac{\sqrt{3}}{2}\right)$ (3) $\left(\frac{\sqrt{2}}{2}, \frac{\sqrt{2}}{2}\right)$

(2) $\left(\frac{\sqrt{3}}{2}, \frac{1}{2}\right)$ (4) $\left(\frac{\sqrt{2}}{2}, \frac{1}{2}\right)$

5. Sketch an angle of 250° in standard position. Express cos 250° as a cosine function of a positive acute angle.

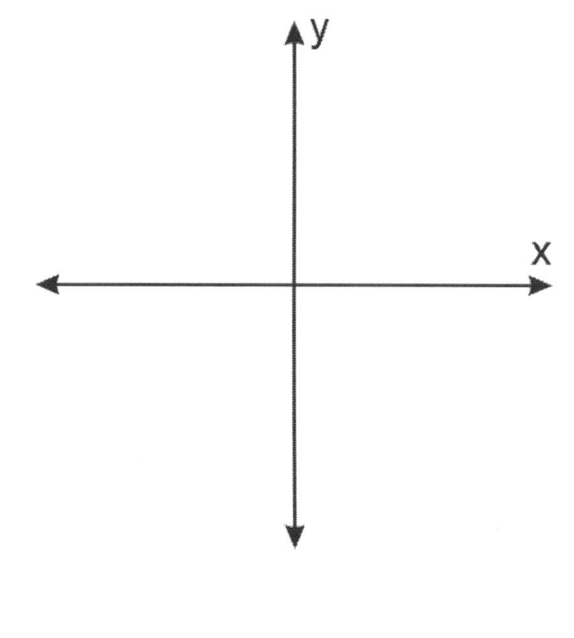

6. Sketch an angle of 310° in standard position. Express sin 310° as a sine function of a positive acute angle.

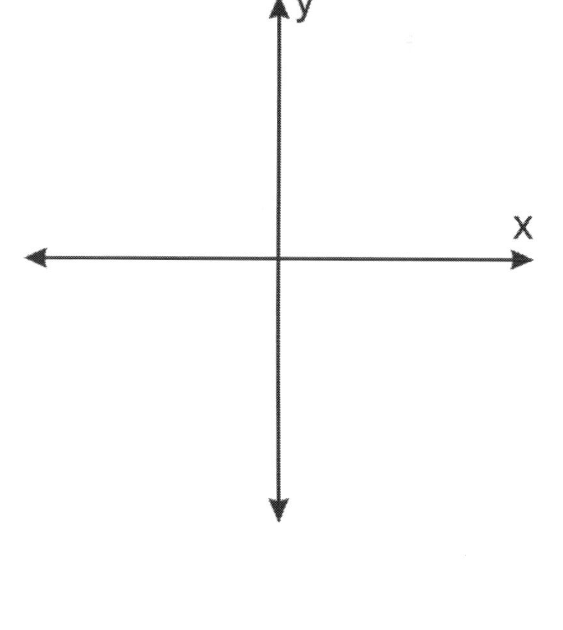

7. Express tan 230° as a tangent function of a positive acute angle.

8. Find the value of sin 135° in radical form.

9. Find the value of $\sin\frac{4\pi}{3}$ in simplest radical form.

10. Find the value of $\sin\frac{3\pi}{2}+\cos\frac{2\pi}{3}$.

11. In the accompanying diagram of a unit circle, the ordered pair (x, y) represents the point where the terminal side of θ intersects the unit circle.

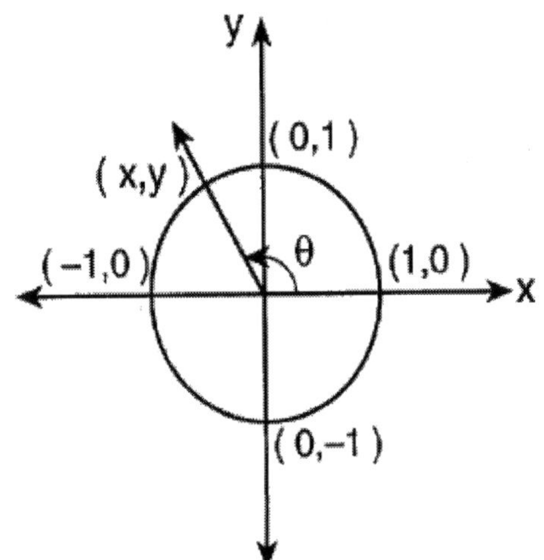

If m$\angle\theta = 120°$, what is the value of x?

12. In the accompanying diagram of a unit circle, the ordered pair (x, y) represents the point where the terminal side of θ intersects the unit circle.

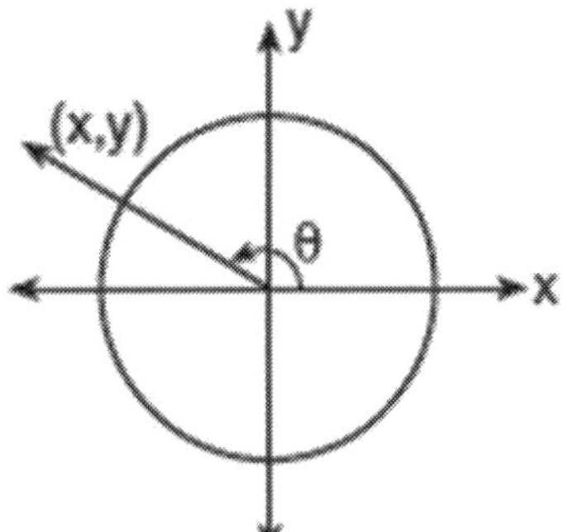

If m$\angle\theta = 145°$, what is the value of y to the *nearest thousandth*?

13. In the unit circle shown in the accompanying diagram, what are the coordinates of (x, y)?

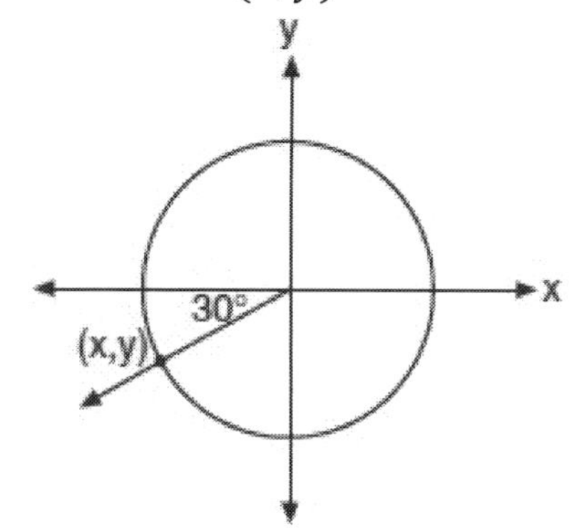

14. The unit circle diagram below shows a 400° angle. Find the coordinates of (x, y), to the *nearest thousandth.*

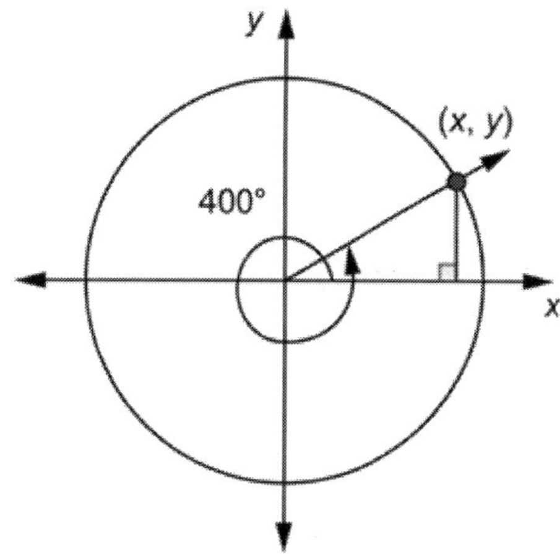

12.4 Solve Simple Trigonometric Equations

Key Terms and Concepts

A trigonometric equation is one where the unknown variable for which we need to solve is an angle measure. There are often multiple solutions to these types of equations, even if we limit the angle measures to between 0° and 360°.

Examples: (a) For $\sin a = \frac{1}{2}$, angle a can be 30° or 150°, as we can see in the unit circle diagram below.

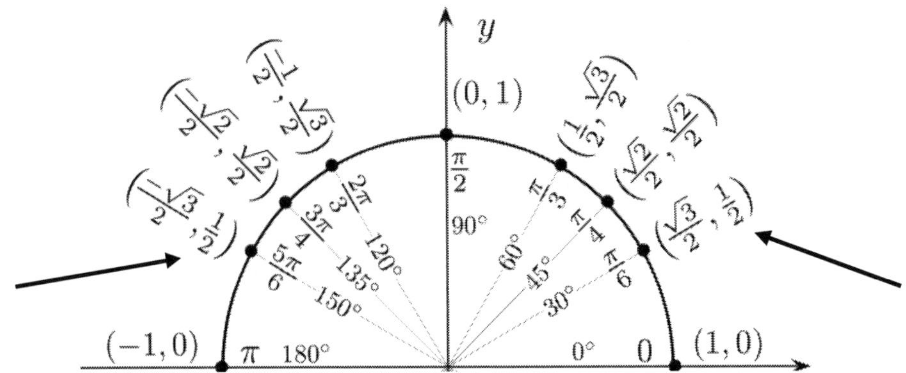

(b) For $\cos b = 0.4$, angle b can be approximately 66.4° or 293.6°.

This is because there are two points on the unit circle where $\sin a$ equals the given y, or $\cos b$ equals the given x.

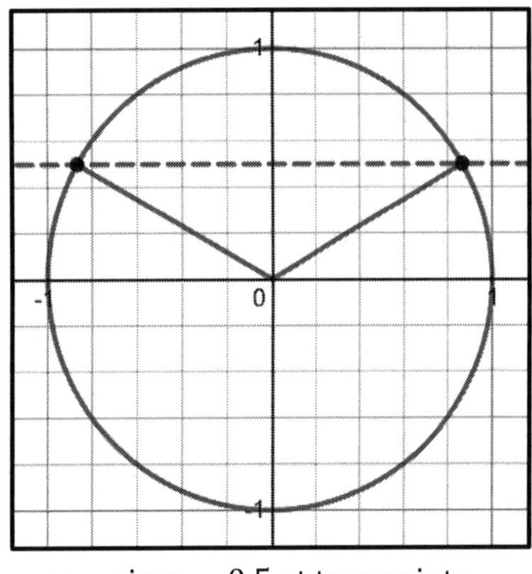

$y = \sin a = 0.5$ at two points

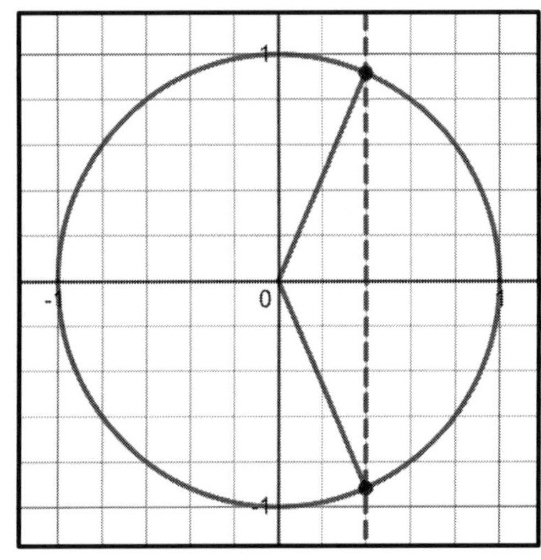

$x = \cos b = 0.4$ at two points

In fact, as long as the value of $\sin a$ or $\cos b$ is between -1 and 1, exclusive, the horizontal or vertical line will cross the unit circle at two points, so there will be two solutions between $0°$ and $360°$. If the value is equal to -1 or 1, then the line will be tangent to the circle, so there will be only one solution between $0°$ and $360°$. For values less than -1 or greater than 1, there are no solutions. (When we look at graphs of trig functions, we'll extend the domain beyond the interval between $0°$ and $360°$.)

To solve a trig equation of the form $\cos a = x$ for a, or $\sin b = y$ for b:

1. Determine in which quadrants the points would be located in a unit circle:
   - For $\sin a = y$, the points are in Quadrants I and II for a positive y or in Quadrants III and IV for a negative y.
   - For $\cos b = x$, the points are in Quadrants I and IV for a positive x or in Quadrants II and III for a negative x.
2. Find the reference angle, which we'll call R, by using the calculator's inverse trig function on the absolute value of the number after the equal sign.
3. Depending on the quadrants of the points in step 1, calculate the angles as follows:
   - For a point in Quadrant I, the angle is the reference angle, R.
   - For a point in Quadrant II, the angle is $180° - R$.
   - For a point in Quadrant III, the angle is $180° + R$.
   - For a point in Quadrant IV, the angle is $360° - R$.

MODEL PROBLEM

On the domain $0° \leq b < 360°$, solve for b to the *nearest tenth of a degree*: $\cos b = -0.6$

Solution:

(A) Points are in quadrants II and III.

(B) Reference angle $R = \cos^{-1} 0.6 \approx 53.1°$.

(C) Solutions are approximately $180° - 53.1° = 126.9°$ and $180° + 53.1° = 233.1°$.

Explanation of steps:

(A) Determine in which quadrants the points would be located in a unit circle.
[For $\cos b = x$, the points are in Quadrants II and III for a negative x.]

(B) Find the reference angle using the calculator's inverse trig function. Remember to always use the absolute value, *not* a negative value.

(C) Depending on the quadrants of the points, calculate the angles.
[For Quadrant II, use $180° - R$, and for Quadrant III, use $180° + R$.]

PRACTICE PROBLEMS

1. In the unit circle diagram below, the terminal point of angle θ is $\left(\frac{1}{2}, \frac{\sqrt{3}}{2}\right)$. Find m$\angle\theta$.

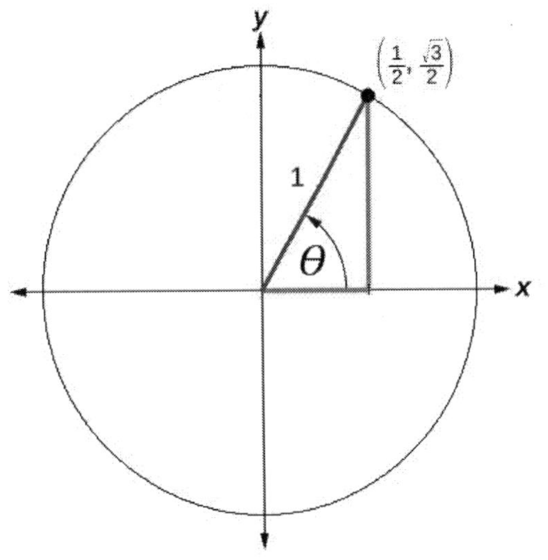

2. In the accompanying diagram of circle O, point O is the origin, $OJ = 1$, and $\overline{TOY}$ is a diameter. If the coordinates of point J are $\left(\frac{\sqrt{2}}{2}, \frac{\sqrt{2}}{2}\right)$, what is m$\angle JOY$?

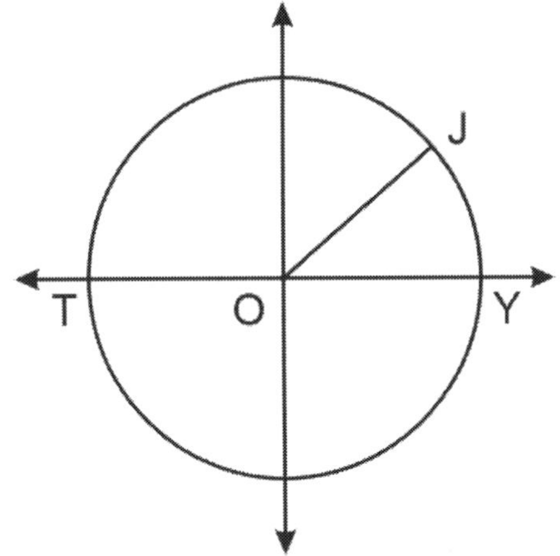

3. In the unit circle diagram below, the terminal point of angle θ is $\left(-\frac{\sqrt{3}}{2}, -\frac{1}{2}\right)$. Find m$\angle\theta$?

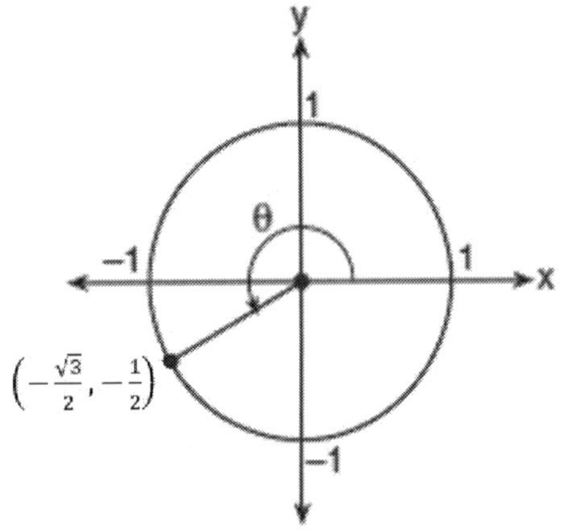

4. In the accompanying diagram, point $P(0.6, -0.8)$ is on unit circle O. What is m$\angle\theta$, to the *nearest degree*?

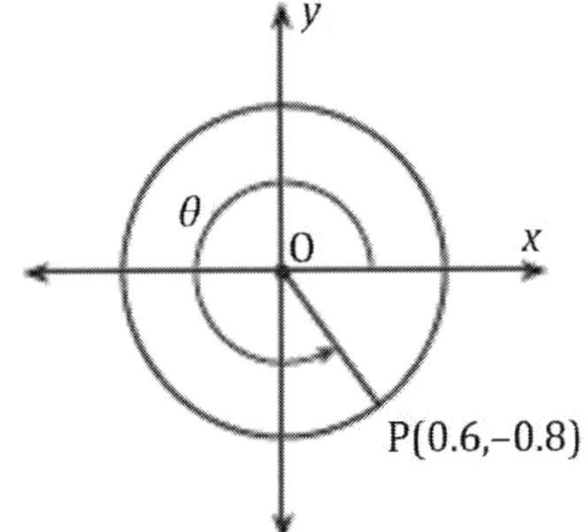

5. Over the domain $0° \leq \theta < 360°$, solve for θ to the *nearest tenth of a degree*: $\cos\theta = 0.25$	6. Over the domain $0° \leq \theta < 360°$, solve for θ to the *nearest tenth of a degree*: $\cos\theta = -0.75$
7. Over the domain $0° \leq \theta < 360°$, solve for θ to the *nearest tenth of a degree*: $\sin\theta = 0.99$	8. Over the domain $0° \leq \theta < 360°$, solve for θ to the *nearest tenth of a degree*: $\sin\theta = -\frac{1}{3}$

12.5 Circles of Any Radius

Key Terms and Concepts

We can enlarge or reduce the unit circle to a circle of any radius. This will allow us to determine the trig functions of an angle given any terminal point on a coordinate grid.

In the right triangle to the right, $\cos\theta = \frac{x}{r}$, $\sin\theta = \frac{y}{r}$, and $\tan\theta = \frac{y}{x}$.

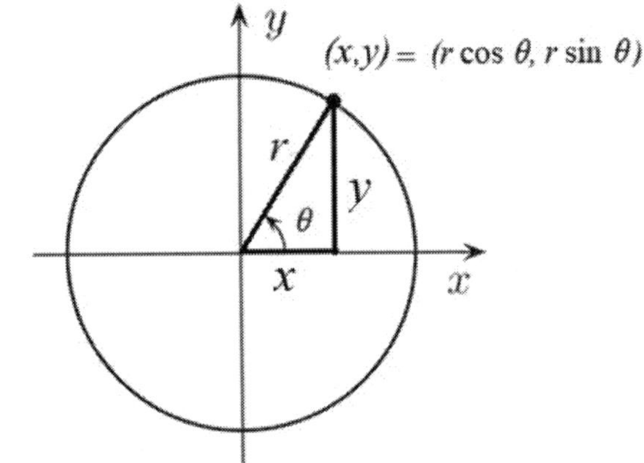

Since $\cos\theta = \frac{x}{r}$, we know $x = r\cos\theta$, which means the x-coordinate of the terminal point now represents $r\cos\theta$.

Likewise, since $\sin\theta = \frac{y}{r}$, we know $y = r\sin\theta$, so the y-coordinate represents $r\sin\theta$.

Given a radius and terminal point, we can find the cosine, sine, and tangent of an angle by using, $\cos\theta = \frac{x}{r}$, $\sin\theta = \frac{y}{r}$, and $\tan\theta = \frac{y}{x}$.

Example: In the diagram below, $m\angle\theta = 120°$, $r = 2$, and the terminal point is $(-1, \sqrt{3})$.

$\sin 120° = \frac{y}{r} = \frac{\sqrt{3}}{2}$, $\cos 120° = \frac{x}{r} = -\frac{1}{2}$, and $\tan 120° = \frac{y}{x} = \frac{\sqrt{3}}{-1} = -\sqrt{3}$.

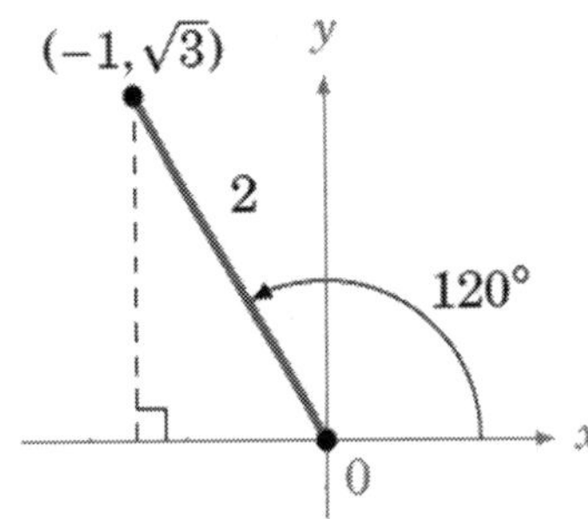

We can find the cosecant, secant, and cotangent of an angle by finding the reciprocals of the sine, cosine, and tangent, respectively.

Example: In the diagram above, $\csc 120° = \frac{r}{y} = \frac{2}{\sqrt{3}} = \frac{2\sqrt{3}}{3}$, $\sec 120° = \frac{r}{x} = \frac{2}{-1} = -2$, and $\cot 120° = \frac{x}{y} = -\frac{1}{\sqrt{3}} = -\frac{\sqrt{3}}{3}$.

If we are not given the radius or one of the coordinates, we can calculate it using the Pythagorean Theorem, $x^2 + y^2 = r^2$, which you should also recognize as the general equation of a circle centered at the origin.

Example: In the previous examples, if we weren't given that the radius was 2, we could have calculated it using the coordinates of the point $\left(-1, \sqrt{3}\right)$.

$$(-1)^2 + \left(\sqrt{3}\right)^2 = r^2$$

$$4 = r^2, \text{so } r = 2 \qquad \text{(the radius must be positive)}$$

If we are not given a terminal point, but we are given the quadrant of an angle and the value of one trigonometric function of that angle, we can use this information to find the exact values of other trigonometric functions of that same angle. We do this by drawing (or imagining) a right triangle with an acute angle in standard position.

Example: Suppose we know $\sin\theta = \frac{1}{2}$ and that θ is in Quadrant I, and we want to find $\cos\theta$. Because $\sin\theta = \frac{y}{r}$, we can draw a right triangle in Quadrant I where $y = 1$ and $r = 2$, as shown below. We can then calculate the corresponding x value by solving $x^2 + 1^2 = 2^2$, or $x = \sqrt{3}$ (in Quadrant I, both x and y are positive). With this information, we can now find $\cos\theta = \frac{x}{r} = \frac{\sqrt{3}}{2}$.

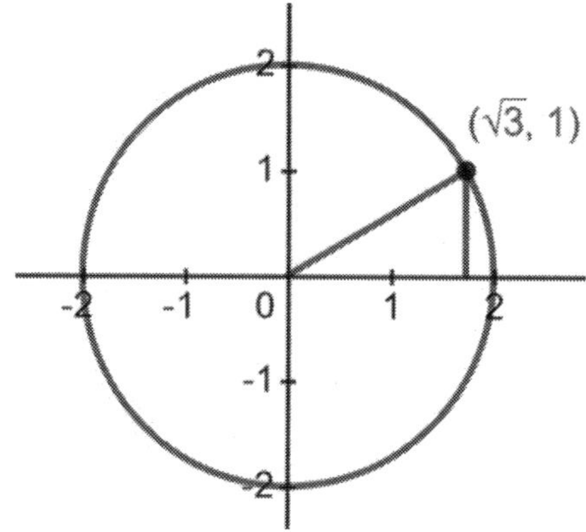

To find the value of a trigonometric function, given the quadrant of an angle and the value of another trigonometric function of that angle:

1. Set the corresponding variables equal to the numerator and denominator of the ratio, using the appropriate signs based on the quadrant.
2. Use the Pythagorean Theorem to find the missing side.
3. Write the value of the function using the appropriate ratio.

CALCULATOR TIP

You can check your answer using the inverse trigonometric functions on the calculator. You'll just need to make sure to use the correct sign, depending on the quadrant.

Example: To check the above example, since $\sin\theta = \frac{1}{2}$, we can calculate θ as $\sin^{-1}\left(\frac{1}{2}\right)$ and then take the cos of the result, as shown to the right. This should give us the same decimal value as the calculation of $\frac{\sqrt{3}}{2}$. Since the cosine is positive in Quadrant I, a positive $\frac{\sqrt{3}}{2}$ is the correct answer.

NORMAL FLOAT AUTO REAL DEGREE MP

$\cos\left(\sin^{-1}\left(\frac{1}{2}\right)\right)$

0.8660254038

$\frac{\sqrt{3}}{2}$

0.8660254038

MODEL PROBLEM 1: GIVEN A POINT

If the terminal side of angle θ passes through point $(-4,3)$, what is the value of $\cos\theta$?

Solution:

(A) $x^2 + y^2 = r^2$

$(-4)^2 + 3^2 = r^2$

$25 = r^2$

$r = 5$

(B) $\cos\theta = \frac{x}{r} = -\frac{4}{5}$

Explanation of steps:

(A) If the radius isn't given, calculate it using the Pythagorean Theorem.

(B) Write the trig function as a ratio and substitute. *[$x = -4$ and $r = 5$]*

PRACTICE PROBLEMS

1. If the terminal side of angle θ passes through point $(-3,-4)$, what is the value of $\cot\theta$?	2. If θ is an angle in standard position and its terminal side passes through the point $(-\sqrt{2},-\sqrt{2})$, what is the value of $\tan\theta$?
3. If θ is an angle in standard position and $(-3,4)$ is a point on the terminal side of θ, what is the value of $\sin\theta$?	4. Angle θ is in standard position and $(-4,0)$ is a point on the terminal side of θ. What is the value of $\sec\theta$?
5. If θ is an angle in standard position and its terminal side passes through the point $(-3,2)$, what is the exact value of $\csc\theta$?	6. If θ is an angle in standard position and its terminal side passes through the point $(-8,5)$, what is the exact value of $\sec\theta$?

7. Find $\sin\theta$, $\cos\theta$, and $\tan\theta$.

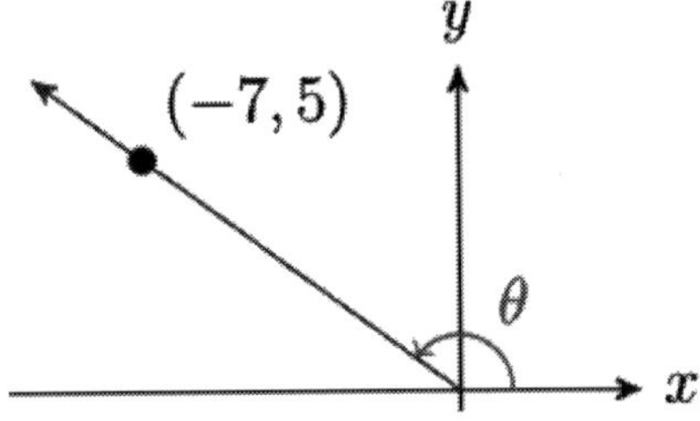

a) $\sin\theta$

b) $\cos\theta$

c) $\tan\theta$

8. Find $\csc\theta$, $\sec\theta$, and $\cot\theta$.

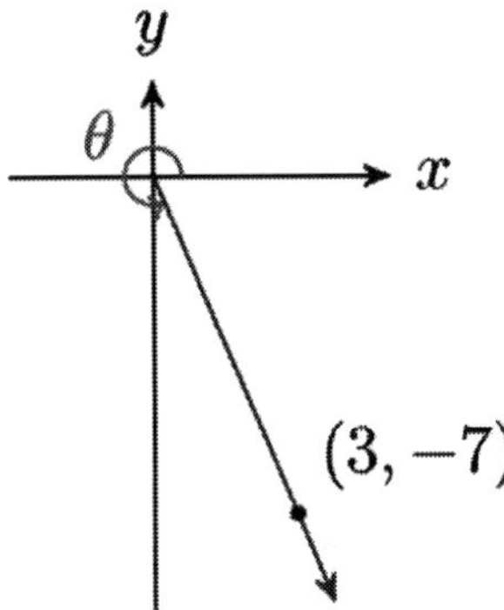

a) $\csc\theta$

b) $\sec\theta$

c) $\cot\theta$

Model Problem 2: given a ratio

If $\sin\theta = -\frac{\sqrt{7}}{4}$ and θ is in Quadrant III, find $\cos\theta$.

Solution:

(A) $\sin\theta = \frac{y}{r}$, so $y = -\sqrt{7}$ and $r = 4$

(B) $x^2 + \left(-\sqrt{7}\right)^2 = 4^2$

$x^2 = 9$

$x = -3$ (reject $x = 3$)

(C) $\cos\theta = \frac{x}{r} = -\frac{3}{4}$

Explanation of steps:

(A) If given a ratio, set the corresponding variables equal to the numerator and denominator, using the appropriate signs based on the quadrant.
[In Quadrant III, x and y are both negative.]

(B) Use the Pythagorean Theorem to find the missing side.
[In Quadrant III, x must be negative, so reject $x = 3$.]

(C) Write the value of the function using the appropriate ratio.

Practice Problems

9. If $\sin\theta = \frac{4}{5}$ and θ is in Quadrant I, find $\cot\theta$.	10. If $\cos\theta = \frac{\sqrt{3}}{2}$ and θ is in Quadrant IV, find $\sin\theta$.

11. If $\sec\theta = -\frac{5}{2}$ and θ is in Quadrant III, find $\sin\theta$.	12. If $\cos\theta = \frac{\sqrt{2}}{3}$ and $\sin\theta < 0$, find $\tan\theta$.
13. If $\tan\theta = \frac{3}{5}$ and θ is in Quadrant III, find $\sec\theta$.	14. If $\cot\theta = -\frac{3\sqrt{2}}{2}$ and θ is in Quadrant II, find $\sin\theta$.

15. Circle O is centered at the origin and has a radius of 2 units. An angle with a measure of $\frac{\pi}{6}$ radians is in standard position. If the terminal side of the angle intersects the circle at point P, what are the coordinates of P?

12.6 Pythagorean Identity

Key Terms and Concepts

Looking at the right triangle in the unit circle below, we know by the Pythagorean Theorem that $x^2 + y^2 = 1$. We also know that $x = \cos\theta$ and $y = \sin\theta$. By substitution and the commutative property of addition, we get the equation $\mathbf{\sin^2\theta + \cos^2\theta = 1}$, which is known as the **Pythagorean identity**. Note that the square of a trig function, such as $(\sin\theta)^2$, is usually written as $\sin^2\theta$.

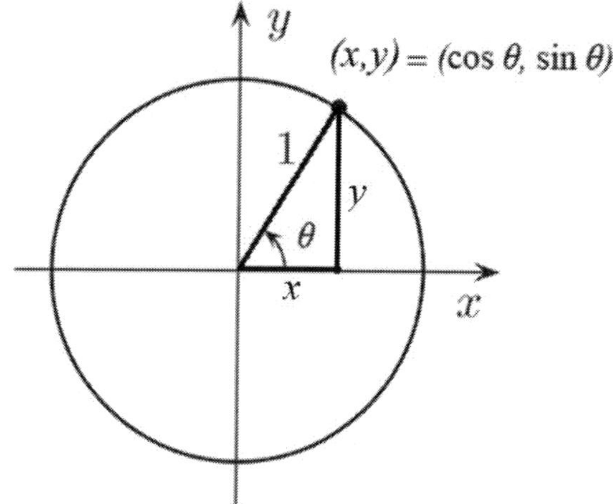

This identity can easily be rewritten as $\mathbf{\sin^2\theta = 1 - \cos^2\theta}$ or as $\mathbf{\cos^2\theta = 1 - \sin^2\theta}$.

If given the sine or cosine of an angle and its quadrant on the unit circle, you can use the Pythagorean identity to find the other trig function's value.

Example: Suppose we know $\cos\theta = -\frac{4}{5}$ and that θ is in Quadrant II.

We can find $\sin\theta$ by:

$\sin^2\theta = 1 - \cos^2\theta$

$\sin^2\theta = 1 - \left(-\frac{4}{5}\right)^2$

$\sin^2\theta = 1 - \frac{16}{25} = \frac{9}{25}$

$\sin\theta = \sqrt{\frac{9}{25}} = \frac{3}{5}$ *[reject $-\frac{3}{5}$ because sine is positive in Quadrant II]*

Of course, we may continue to solve these problems using the method we learned earlier.

Example: For the above problem, where $\cos\theta = -\frac{4}{5}$ and θ is in Quadrant II, we can find $\sin\theta$ as follows.

In Quadrant II, $x < 0$ and $y > 0$

$\cos\theta = \frac{x}{r}$, so we can use $x = -4$ and $r = 5$

$(-4)^2 + y^2 = 5^2$

$y^2 = 9$, so $y = 3$ *[reject $y = -3$]*

$\sin\theta = \frac{y}{r} = \frac{3}{5}$

MODEL PROBLEM

Given $\sin\theta = \frac{1}{4}$ and θ is in Quadrant II, find $\cos\theta$ using the Pythagorean identity.

Solution:	**Explanation of steps:**
(A) $\cos^2\theta = 1 - \sin^2\theta$	(A) Write the appropriate form of the Pythagorean identity. *[Since we are trying to find the cosine, use the form that has cosine already isolated.]*
(B) $\cos^2\theta = 1 - \left(\frac{1}{4}\right)^2$	(B) Substitute the given value.
(C) $\cos^2\theta = 1 - \frac{1}{16} = \frac{15}{16}$ $\cos\theta = -\frac{\sqrt{15}}{4}$	(C) Solve. Use the correct sign for your answer *[in Quadrant II, cosine is negative]*.

PRACTICE PROBLEMS

1. Given $\cos\theta = \frac{1}{2}$ and θ is in Quadrant IV. Find $\sin\theta$ using the Pythagorean identity.	2. Given $\sin\theta = -\frac{\sqrt{2}}{2}$ and θ is in Quadrant III. Find $\cos\theta$ using the Pythagorean identity.

3. Given $\cos\theta = \frac{1}{3}$ and θ is in Quadrant I. Find $\sin\theta$ using the Pythagorean identity.	4. Given $\sin\theta = \frac{\sqrt{3}}{4}$ and θ is in Quadrant II. Find $\cos\theta$ using the Pythagorean identity.
5. Given $\sec\theta = 3$ and θ is in Quadrant IV. Find $\csc\theta$ using the Pythagorean identity.	6. Given $\csc\theta = -\sqrt{5}$ and θ is in Quadrant III. Find $\tan\theta$ using the Pythagorean identity.
7. Point $\left(k, \frac{9}{10}\right)$ is located in Quadrant I on the unit circle. Determine the exact value of k using the Pythagorean identity.	8. Point $\left(\frac{4}{9}, k\right)$ is located in Quadrant IV on the unit circle. Determine the exact value of k using the Pythagorean identity.

12.7 Simplify Trigonometric Expressions

KEY TERMS AND CONCEPTS

A **trigonometric expression** is one that involves trigonometric functions. Like any algebraic expression, a trigonometric expression can often be simplified by rewriting it as an equivalent expression using as few terms (and as few trig functions) as possible, preferably without any fractions.

Note that when trig functions are written next to each other with no operation symbol between them, then multiplication is implied.
Example: $\sin\theta\cos\theta$ means $(\sin\theta) \times (\cos\theta)$.

To simplify a trigonometric expression:

1. Rewrite each function as an equivalent expression involving only sin and/or cos.
2. If possible, simplify binomial expressions involving $\sin^2\theta$ or $\cos^2\theta$ using the Pythagorean identity:
   - Rewrite $1 - \sin^2\theta$ as $\cos^2\theta$
   - Rewrite $1 - \cos^2\theta$ as $\sin^2\theta$
   - Rewrite $\sin^2\theta + \cos^2\theta$ as 1
3. Rewrite complex fractions as the product of the numerator times the reciprocal of the denominator.
4. Simplify fractions by cancelling common factors.
5. If possible, eliminate fractions by using reciprocal functions (e.g., $\frac{1}{\sin x} = \csc x$).

Example: To express $\frac{\tan x}{\sec x}$ as a single trigonometric function in simplest form,

$$\frac{\tan x}{\sec x} = \frac{\frac{\sin x}{\cos x}}{\frac{1}{\cos x}} = \frac{\sin x}{\cancel{\cos x}} \cdot \frac{\cancel{\cos x}}{1} = \sin x.$$

MODEL PROBLEM 1: *REWRITE TRIG FUNCTIONS AS SIN OR COS*

Express as a single trigonometric function in simplest form: $\frac{\sin x \cos x}{\cot x}$.

Solution:

$$\frac{\sin x \cos x}{\cot x} \overset{(A)}{=} \frac{\sin x \cos x}{\frac{\cos x}{\sin x}} \overset{(B)}{=} \frac{\sin x \cancel{\cos x}}{1} \cdot \frac{\sin x}{\cancel{\cos x}} \overset{(C)}{=} \sin^2 x$$

Explanation of steps:

(A) Rewrite each function as an equivalent expression involving only sin and/or cos.

(B) For clarity, rewrite complex fractions as the product of the numerator times the reciprocal of the denominator.

(C) Simplify by cancelling any factors that are common to both the numerator and denominator.

PRACTICE PROBLEMS

1. Simplify the expression: $\sec x \cot x$	2. Simplify the expression: $\tan x \csc x$
3. Simplify the expression: $\sin x \cos x \tan x$	4. Simplify the expression: $\frac{\sin \theta}{\csc \theta}$

5. Simplify the expression: $\sin^2\theta\csc\theta$	6. Simplify the expression: $\dfrac{\tan\theta}{\sin\theta}$
7. Simplify the expression: $\dfrac{\cot x}{\csc x}$	8. Simplify the expression: $\dfrac{\cot x\sin x}{\sec x}$

MODEL PROBLEM 2: ***SIMPLIFY USING THE PYTHAGOREAN IDENTITY***

Simplify $\cot\theta\,(1-\cos^2\theta)$.

Solution:

(A) (B) (C)

$$\cot\theta\,(1-\cos^2\theta)=\frac{\cos\theta}{\sin\theta}\cdot\sin^2\theta=\cos\theta\sin\theta$$

Explanation of steps:

(A) Rewrite each function as an equivalent expression involving only sin and/or cos. $[\cot\theta=\dfrac{\cos\theta}{\sin\theta}]$

(B) If possible, simplify binomial expressions involving $\sin^2\theta$ or $\cos^2\theta$ using the Pythagorean identity. $[1-\cos^2\theta=\sin^2\theta]$

(C) Simplify fractions by cancelling common factors. $[\dfrac{\sin^2\theta}{\sin\theta}=\sin\theta]$

PRACTICE PROBLEMS

9. Simplify the expression: $\dfrac{1-\sin^2\theta}{\cos\theta}$	10. Simplify the expression: $\dfrac{1-\sin^2 x}{\cos^2 x}$
11. Simplify the expression: $\dfrac{\sin^2 x+\cos^2 x}{1-\sin^2 x}$	12. Simplify the expression: $\cos x\,(\cos x+1)+\sin^2 x$
13. Simplify the expression: $\cos x\,(\sec x-\cos x)$	14. Simplify the expression: $\sin^2 x\,(1+\cot^2 x)$

15. Simplify the expression: $\dfrac{2 - 2\sin^2 x}{\cos x}$	16. Simplify the expression: $\sec x - \tan x \sin x$
17. Simplify the expression: $\csc^2 x\,(1 + \sin x)(1 - \sin x)$	18. Simplify the expression: $\dfrac{\tan^2 x - \sec^2 x}{\cot^2 x - \csc^2 x}$

12.8 Graphs of Parent Trig Functions

Key Terms and Concepts

Let's take another look at the unit circle below.

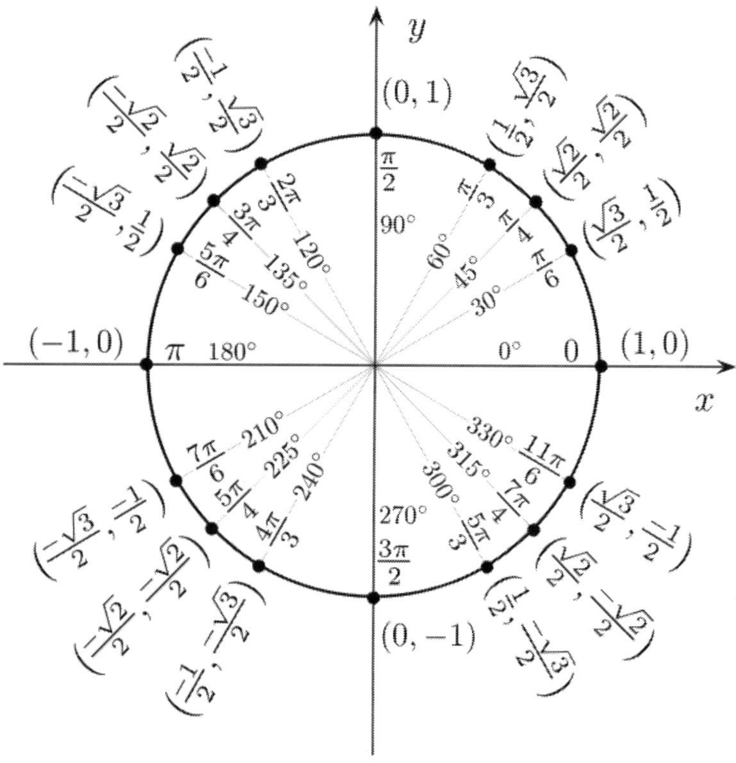

Suppose we create a coordinate graph where the x-axis represents the measure of angle θ. Here we mark along this axis the measures of the angles shown in the unit circle.

Now, to graph the values of $\sin\theta$, where $0 \leq \theta \leq 2\pi$ (or $0° \leq \theta \leq 360°$) we can use the y-coordinates of the terminal points $(\cos\theta, \sin\theta)$ in the unit circle.

Example: At 90° (or $\frac{\pi}{2}$ radians), we can graph $y = \sin 90° = 1$ at the point $\left(\frac{\pi}{2}, 1\right)$.

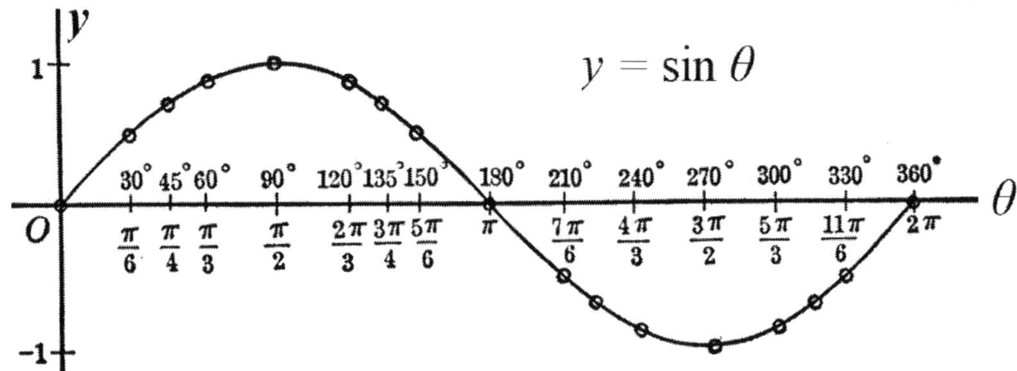

All of the points that make up the unit circle can be graphed as a continuous function. The diagram below shows two such points, at angle θ in Quadrant I and its reflection over the y-axis in Quadrant II. We can see how the points in these two quadrants form a kind of bell-shaped curve between 0 and π, with $\theta = \frac{\pi}{2}$ as a line of symmetry.

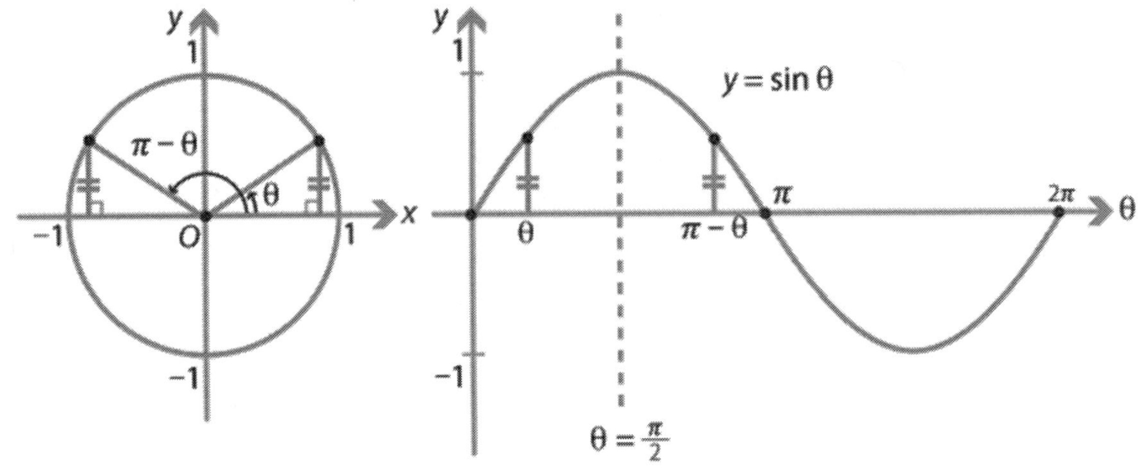

If we replace the number of radians in θ with the variable x, we have a function, $f(x) = \sin x$, graphed below. This is the *parent function* for the family of sine functions. It intersects the x-axis at every multiple of π: at $x = \pi \approx 3.14$, at $x = 2\pi \approx 6.28$, etc. The domain is all real numbers and the range is $-1 \leq y \leq 1$.

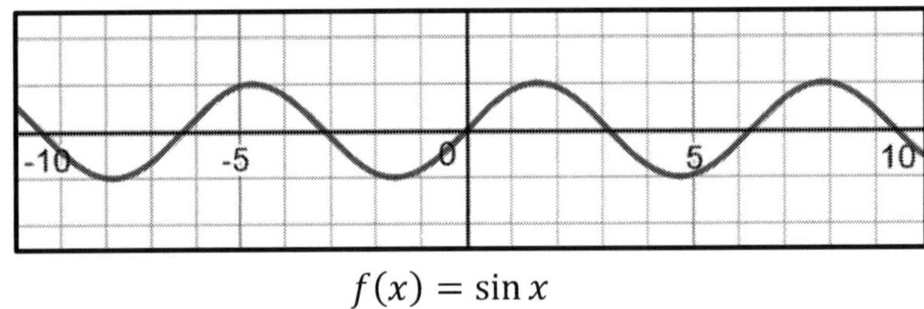

$f(x) = \sin x$

Or, if we prefer, we can label the x-axis with radian measures.

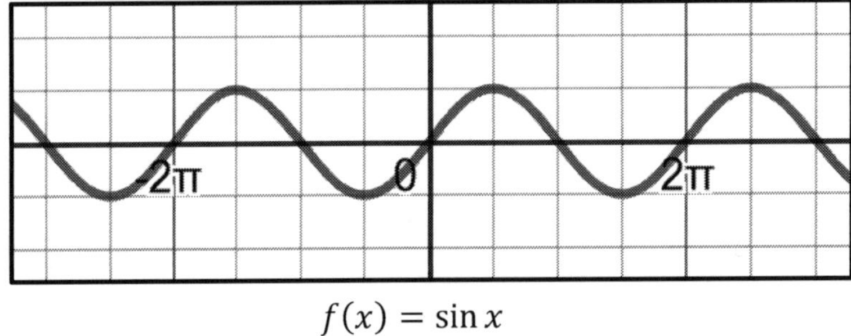

$f(x) = \sin x$

A sine graph, like the one above, is sometimes called a **sinusoid**.

CALCULATOR TIP

If you want to graph a trigonometric function on the calculator using radian measures along the x-axis, use the Trig mode. By default. this will display the graph for $-2\pi \leq x \leq 2\pi$ radians, with tick marks on the x-axis that are $\frac{\pi}{2}$ apart.

You can view a graph in Trig mode by entering the equation and then pressing ZOOM 7.

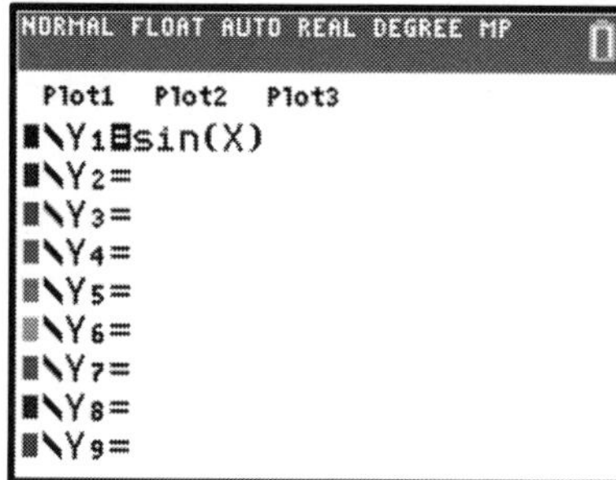

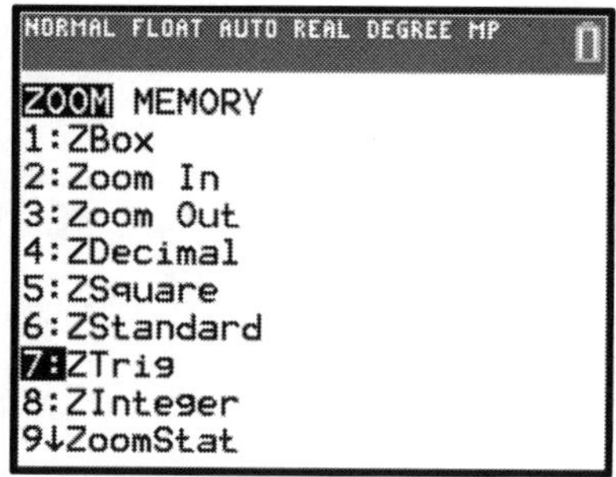

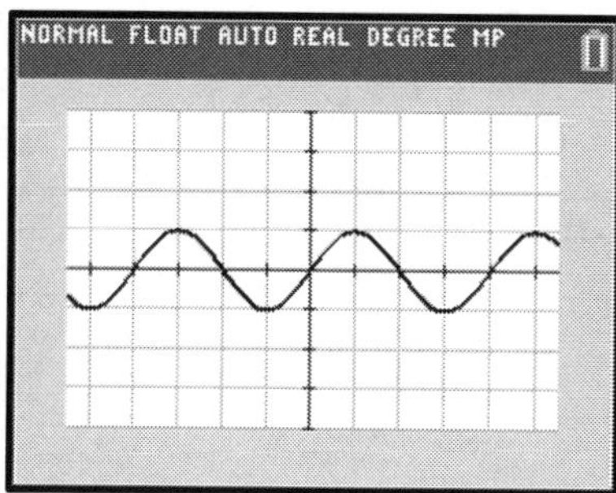

After viewing the graph in Trig mode, you may want to press WINDOW to adjust the minimum and maximum y-values (Ymin and Ymax) of the graph. The Xscl value, which specifies how apart the horizontal tick marks are, should stay set at $\frac{\pi}{2} \approx 1.570796 \ldots$. However, the Yscl value, which specifies how far apart the vertical tick marks are, may be adjusted depending on the amplitude of your function.

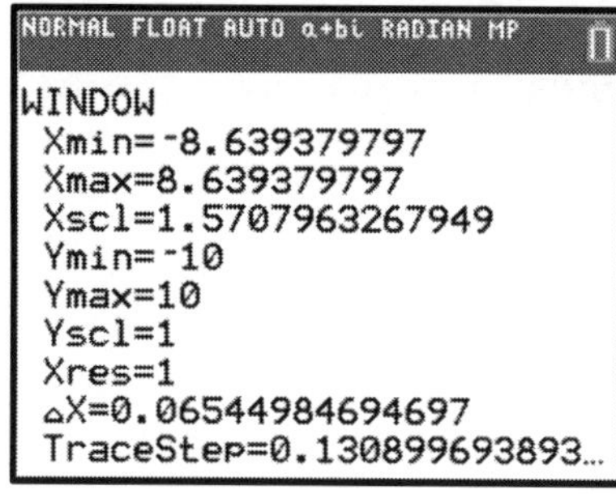

The **sine function** $f(x) = \sin x$ is an example of a periodic function. A **periodic function** is one that repeats a pattern of y-values in regular cycles along the x-axis. The horizontal length of each **cycle**, or one complete pattern, is called a **period**. We can therefore graph the function over the interval $0 \leq x \leq 2\pi$ and then repeat the cycle over and over, both to the left and right of the origin.

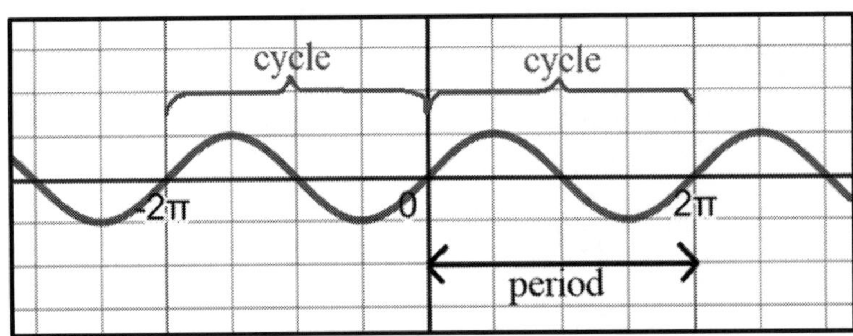

More formally, $f(x)$ is periodic if, for some constant $p > 0$, $f(x + p) = f(x)$ for all x in the domain. For $f(x) = \sin x$, the period p is 2π.

The **frequency** of a graph is the number of cycles that fit within the interval $0 \leq x \leq 2\pi$. For $f(x) = \sin x$, the frequency is 1.

Some periodic functions, such as $f(x) = \sin x$, **oscillate** around a central value. As we see above, $f(x) = \sin x$ oscillates around a central value of $y = 0$, and it reaches as high as $y = 1$ and as low as $y = -1$. Therefore, it has a **range** of $[-1,1]$.

The **midline** is the horizontal line about which the graph oscillates.

Example: In $f(x) = \sin x$, the midline is $y = 0$.

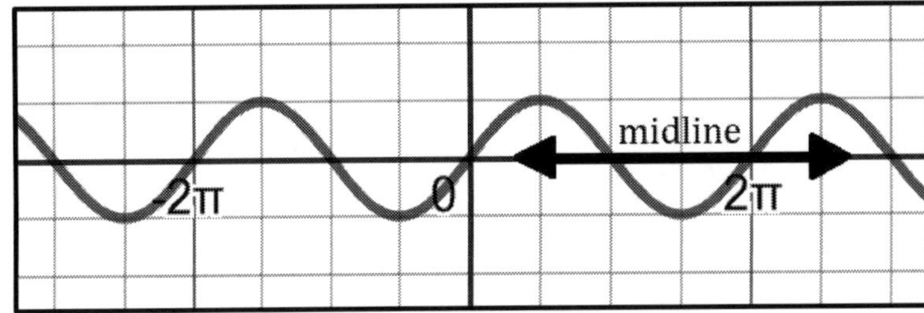

The **amplitude** is the distance from the graph's midline to its maximum (or half the distance between its minimum and maximum values).

Example: In $f(x) = \sin x$, the midline is at $y = 0$ and its maximum y-value is 1, so the amplitude is 1.

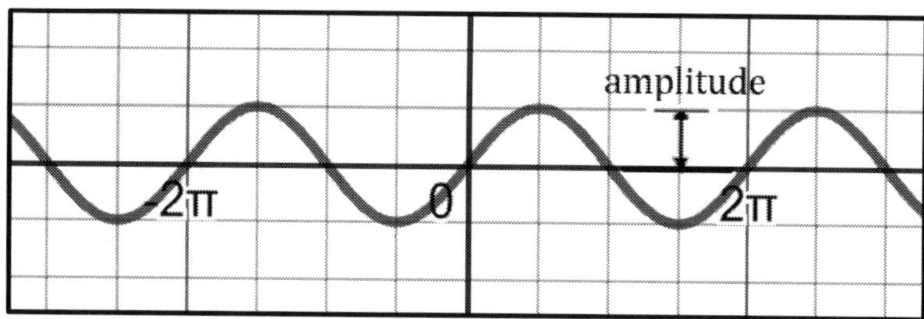

The graph of the **cosine function** $f(x) = \cos x$, shown below, looks very similar to the sine graph. This is the *parent function* for the family of cosine functions. Just like the sine graph, the period is 2π, the midline is $y = 0$, and the amplitude is 1. However, look at the y-intercepts of the two graphs; in the sine graph, the point is (0,0), but in the cosine graph, the point is (0,1). While the sine graph has x-intercepts at $0, \pi, 2\pi$, etc., the cosine graph has x-intercepts at $\frac{\pi}{2}, \frac{3\pi}{2}$, etc.

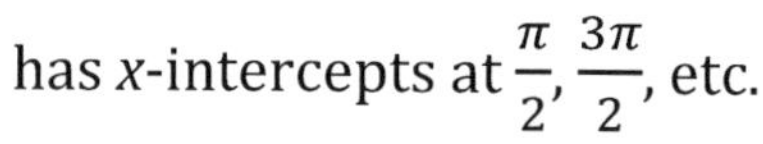

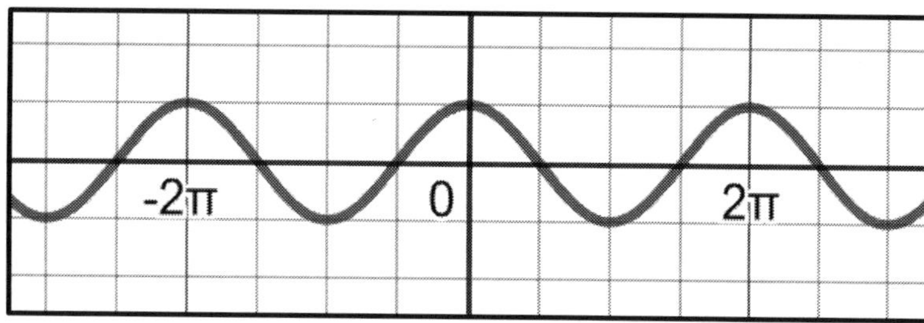

$f(x) = \cos x$

The graph shown below for $f(x) = \tan x$, the *parent function* for the family of tangent functions, looks very different. Rather than being a continuous curve, the tangent graph increases infinitely as it approaches vertical asymptotes from the left, or decreases infinitely as it approaches the vertical asymptotes from the right.

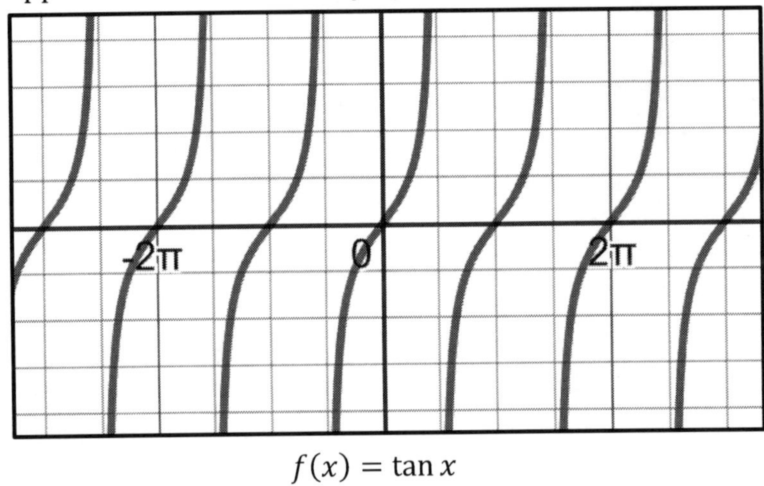

$f(x) = \tan x$

As you may recall from Section 9.8, a **vertical asymptote** is a line that a curve approaches more and more closely from the left or right without ever touching the asymptote.

Example: In the function $f(x) = \tan x$, there are asymptotes at $x = \frac{\pi}{2}, x = \frac{3\pi}{2}$, and so on. They occur at regular intervals that are π units apart. Therefore, the domain of $f(x) = \tan x$ is all real numbers except the odd multiples of $\frac{\pi}{2}$.

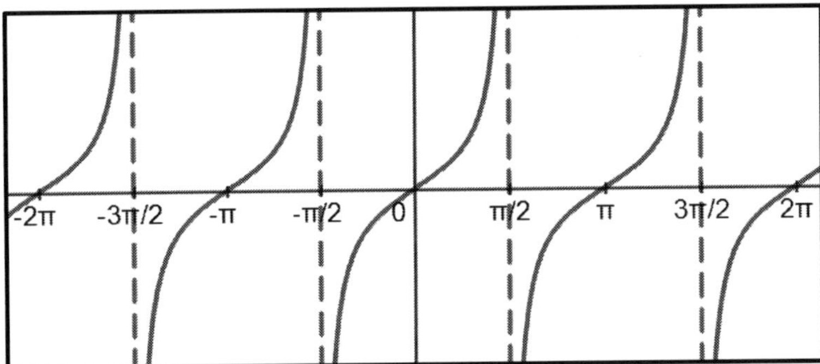

Why is the tangent function undefined at the asymptotes? Consider that $\tan x = \frac{\sin x}{\cos x}$. Therefore, $\tan x$ is an undefined fraction when $\cos x = 0$. As we saw in the cosine graph, $\cos x = 0$ at $\frac{\pi}{2}, \frac{3\pi}{2}$, etc. The domain of $f(x) = \tan x$ includes all real x except where $\cos x = 0$.

Unlike the sine and cosine functions above, the period of $f(x) = \tan x$ is π units. The function has no maximum or minimum values, so it has no amplitude; the range is all real numbers. It has the same x-intercepts as the function $f(x) = \sin x$, at $0, \pi, 2\pi$, etc.

In the unit circle, the tangent of the angle is the slope of the terminal side.

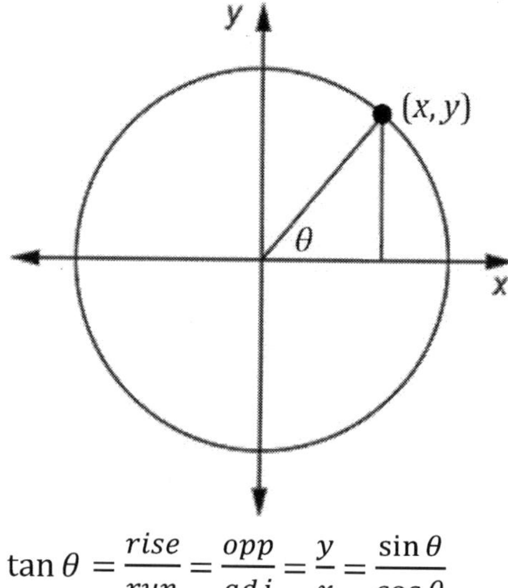

$$\tan\theta = \frac{rise}{run} = \frac{opp}{adj} = \frac{y}{x} = \frac{\sin\theta}{\cos\theta}$$

For $\theta = 0°$, the slope is zero, so $\tan 0 = 0$. As θ increases towards $90°$ or $\frac{\pi}{2}$ rad, the slope (and therefore $\tan\theta$) increases. But at $\theta = 90°$, the slope and $\tan\theta$ are undefined. When the terminal side is located in Quadrant II where $90° < \theta < 180°$ (or $\frac{\pi}{2} < \theta < \pi$ radians), the slope is negative, and so $\tan\theta$ is negative.

A summary of key features of these three parent functions is provided in the table below.

	Sine	**Cosine**	**Tangent**
parent function	$y = \sin x$	$y = \cos x$	$y = \tan x$
domain	$(-\infty, \infty)$	$(-\infty, \infty)$	$(-\infty, \infty)$ except $n \cdot \frac{\pi}{2}$ for odd integers n
range	$[-1,1]$	$[-1,1]$	$(-\infty, \infty)$
asymptotes	none	none	vertical asymptotes at $x = n \cdot \frac{\pi}{2}$ for odd integers n
midline	$y = 0$	$y = 0$	$y = 0$
amplitude	1	1	no amplitude
period	2π	2π	π
frequency	1	1	2
y-intercept	0	1	0
x-intercepts	$n\pi$ for integers n	$n \cdot \frac{\pi}{2}$ for odd integers n	$n\pi$ for integers n

Model Problem

$P\left(\frac{\pi}{4}, y\right)$ is a point on the graph of $f(x) = \tan x$.

a) Find the value of y.

b) What is the image of point P after a 180° rotation around the origin?

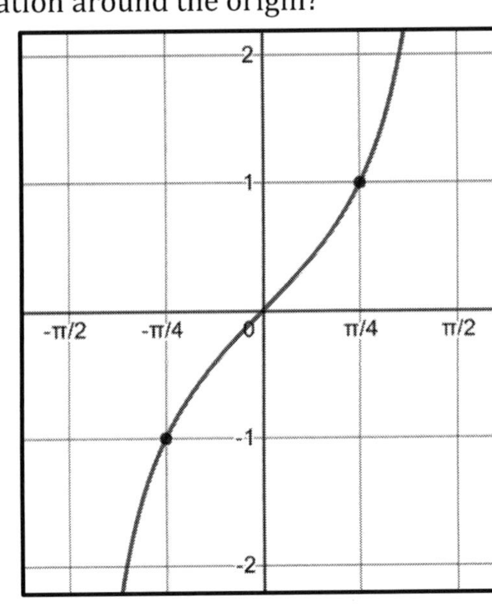

Solution:

(A) $y = \tan\frac{\pi}{4} = 1$

(B) $\left(-\frac{\pi}{4}, -1\right)$

Explanation of steps:

(A) Calculate $y = f(x)$. *[*$f\left(\frac{\pi}{4}\right) = \tan\frac{\pi}{4} = 1$*]*

(B) 180° rotations around the origin will map $(x, y) \rightarrow (-x, -y)$. *[Point* $P\left(\frac{\pi}{4}, 1\right) \rightarrow P'\left(-\frac{\pi}{4}, -1\right)$*, as shown to the right.]*

PRACTICE PROBLEMS

1. In the interval $0° \leq x \leq 360°$, $\tan x$ is undefined when x equals

   (1) 0° and 90° (3) 180° and 270°

   (2) 90° and 180° (4) 90° and 270°

2. The graph of the equation $y = |\sin x|$ will contain *no* points in Quadrants

   (1) I and II (3) III and IV

   (2) II and III (4) I and IV

3. As θ increases from π to $\frac{3\pi}{2}$, which statement is true?

   (1) $\sin\theta$ increases from -1 to 0.

   (2) $\sin\theta$ decreases from 1 to 0.

   (3) $\cos\theta$ increases from -1 to 0.

   (4) $\cos\theta$ decreases from 0 to -1.

4. As x increases from π to 2π, the value of $\sin x$

   (1) increases, only

   (2) decreases, only

   (3) increases, then decreases

   (4) decreases, then increases

5. Which type of symmetry does the equation $y = \cos x$ have?

   (1) line symmetry with respect to the x-axis

   (2) line symmetry with respect to the y-axis

   (3) line symmetry with respect to $y = x$

   (4) point symmetry with respect to the origin

6. Which type of symmetry does the equation $y = \tan x$ have?

   (1) line symmetry with respect to the x-axis

   (2) line symmetry with respect to the y-axis

   (3) line symmetry with respect to $y = x$

   (4) point symmetry with respect to the origin

12.9 Trigonometric Transformations

Key Terms and Concepts

Now that we have seen the parent functions for sine, cosine and tangent, let's take a look at how we can transform the graphs of these functions.

Changing the amplitude

For the graphs of $y = a\sin x$ and $y = a\cos x$, the amplitude is $|a|$.
(Remember that the tangent function has no amplitude.)

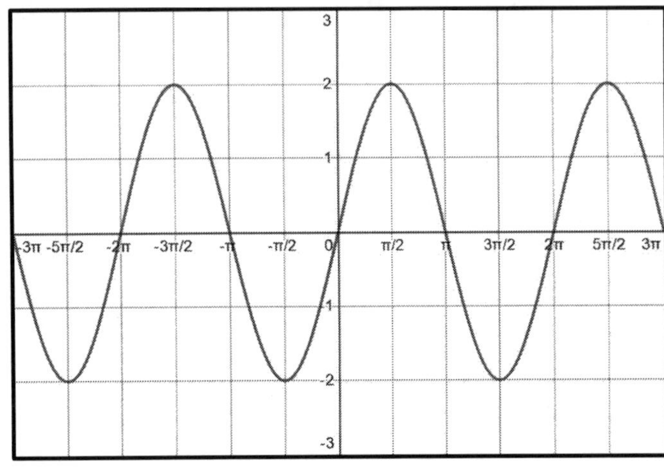

$y = 2\sin x$

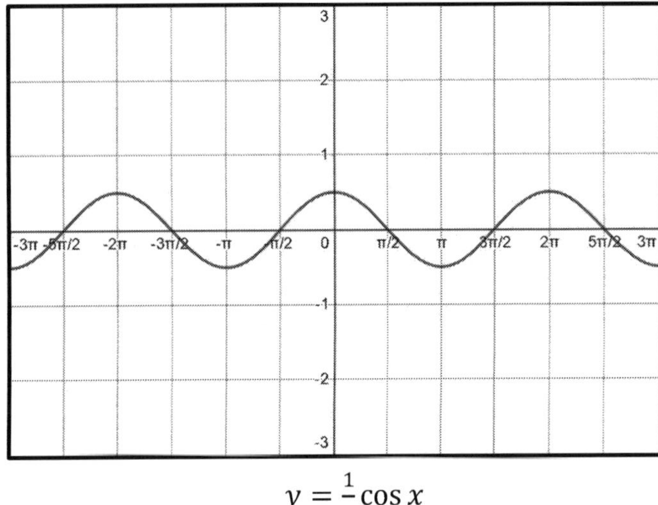

$y = \frac{1}{2}\cos x$

Note that if a is negative, we would have a reflection of the curve over the x-axis.

Changing the frequency

For the graphs of $y = \sin bx$ and $y = \cos bx$, the frequency is $|b|$. Given the frequency, we can calculate $period = \frac{2\pi}{frequency} = \frac{2\pi}{|b|}$. If given the period, $frequency = \frac{2\pi}{period}$. For a sine or cosine graph, a period is sometimes called a **wavelength**.

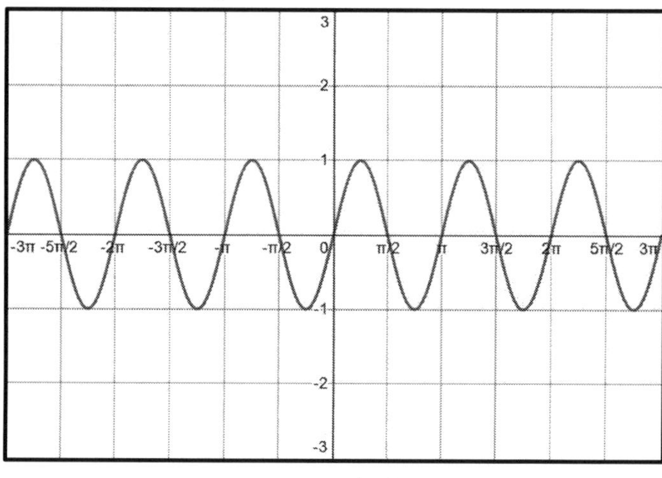

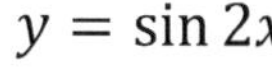
$y = \sin 2x$

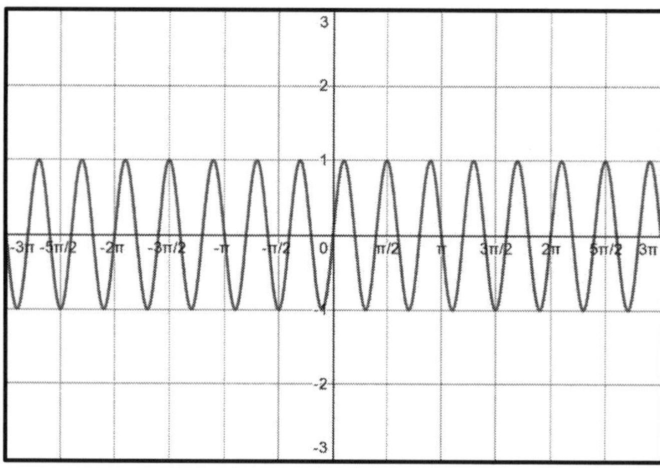

$y = \sin 5x$

Note that in science, frequency is sometimes defined as the number of cycles between 0 and 1 rather than between 0 and 2π. For example, hertz is a measure of sound frequency and is equal to the number of cycles between 0 and 1 second. However, if we multiply the frequency by 2π, this tells us the frequency for the corresponding sine or cosine graph. We will take a closer look at this distinction in the next section.

For the graph of $y = \tan bx$, the frequency is $|2b|$. Given the frequency, we can calculate $period = \frac{2\pi}{frequency} = \frac{\pi}{|b|}$. If given the period, $frequency = \frac{2\pi}{period}$.

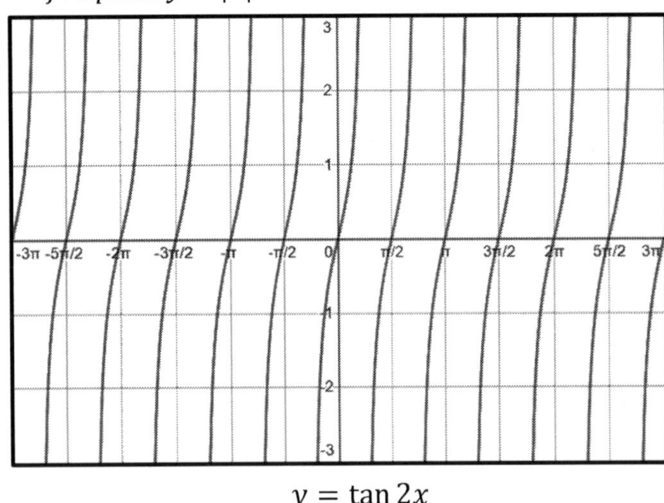

$y = \tan 2x$

In a unshifted tangent graph, the vertical asymptotes appear at every other multiple of $\frac{\pi}{2b}$.

Example: Below is a graph of $y = \tan 3x$. Asymptotes appear at $\frac{\pi}{6}$ and $\frac{\pi}{2}$, as well as at $-\frac{\pi}{6}$ and $-\frac{\pi}{2}$.

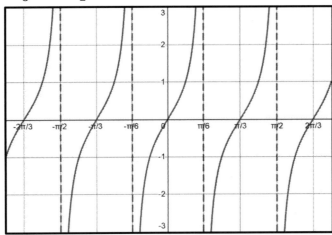

In any of the trigonometric functions, negating b will reflect the graph over the y-axis.

Remember that for sin and cos functions, $f = |b|$, but for tan functions, $f = |2b|$.

In either case, $f = \frac{2\pi}{p}$ and $p = \frac{2\pi}{f}$.

Horizontal shifts

The graph of $y = \sin(x + c)$ will shift the graph of $y = \sin x$ horizontally by c units to the left if $c > 0$, or by c units to the right if $c < 0$. A horizontal shift of a periodic function is often called a **phase shift**. The same phase shifts occur for the graphs of $y = \cos(x + c)$ and $y = \tan(x + c)$.

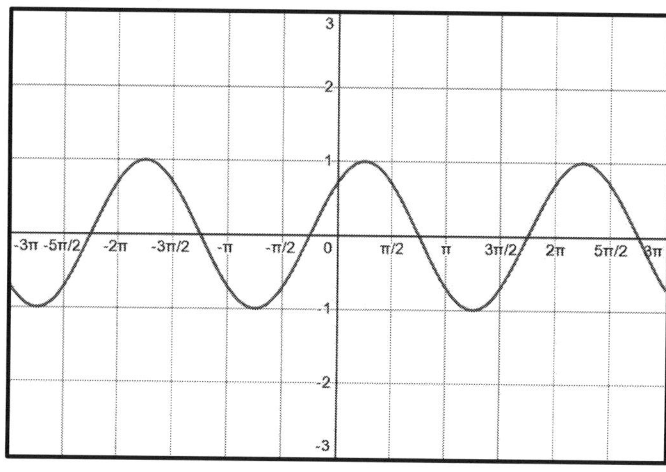

$y = \sin\left(x + \frac{\pi}{4}\right)$

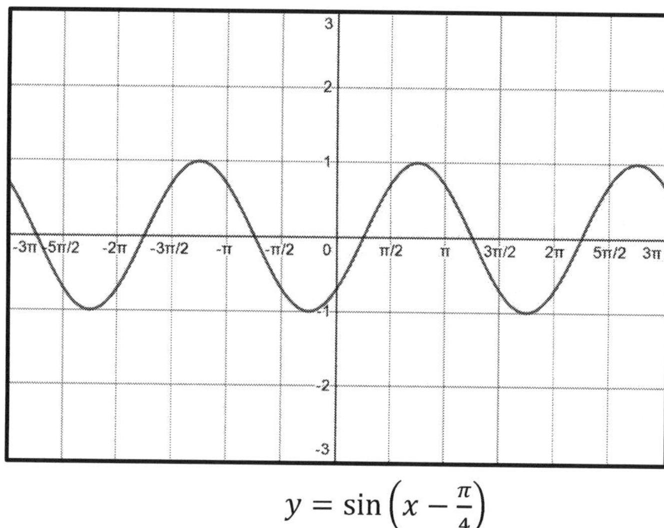

$y = \sin\left(x - \frac{\pi}{4}\right)$

To find the left and right endpoints of one cycle, we can set the expression in parentheses equal to 0 and the period (in this case, 2π).

Example: For the first graph above, solve for $x + \frac{\pi}{4} = 0$ and $x + \frac{\pi}{4} = 2\pi$ to get $x = -\frac{\pi}{4}$ and $x = \frac{7\pi}{4}$. As shown, these are the left and right x-values of one cycle.

Note that if the function is written as $y = \sin(bx + c)$, where $b \neq 1$, you'll need to divide c by b to determine the phase shift. This is because $\sin(bx + c) = \sin\left(b\left(x + \frac{c}{b}\right)\right)$.

Example: The horizontal shift for $y = \sin\left(3x - \frac{\pi}{2}\right)$ is $\frac{\pi}{2} \div 3 = \frac{\pi}{6}$ units to the right.

Vertical shifts

The graph of $y = \sin x + d$ will shift the graph of $y = \sin x$ vertically by d units up if $d > 0$ or by d units down if $d < 0$. The same vertical shifts occur for $y = \cos x + d$ and $y = \tan x + d$. In each case, the equation of the midline is $y = d$.

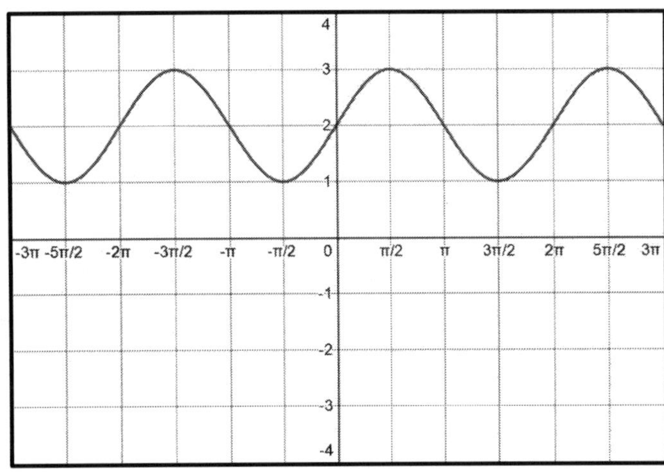

$y = \sin x + 2$

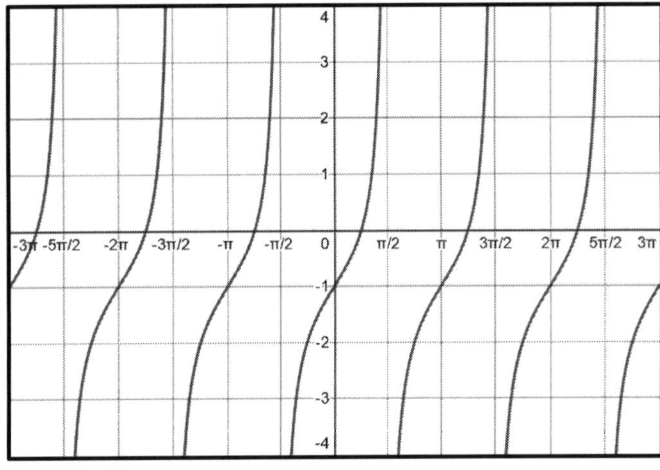

$y = \tan x - 1$

Combining transformations

Of course, multiple transformations can be applied to a trigonometric function. You may use the following list of rules as a guide (replace sin with cos or tan as needed).

In the graph of $y = a\sin(b(x + c)) + d$,

- a change in a affects the amplitude *(or vertically stretches* tan*)*
- a change in b affects the frequency (and period)
- a change in c causes a horizontal shift (left if $c > 0$)
- a change in d causes a vertical shift (up if $d > 0$)
- negating a reflects the graph over the x-axis
- negating b reflects the graph over the y-axis

Determining the equation of a graph

Given a graph, we can determine the equation of the function by examining the amplitude, frequency, and any shifts.

Example: The following graph seems to represent a transformed cosine graph, so we'll use the form $f(x) = a\cos\big(b(x + c)\big) + d$. We start with cosine rather than sine because there is a maximum on the y-axis, but see the note below. Since the maximum is 2 and the midline is $y = -1$, the amplitude is $2 - (-1) = 3$. Also, the graph is not reflected over the x-axis (a is positive), so $a = 3$. To find b, we notice that there is a complete cycle from $x = 0$ to $x = \pi$, so the period is π. Therefore, the frequency $b = \dfrac{2\pi}{\pi} = 2$. There is no horizontal shift, so $c = 0$. There is, however, a vertical shift down 1, since the midline is at $y = -1$, so $d = -1$. By substitution, our equation is $f(x) = 3\cos(2x) - 1$

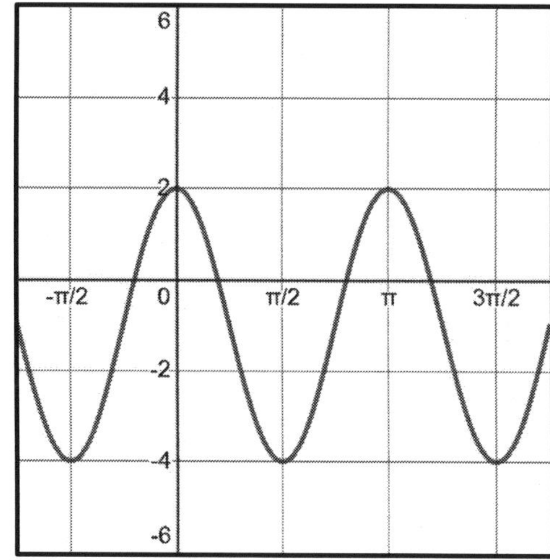

Note: A cosine graph looks just like a sine graph that has been horizontally shifted, so it would not be incorrect to write a sin function for the graph above, as long as a correct value of c is used (such as $\frac{\pi}{4}$ in this case, phase shifting the sin graph $\frac{\pi}{4}$ units left). Therefore, the equation for the above graph could have been written as $f(x) = 3\sin\left(2\left(x + \frac{\pi}{4}\right)\right) - 1$, or more simply as $f(x) = 3\sin\left(2x + \frac{\pi}{2}\right) - 1$.

Finding the maximum or minimum

To find the maximum or minimum value of a sine or cosine graph, we can use the facts that the maximum values occur when $\sin x = 1$ or $\cos x = 1$ and the minimum values occur when $\sin x = -1$ or $\cos x = -1$. Only the amplitude ($|a|$) and vertical shift (or midline, d) may affect the maximum or minimum, so we don't need to be concerned about the graph's frequency (b) or horizontal shifts (c). For any sin or cos function, we can start at the midline and add the amplitude to get the maximum, or start at the midline and subtract the amplitude to get the minimum. Therefore, the maximum is always $d + |a|$ and the minimum is always $d - |a|$.

Example: For the function $f(x) = 6\sin(3x) + 10$, graphed below, the maximum is $d + |a| = 10 + 6 = 16$ and the minimum is $d - |a| = 10 - 6 = 4$.

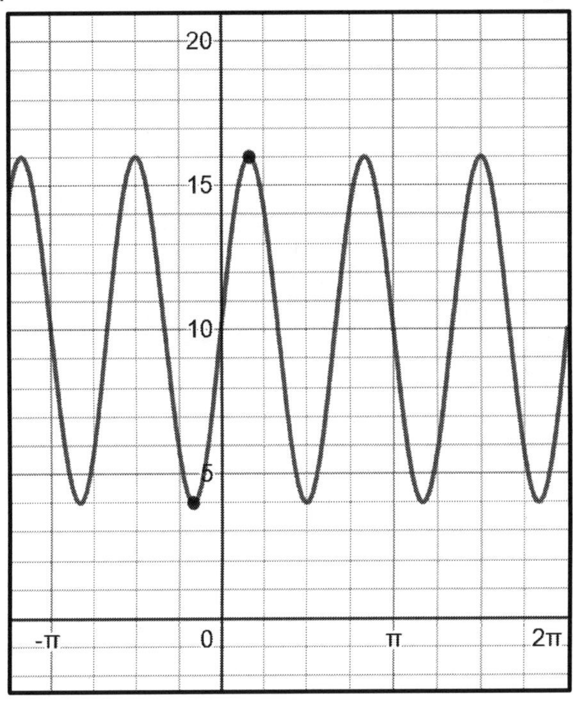

If we want to know the x-values at the maximum or minimum points, we can find one such point by setting the sin or cos equal to 1 or -1 (ignoring the amplitude and midline values of a and d) and solving for x in radians. (Be sure to set your calculator to Radian mode.)

Example: For the function $f(x) = 6\sin(3x) + 10$ which we saw above, we can set $\sin(3x) = 1$ and solve for x. This gives us $3x = \sin^{-1}(1) = \frac{\pi}{2}$, so $x = \frac{\pi}{6}$ at the maximum point $\left(\frac{\pi}{6}, 16\right)$. To find a minimum, set $\sin(3x) = -1$, giving us $3x = \sin^{-1}(-1) = -\frac{\pi}{2}$, so $x = -\frac{\pi}{6}$ at the minimum point $\left(-\frac{\pi}{6}, 4\right)$.

When solving for x in these problems, it is helpful to remember (with results in radians):

$$\sin^{-1}(1) = \frac{\pi}{2} \qquad \sin^{-1}(-1) = -\frac{\pi}{2}$$

$$\cos^{-1}(1) = 0 \qquad \cos^{-1}(-1) = \pi$$

Then, since there are an infinite number of maximum and minimum points on the graph, each one period apart, the x-values of the other points can be determined by adding or subtracting any multiple of the period to x.

Example: For the function $f(x) = 6\sin(3x) + 10$, since the frequency is 3, the period is $\frac{2\pi}{3}$, which is equivalent to $\frac{4\pi}{6}$. So, we can find the x-values of other maximum points by adding or subtracting any multiple of $\frac{4\pi}{6}$ to $\frac{\pi}{6}$, such as $\frac{5\pi}{6}$, or $\frac{9\pi}{6}$, or $-\frac{3\pi}{6}$, etc. We can also find the x-values of other minimum points by adding or subtracting any multiple of $\frac{4\pi}{6}$ to $-\frac{\pi}{6}$, such as $\frac{3\pi}{6}$, or $\frac{7\pi}{6}$, or $-\frac{5\pi}{6}$, etc.

Model Problem

Write an equation for the cosine graph below.

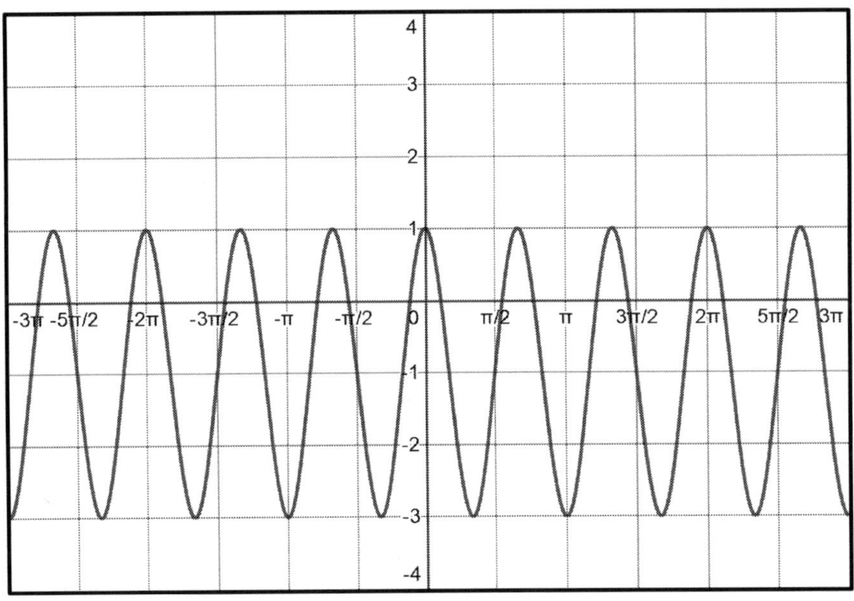

Solution:

(A) (B) (C)

$$y = 2\cos 3x - 1$$

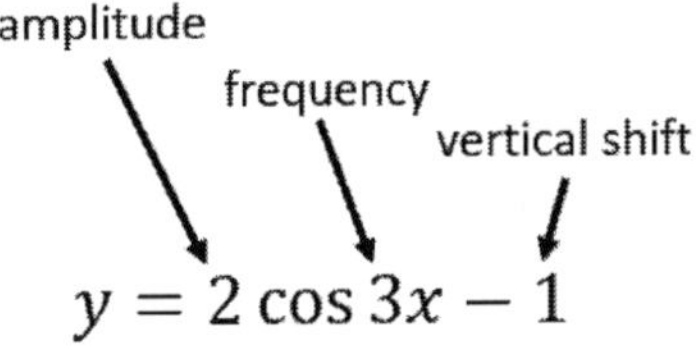

Explanation of steps:

(A) The absolute value of a represents the amplitude. [This graph ranges from $y = -3$ to $y = 1$, which are 4 units apart. The amplitude is half of this distance, so $a = 2$.]

(B) The coefficient of x, b, represents the frequency. [Counting the cycles from $x = 0$ to $x = 2\pi$, the frequency is 3, so $b = 3$.]

(C) The constant value, k, represents the vertical shift, which we can determine by finding the midline. [The midline is at $y = -1$, so $k = -1$.]

Note that there is no horizontal shift in this graph, which we can tell by the fact that the y-intercept is the maximum of the curve, as it would be for the parent function $y = \cos x$.

PRACTICE PROBLEMS

1. What are the amplitude and period for the graph below?

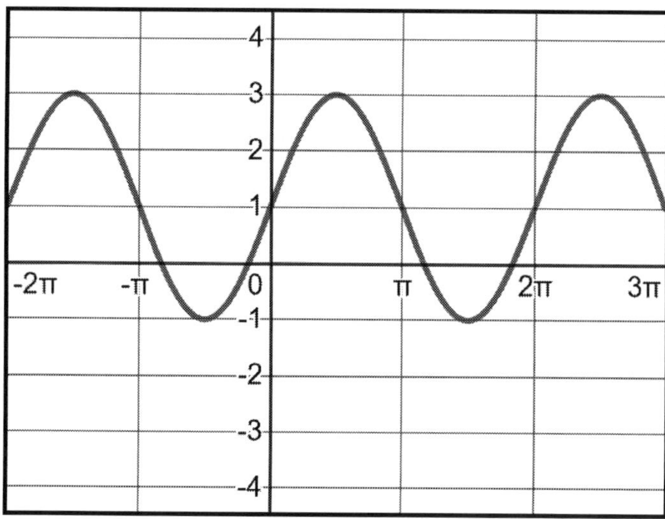

2. What is the amplitude of the function shown in the graph below?

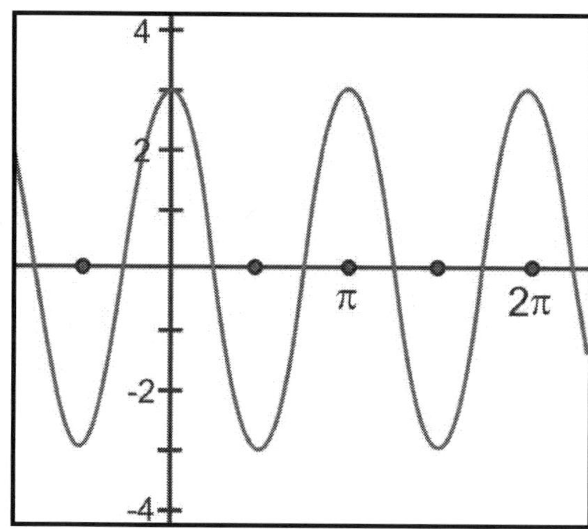

What is the frequency of this graph and how long is each period?

3. Which graph represents an equation of the form $y = a \sin bx$ with a period of π?

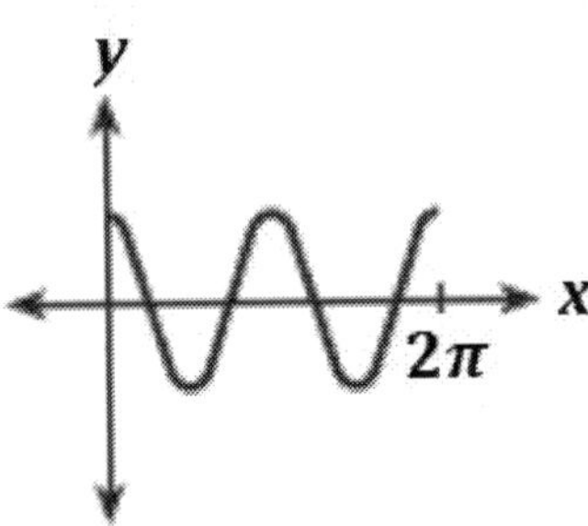

(1)

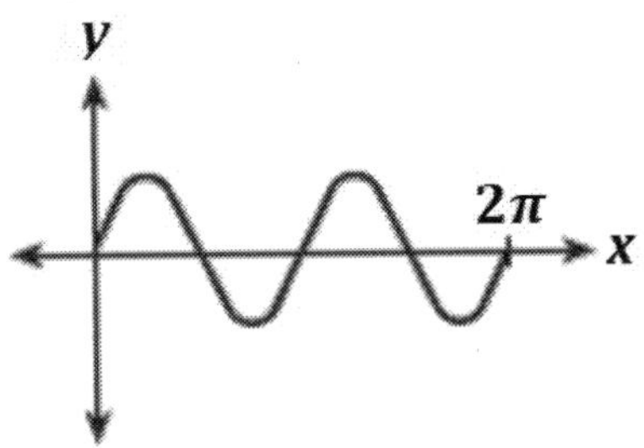

(3)

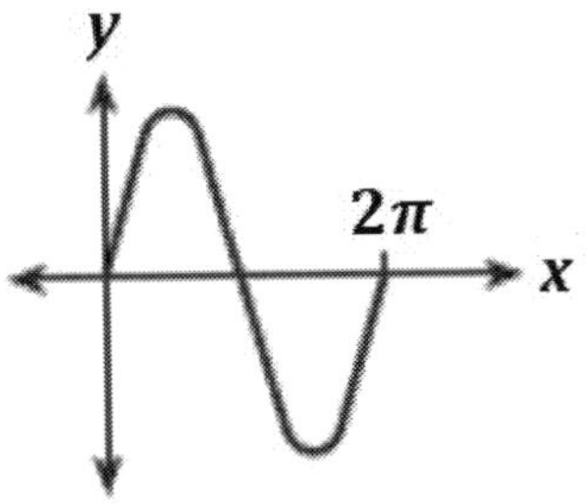

(2)

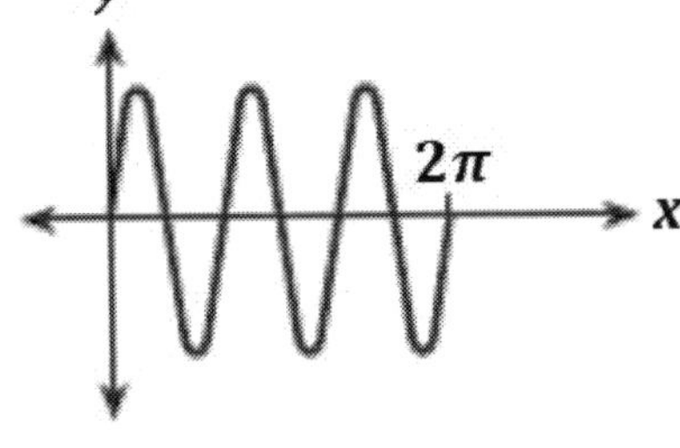

(4)

4. Which graph represents the function $f(x) = -\sin x$ in the interval $-\pi \leq x \leq \pi$?

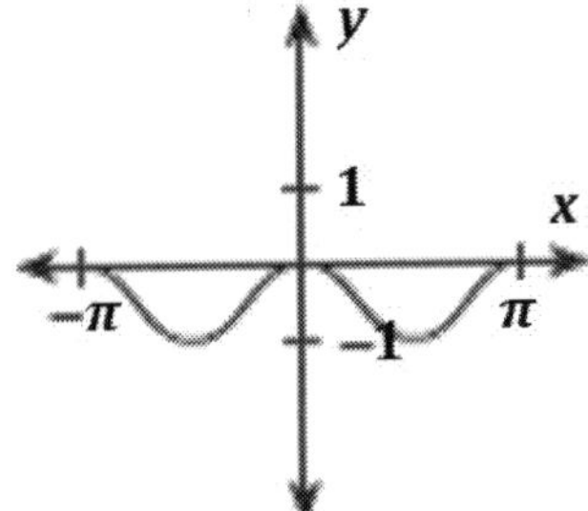

(1)

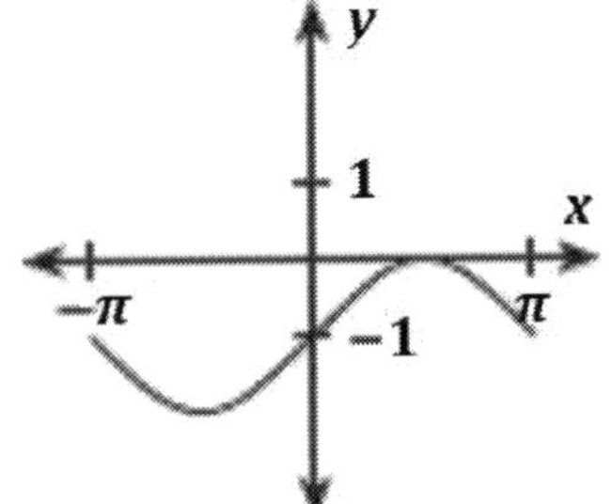

(3)

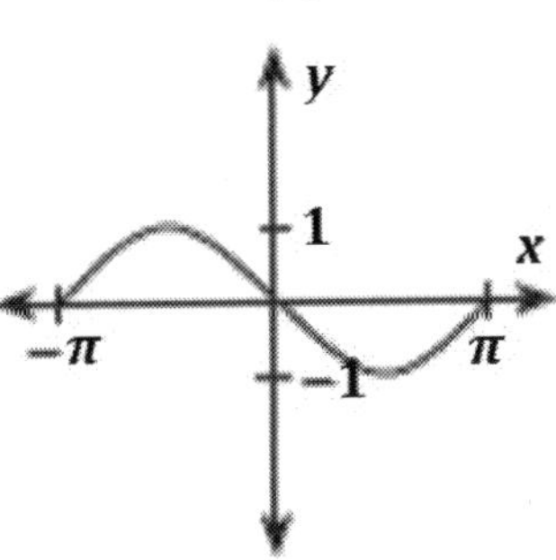

(2)

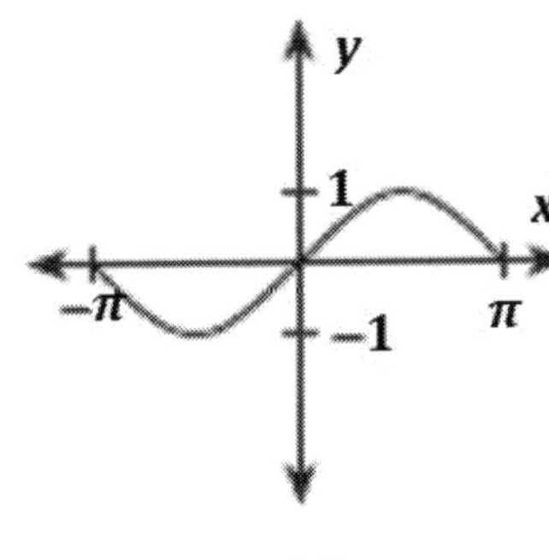

(4)

5. What is the period for $f(x) = \sin 4x$?	6. What is the frequency for a sine function with a period of π?
7. What is the period for $f(x) = \tan(-4x)$?	8. What is the frequency for a tangent function with a period of $\frac{\pi}{6}$?
9. What is the amplitude of the function $f(x) = \frac{2}{3}\cos 4x$? What is the function's frequency and the length of its period?	10. What is the midline of the function $g(x) = 2\tan 3x + 4$? What is the function's frequency and the length of its period?
11. What is the maximum value of the function $f(x) = 3\sin 2x$?	12. What is the minimum value of the function $f(x) = \frac{1}{3}\sin 5x$?

13. What is the maximum value of $f(x) = 2\sin 3x + 4$?	14. What is the minimum value of $f(x) = -2\cos x - 3$?
15. For the function $y = 4\sin(2x - \pi)$, state the midline, amplitude, period, and phase shift.	16. For the function $y = \frac{1}{2}\cos\left(3x + \frac{\pi}{2}\right) - 5$, state the midline, amplitude, period, and phase shift.
17. As angle x increases from 180° to 270°, the value of $\cos x$ will (1) increase from 0 to 1 (2) increase from −1 to 0 (3) decrease from 0 to −1 (4) decrease from 1 to 0	18. Given $f(x) = \sin(x + h)$ and $g(x) = \cos x$, what value of h will make the graph of $f(x)$ coincide with the graph of $g(x)$?

12.10 Graph Trigonometric Functions

KEY TERMS AND CONCEPTS

To graph a trigonometric function, write it in $f(x) = a\sin(b(x + c)) + d$ form (replacing sin with cos or tan as needed) to determine the function's properties.

In the graph of $y = a\sin(b(x + c)) + d$,

- a change in a affects the amplitude *(or vertically stretches* tan*)*
- a change in b affects the frequency (and period)
- a change in c causes a horizontal shift (left if $c > 0$)
- a change in d causes a vertical shift (up if $d > 0$)
- negating a reflects the graph over the x-axis
- negating b reflects the graph over the y-axis

Based on the values of a, b, c, and d, we should be able to sketch a graph of any function manually. However, we have a graphing calculator for just that purpose, so it's best to start by graphing the function on the calculator. Before copying from the calculator, check that the attributes of the graph – the midline, amplitude, frequency, reflections, etc. – are as expected based on the values of a, b, c, and d. Then, carefully copy the graph to the grid.

CALCULATOR TIP

When graphing a trigonometric function on the calculator, it is often best to use the ZTrig Zoom mode. By default. this will display the graph for $-2\pi \leq x \leq 2\pi$ radians, with tick marks on the x-axis that are $\frac{\pi}{2}$ apart. *[Look back at Section 12.8 for steps on how to do this.]*

MODEL PROBLEM

Sketch the graph of $y = 3\cos 2x - 1$.

Solution:

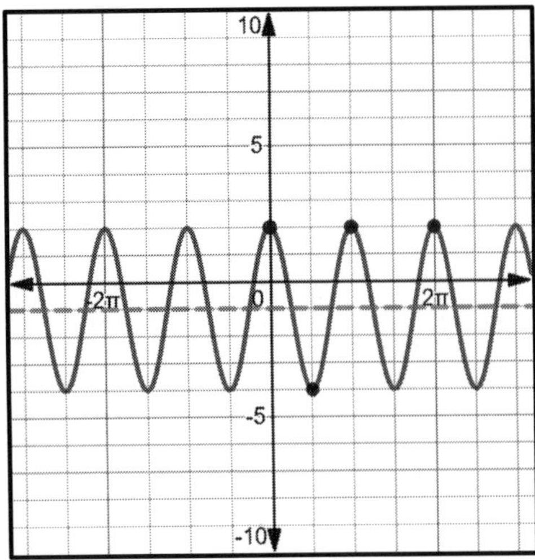

Explanation of steps:

The function is written in the form $y = a\cos(b(x + c)) + d$ [*where* $a = 3$, $b = 2$, $c = 0$, *and* $d = -1$]. For cos graphs, $|a|$ represents the amplitude, $|b|$ represents the frequency, and c and d shift the graph horizontally and vertically, respectively.

[Method 1: Draw the midline at $y = -1$ *since* $d = -1$. *There is no horizontal shift, so the cosine graph will have a maximum on the y-axis. Since* $a = 3$, *that point will be 3 units above the midline, at* $(0,2)$. *The period is* $\frac{2\pi}{2} = \pi$, *so there will be additional maximums at* $(\pi, 2)$ *and* $(2\pi, 2)$, *and at every interval* π *units apart. There will be a minimum 3 units below the midline, half-way between 0 and* π *at* $\left(\frac{\pi}{2}, -4\right)$. *Plot these points and sketch the graph.]*

[Method 2: Graph the function on the calculator and carefully copy it onto the grid.]

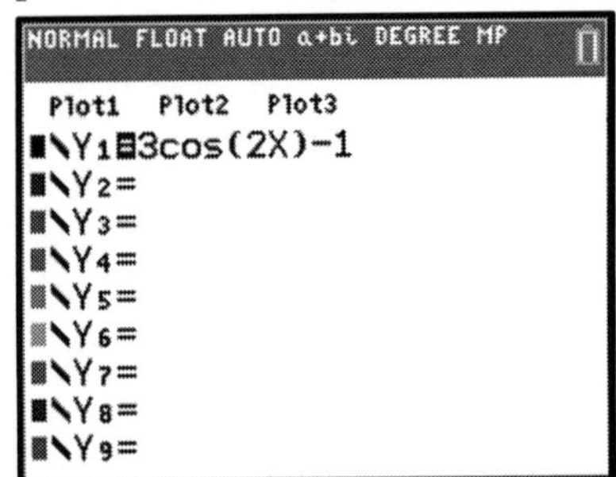

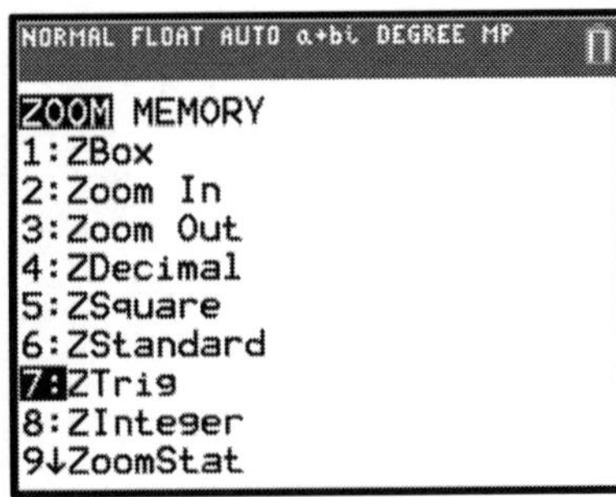

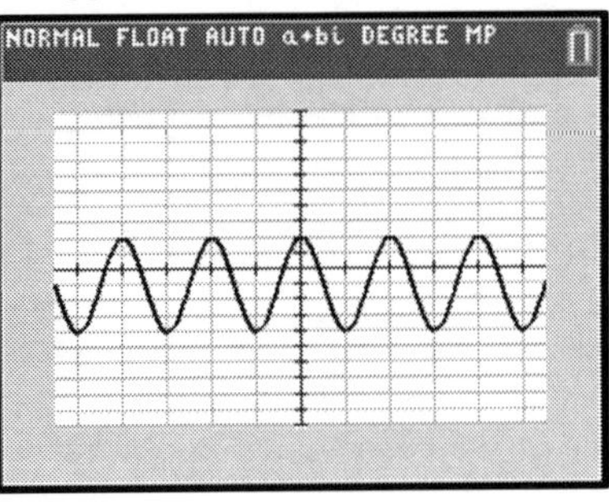

PRACTICE PROBLEMS

1. The graph of an unshifted sine function has an amplitude of 3 and a period of π. Sketch a graph of the curve over the interval 0 to 2π.

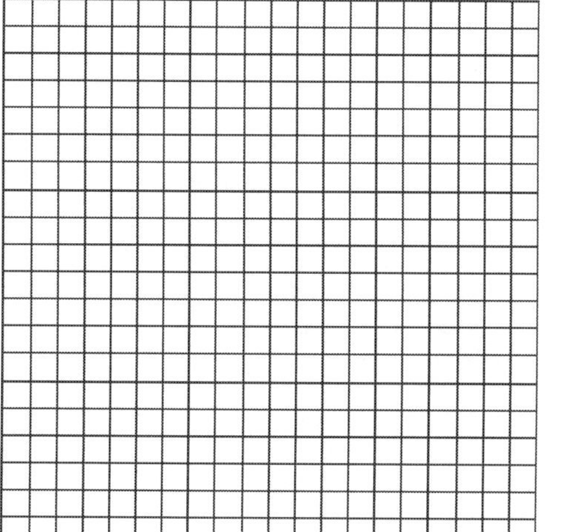

Write an equation for your graph.

2. The graph of a cosine function has a midline at $y = -2$, an amplitude of 4, a period of 2π, and no phase shift. Sketch the graph over the interval 0 to 2π.

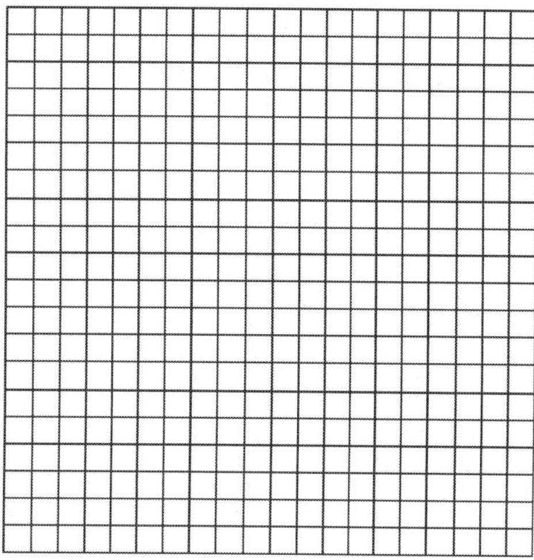

Write an equation for your graph.

3. Sketch the graph of $y = 2\sin\frac{1}{2}x$ in the interval $-2\pi \leq x \leq 2\pi$.

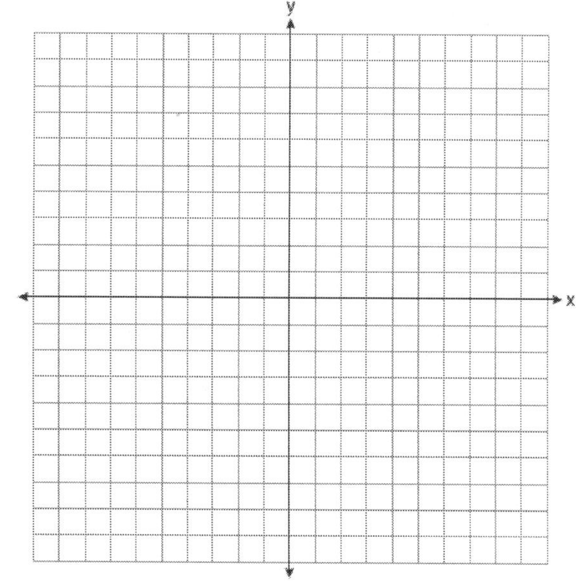

4. Sketch the graph of $y = 2\tan\frac{1}{2}x$ in the interval $-2\pi \leq x \leq 2\pi$.

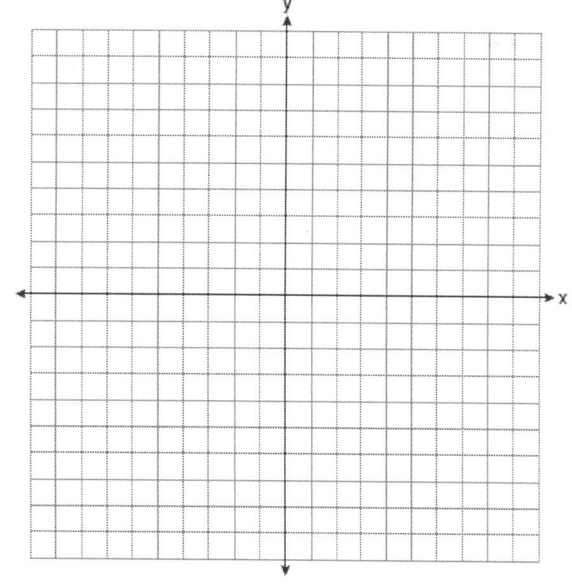

5. On the grid below, sketch the graphs of $f(x) = 2\cos x$ and $g(x) = -2\cos x$ in the interval $-\pi \le x \le \pi$.

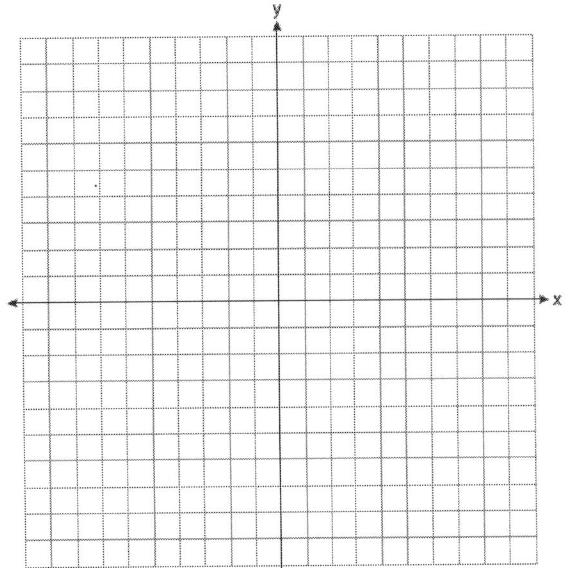

How many points of intersection do the graphs have in this interval?

What is the value of $g\left(\frac{\pi}{6}\right)$?

6. Graph the functions $y = 4\cos x$ and $y = 2$ over the domain $-\pi \le x \le \pi$.

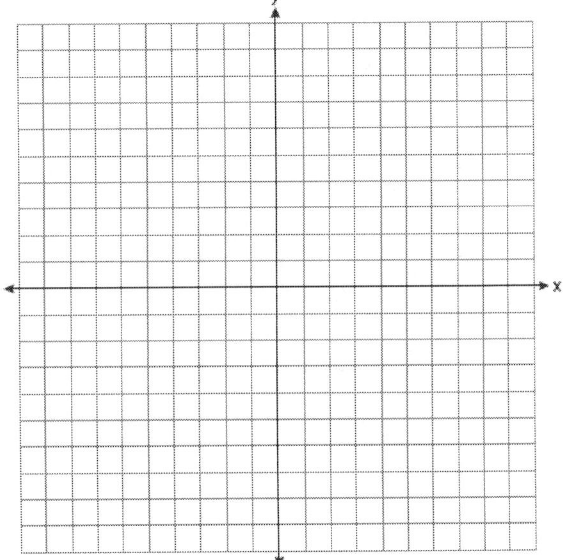

Express, in terms of π, the interval for which $4\cos x \ge 2$.

12.11 Model Trigonometric Functions

Key Terms and Concepts

There are many applications of trigonometric functions to real world situations. For examples, sound and light are represented by sine waves. Any situation that involves a continuous ebb and flow, such as the changes in water levels from high tide to low tide, could be represented by a cosine or sine function.

The swinging of a pendulum can be represented by a cosine or sine function that calculates the displacement of the pendulum from its lowest point. We could represent the displacement to the right as positive and to the left as negative.

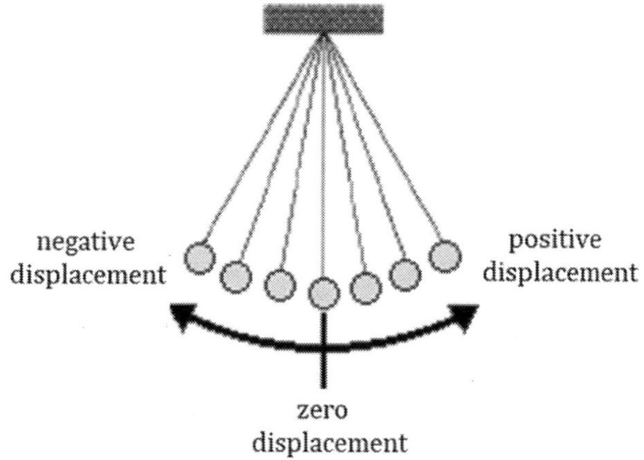

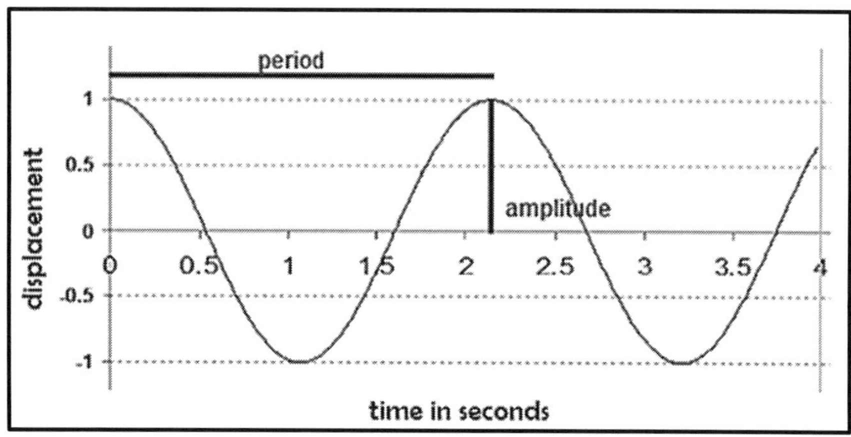

If we start the pendulum at the position of maximum displacement, the function $p(t) = a\cos(bt)$ could be used as a template.

The amplitude – the maximum amount of displacement – would be used for a. The time that it takes to swing from the maximum positive position back to the same position would represent one period. The frequency (b) is calculated as 2π divided by the period.

The height of a point on a turning wheel could also be represented by a sine or cosine function. For example, a function could calculate the altitude of a rider on a Ferris wheel as it turns, with the altitude at the base representing the minimum and the altitude at the top as the maximum. The amplitude of the function would be the radius of the Ferris wheel.

The function $h(t) = -a\cos(bt) + d$, where t represents time and h represents height, could be used as a template. The amplitude, a, is the radius of the wheel. However, since we start the ride at the lowest height, we would negate the function. Remember, the function $\cos t$ normally starts at its maximum value when $t = 0$, so by negating the function, we reflect it over the midline so that it starts at its minimum value instead.

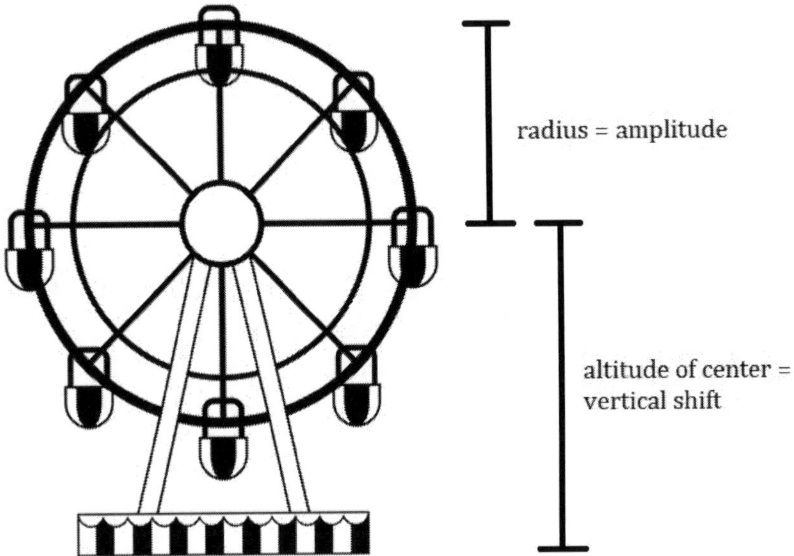

The amount of time it takes to travel from the starting position, the base of the Ferris wheel, back to the same position represents one period. The frequency (b) is calculated as 2π divided by the period.

The vertical shift, d, would be the altitude of the center of the wheel, since this is the distance above ground, represented by the midline of the graph.

We can also model sound waves as sine graphs. The amplitude of the sine graph represents relative **volume** of the sound, which is why we call a device that make sound louder an amplifier. Volume may be measured in linear units of air pressure such as pascals (Pa). Although volume is more often measured in decibels (dB), the decibel scale is logarithmic, not linear, so you are not likely to see it used for problems in this course.

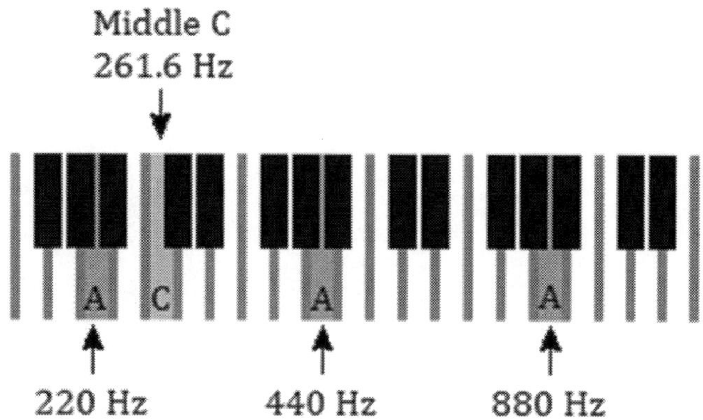

For sound waves, the **pitch** of a sound depends on its frequency; the higher the frequency, the higher the pitch. However, the frequency of a sound is related to, but not exactly the same, as the frequency of the sine graph that represents it. The **frequency** of a sound is usually measured in hertz (Hz), which is the number of cycles per second, or in kilohertz (kHz), which is equal to 1,000 cycles per second. To determine the frequency for the corresponding sine function and its graph, multiply the number of hertz by 2π.

Humans can hear sounds in a range from about 20 Hz to 20 kHz. If we select a sound near the middle of that range, say 10 kHz, which equals 10,000 Hz, then the frequency of the corresponding sine graph would be $10{,}000 \times 2\pi = 20{,}000\pi$. We can model this sound using the function $y = \sin 20{,}000\pi\, t$, where t is the number of seconds, assuming an amplitude of 1 unit. Of course, if we increase the loudness (i.e., amplitude) of the sound by a factor of 5, the function would be represented by $y = 5 \sin 20{,}000\pi\, t$.

MODEL PROBLEM

A Ferris wheel has a radius of 25 feet and makes one complete revolution counterclockwise every 20 seconds. The center of the Ferris wheel is 30 feet above the ground. Graph a function representing the height of a rider on the Ferris wheel over time for at least two complete revolutions.

Solution:

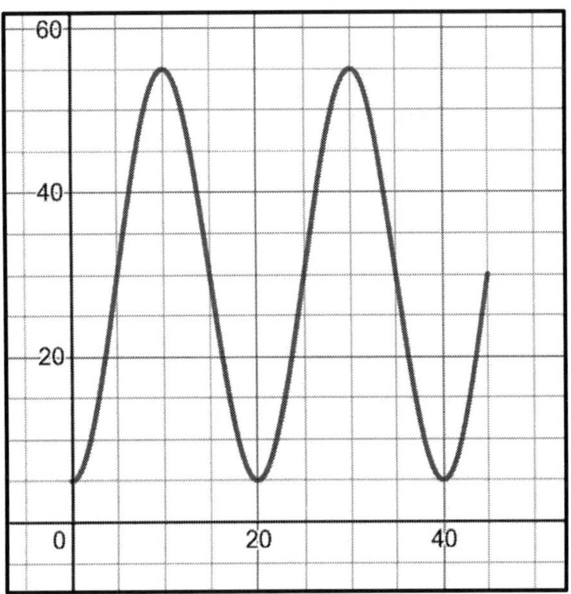

Explanation of steps:

The function is represented by a negated cosine graph, starting at a minimum point when $t = 0$ since the rider starts at the lowest point on the wheel. The midline is the center of the Ferris wheel *[30 feet]*. The altitude is the radius of the wheel *[25 feet, so the curve oscillates between 5 and 55 feet]*. Each period is the time it takes for a complete revolution *[20 seconds]*.

[Note: The equation of the graph is $h(t) = -25\cos\left(\frac{\pi}{10}t\right) + 30$.
The frequency $\frac{\pi}{10}$ *is calculated as* 2π *divided by the period, 20.]*

PRACTICE PROBLEMS

1. A sine function increasing through the origin can be used to model sound waves. A certain note produces a wavelength of 440 hertz. Over which interval, in hertz, is the height of the wave *increasing*, only?

   (1) (0,220) (3) (220,330)
   (2) (110,220) (4) (330,440)

2. A roller coaster travels along a path modeled by the equation $y = 27\sin 0.05x + 30$, with x and y measured in feet. What is the maximum altitude of the roller coaster?

3. The Half Moon, a pendulum ride at an amusement park, rocks back and forth from its starting position according to the function $h(t) = 10\sin\left(\frac{\pi}{3}t\right)$, where t represents time, in seconds. How many seconds does the ride take to complete one full cycle?	4. The altitude, in feet, of a rider on a continuously moving Ferris wheel is modeled by the function $f(t) = -20\cos\left(\frac{\pi}{5}t\right) + 24$, where t represents time, in seconds. a) How many complete revolutions will the rider complete on a 40 second ride? b) What is the radius of the Ferris wheel? c) What is the altitude of the rider at the lowest point on the Ferris wheel?

5. A helicopter is circling a 2-mile stretch of a highway, starting at point A on the diagram to the right. Traveling counterclockwise at a constant speed and altitude, it takes 2π minutes for the helicopter to make a complete revolution and return to point A.

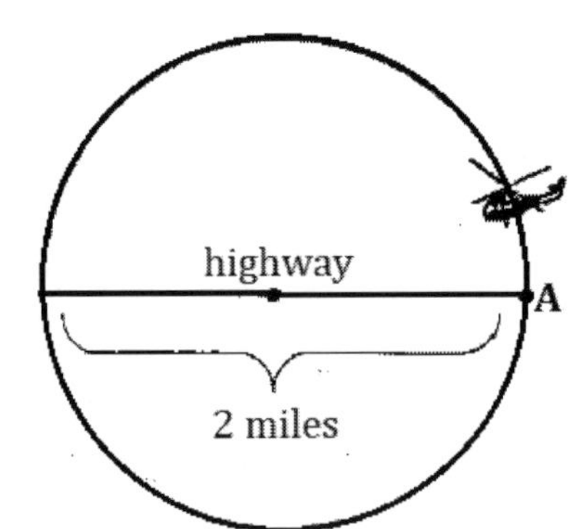

Sketch a graph of $d(t)$, where d is the distance between the helicopter and the highway in miles (disregarding altitude), and t is the time in minutes. If the helicopter is north of the highway, d is positive; if it is south of the highway, d is negative.

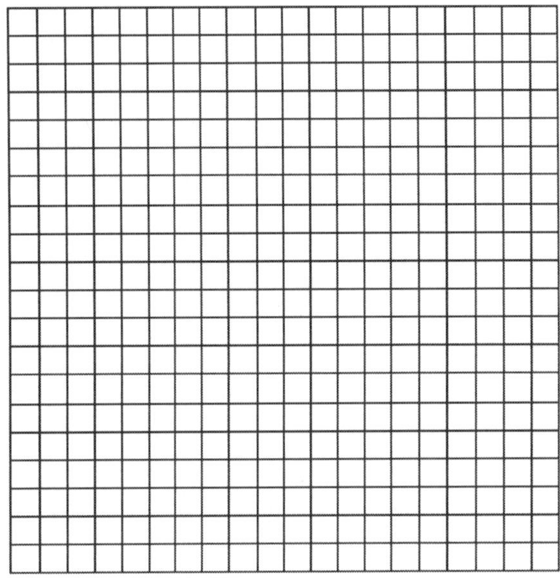

What is the equation of this graph?

6. The price of a share of a certain stock has fluctuated since 1980 according to the function $P(t) = 10\cos\left(\frac{\pi}{5}t\right) + 20$, where t represents the number of years since 1980 and P represents the price of a share in dollars. Graph the function on the grid below for the years between 1980 and 2010.

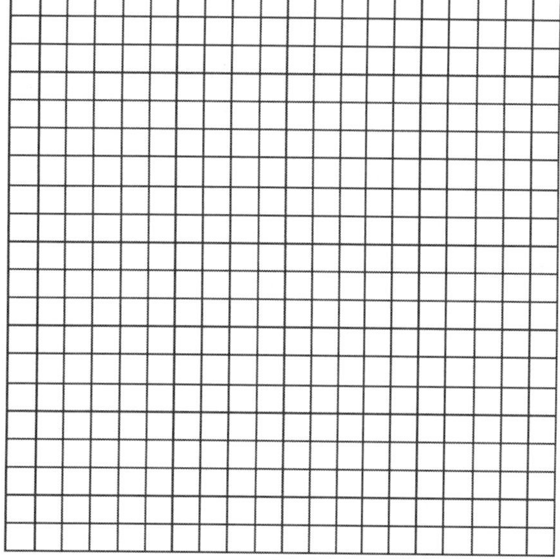

At what years did the stock reach its lowest price, and what was the price?

7. The temperature in a building varies throughout the course of a day according to the function $T(t) = 8\cos t + 78$, where t is the time in hours and T is the temperature. On the grid below, sketch a graph of the function for the interval $6 \leq t \leq 17$.

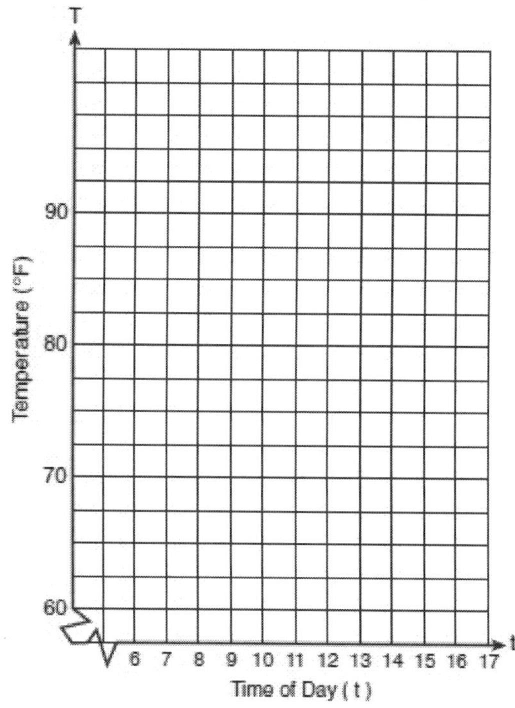

To the *nearest tenth of an hour*, at what times of day within this interval is the temperature at $80°F$?

Chapter 13. Examine Function Graphs

13.1 Even and Odd Function Graphs

Key Terms and Concepts

A function is considered an **even function** if its graph is symmetrical over the y-axis; that is, reflecting the graph over the y-axis will map the graph onto itself.

Example: The graph of $f(x) = x^4 - 3x^2 + 4$ to the right is an even function.

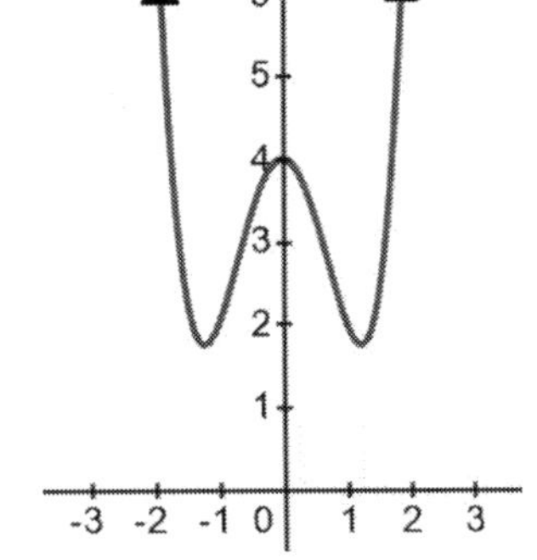

A polynomial function of x will be even if all of its terms have an even power of x, including a possible constant term (cx^0).

Example: Note that $f(x) = x^4 - 3x^2 + 4$ has all even powers of x.

A function is considered an **odd function** if its graph has origin symmetry; that is, rotating the graph 180° around the origin will map the graph onto itself.

Example: The graph of $f(x) = x^5 - 3x^3$ to the right is an odd function.

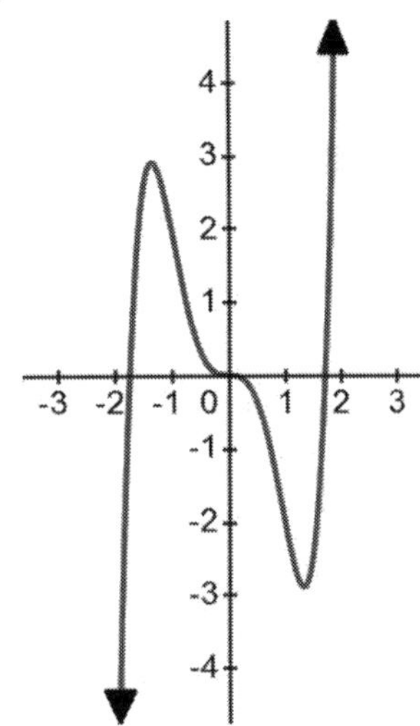

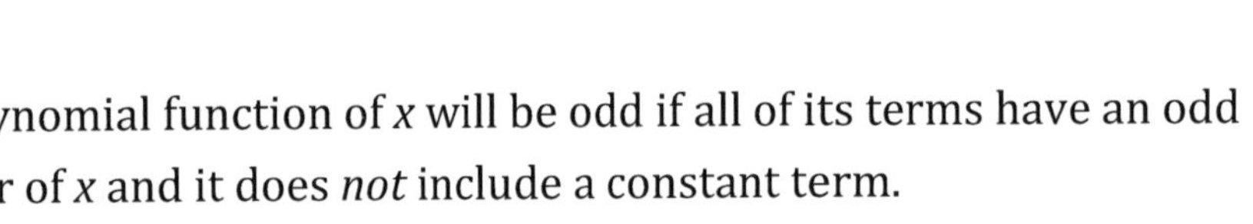

A polynomial function of x will be odd if all of its terms have an odd power of x and it does *not* include a constant term.

Example: Note that $f(x) = x^5 - 3x^3$ has all odd powers of x.

Many functions are *neither even nor odd.*

Example: The function $f(x) = 2x^4 + 4x^3$, shown below, is neither even nor odd.

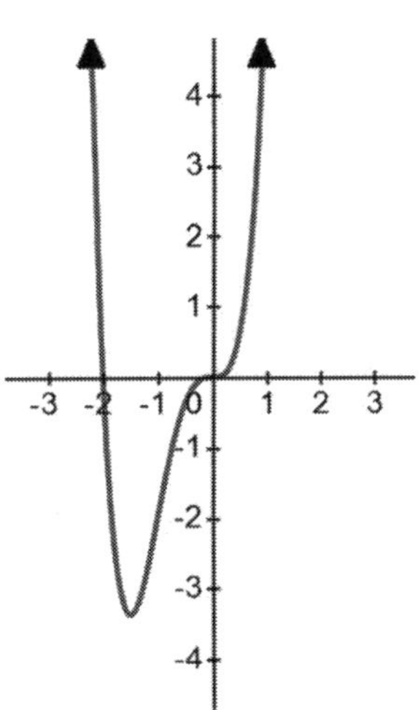

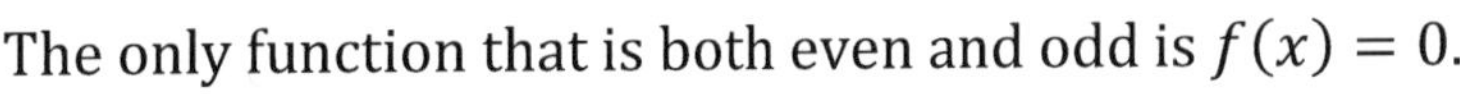

The only function that is both even and odd is $f(x) = 0$.

Among the **trigonometric functions**, $y = \cos x$ is an even function and both $y = \sin x$ and $y = \tan x$ are odd functions. When you look at the function graphs, the graph of $y = \cos x$ has y-axis symmetry, a property of even functions, and the graphs of $y = \sin x$ and $y = \tan x$ have rotational symmetry around the origin, a property of odd functions.

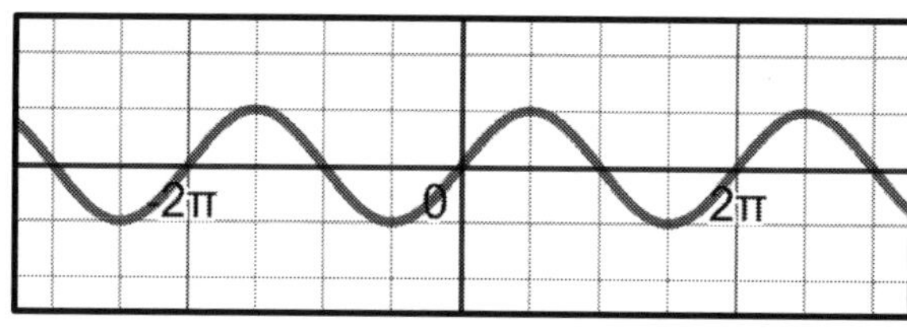

$y = \sin x$

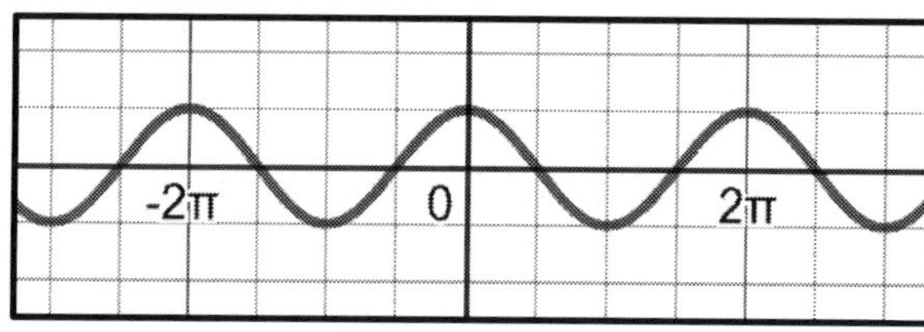

$y = \cos x$

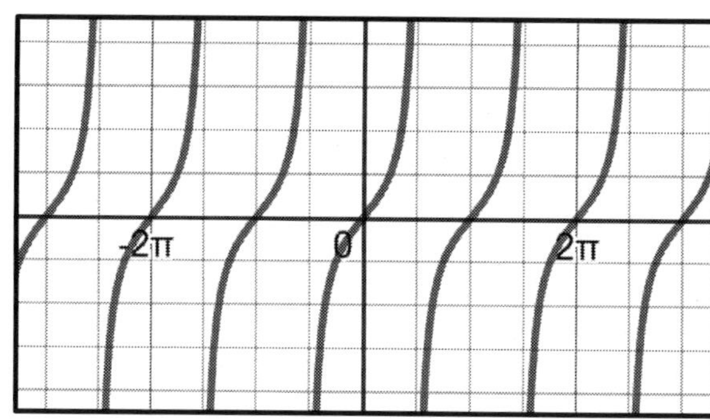

$y = \tan x$

When we **add or subtract two functions**, the sum/difference of two even functions is even and the sum/difference of two odd functions is odd. The sum/difference of an even function and an odd function is neither even nor odd (unless one of them is zero).

Example: $f(x) = x^3$ and $g(x) = 3x$ are both odd functions, so adding them will result in an odd function: $f(x) + g(x) = x^3 + 3x$. Adding two monomial functions yields a binomial function of the same two terms with no change to the exponents, which validates the rules.

When we **multiply or divide two functions**, the product/quotient of two even functions is even and the product/quotient of two odd functions is even. However, the product/quotient of an even function and an odd function is odd.

Example: $f(x) = x^3$ and $g(x) = 3x$ are both odd functions, so multiplying them will result in an even function: $f(x) \cdot g(x) = 3x^4$. Multiplying two monomial functions yields a monomial function in which the exponents of the variable are added, which validates the rules.

If a function is odd, the absolute value of that function is an even function.

Model Problem

State whether the function $f(x) = x^2 - 2\sin x$ is even, odd, or neither.

Solution:

The function $y = x^2$ is even and the function $y = -2\sin x$ is odd, so $f(x)$ is the sum of an even function and an odd function. Therefore, $f(x)$ is neither even nor odd.

Explanation of steps:

It is sometimes easier to look at a function as a sum or product of two functions, and then use the rules for whether the sum or product of functions are odd, even, or neither.

[$y = x^2$ is even because it is a polynomial with only an even degree term, and $y = -2\sin x$ is odd because $y = \sin x$ has point symmetry with respect to the origin, which does not change if we reflect the graph over the x-axis or change its amplitude (i.e., multiply by −2). The sum of an even function and an odd function is neither even nor odd.]

PRACTICE PROBLEMS

1. State whether $f(x)$ below is even, odd, or neither:
   $$f(x) = x^4 - 3x^2 + 7$$

2. State whether $f(x)$ below is even, odd, or neither:
   $$f(x) = x^5 - 3x^3 + 7$$

3. From the graph, state whether the function appears to be even, odd, or neither.

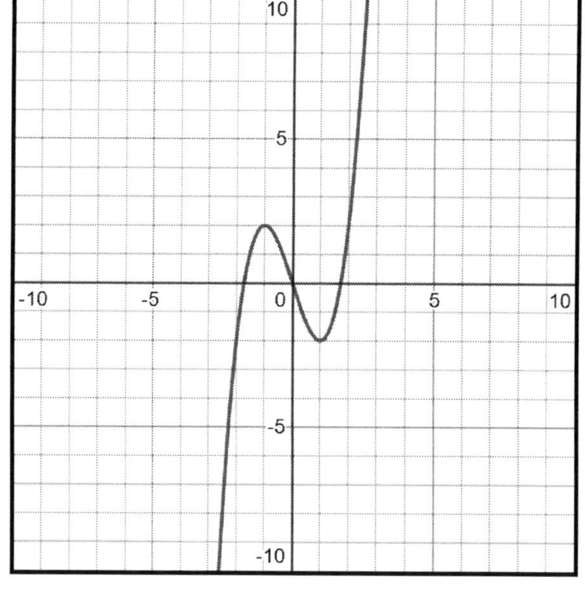

4. From the graph, state whether the function appears to be even, odd, or neither.

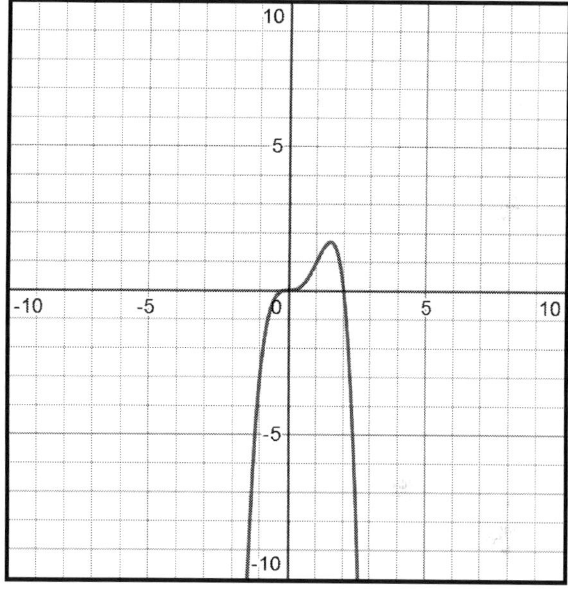

5. From the graph, state whether the function appears to be even, odd, or neither.

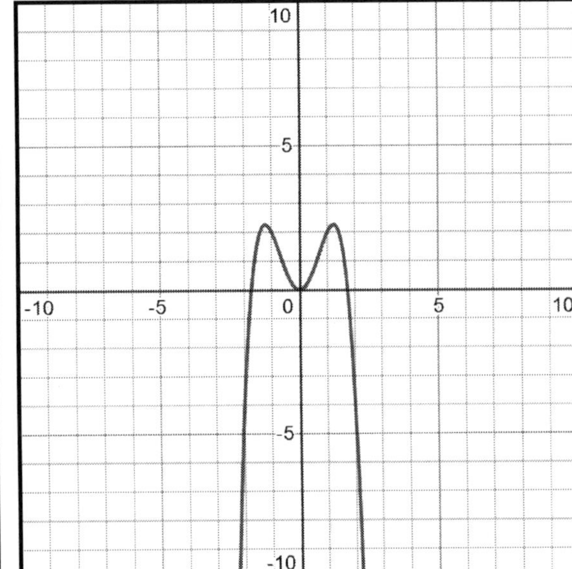

6. From the graph, state whether the function appears to be even, odd, or neither.

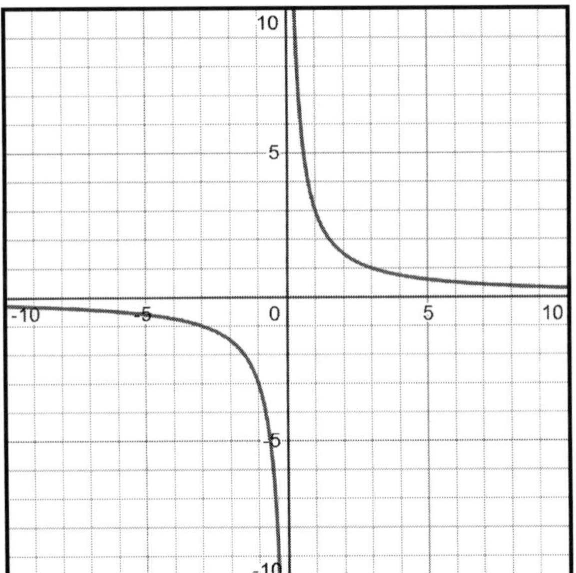

7. Which function is even?

(1) $y = \sin x$ (3) $y = \tan x$

(2) $y = \cos x$ (4) $y = x$

8. If $f(x)$ is an odd function, which function must also be odd?

(1) $g(x) = 2 \cdot f(x)$

(2) $g(x) = f(x) + 2$

(3) $g(x) = x \cdot f(x)$

(4) $g(x) = f(x + 2)$

13.2 Algebraically Determine Even or Odd [CC]

KEY TERMS AND CONCEPTS

$f(x)$ is an even function if, for every x in its domain, $f(-x) = f(x)$. We can use this rule to determine whether a function is an even function.

Example: For the even function $f(x) = x^4 - 3x^2 + 4$, $f(-x) = (-x)^4 - 3(-x)^2 + 4$, which simplifies to $f(-x) = x^4 - 3x^2 + 4$, the same expression as $f(x)$.

$f(x)$ is an odd function if, for every x in its domain, $f(-x) = -f(x)$. We can use this rule to determine whether a function is an odd function.

Example: For the odd function $f(x) = x^5 - 3x^3$, $f(-x) = (-x)^5 - 3(-x)^3$, which can be written as $f(-x) = -(x^5 - 3x^3)$, so it is equivalent to $-f(x)$.

We saw that, among the trigonometric functions, $y = \cos x$ is an even function and both $y = \sin x$ and $y = \tan x$ are odd functions. This is confirmed, according to the rules above, by looking back at the *negative angle identities* in Section 12.3. In that section, we saw that $\cos(-x) = \cos x$, so $y = \cos x$ is an even function. Also, since $\sin(-x) = -\sin x$ and $\tan(-x) = -\tan x$, we know that $y = \sin x$ and $y = \tan x$ are odd functions.

MODEL PROBLEM

State whether the function is even, odd, or neither.

a) $a(x) = \ln|x|$
b) $b(x) = \sin(2x)$
c) $c(x) = \sin^2(4x)$

Solution:

a) $a(-x) = \ln|-x| = \ln|x| = a(x)$, so a is even.
b) $b(-x) = \sin(-2x) = -\sin(2x) = -b(x)$, so b is odd.
c) $c(-x) = \sin^2(-4x) = [\sin(-4x)]^2 = [-\sin(4x)]^2 = \sin^2(4x) = c(x)$, so c is even.

Explanation of steps:

Write $f(-x)$ by substituting $-x$ for x, and determine whether the expression is equivalent to $f(x)$, which means the function is even; $-f(x)$, which means it is odd; or neither.
[For b) and c), remember that $\sin(-\theta) = -\sin\theta$.]

Practice Problems

1. Algebraically determine whether $f(x)$ below is even, odd, or neither: $f(x) = x^4 - 3x^2 + 7$	2. Algebraically determine whether $f(x)$ below is even, odd, or neither: $f(x) = x^5 - 3x^3 + 7$
3. Algebraically determine whether $f(x)$ below is even, odd, or neither: $f(x) = \frac{1}{x}$	4. Algebraically determine whether $f(x)$ below is even, odd, or neither: $f(x) = \left\|\frac{1}{x}\right\|$
5. Algebraically determine whether $f(x)$ below is even, odd, or neither: $f(x) = 2^x - 1$	6. Algebraically determine whether $f(x)$ below is even, odd, or neither: $f(x) = \frac{x^2 + 4}{x^3 - x}$

13.3 Inverse Functions

Key Terms and Concepts

The inverse of a function is usually denoted by a superscript –1, as in $f^{-1}(x)$. The inverse of a function can be defined as follows: if $f(a) = b$, then $f^{-1}(b) = a$.

For basic functions involving one operation, the inverse function will have the opposite operation.

Examples: (1) If $f(x) = x + 2$, then $f^{-1}(x) = x - 2$.
(2) If $f(x) = 2x$, then $f^{-1}(x) = \frac{x}{2}$.
(3) If $f(x) = x^2$ for $x \geq 0$, then $f^{-1}(x) = \sqrt{x}$.
(4) If $f(x) = 2^x$, then $f^{-1}(x) = \log_2 x$.

In each case above, the reverse is also true. For example, with respect to statement (1), if $f(x) = x - 2$, then $f^{-1}(x) = x + 2$, and so on.

Note that the inverse of the exponential function $f(x) = b^x$ (where $b > 0$ and $b \neq 1$) is the logarithmic function, $f^{-1}(x) = \log_b x$. This also means that the inverse of $f(x) = e^x$ is $f^{-1}(x) = \ln x$.

We've also seen and used inverse trigonometric functions. For example, the inverse of $\sin x$, if we restrict the domain of x to $[-90°, 90°]$, is $\sin^{-1} x$ (often called the *arcsine*), which is a function that is included on the calculator.

Example: For the function $f(x) = \sin x$, we know $f(30°) = \frac{1}{2}$. Therefore, for the inverse function, $f^{-1}(x) = \sin^{-1} x$, we could say that $f^{-1}\left(\frac{1}{2}\right) = 30°$. We've used this function to solve equations such as $\sin x = \frac{1}{2}$.

Essentially, an inverse function will "undo" anything that the original function does.

Example: If $f(x) = 3x + 4$, the function multiplies the input by 3 and adds 4. Its inverse will undo this – subtract 4 and then divide by 3. So, $f^{-1}(x) = \frac{x-4}{3}$.

Not all functions have inverse functions. Specifically, only **one-to-one functions** (*also known as injective functions*) will have inverse functions. A function f is one-to-one when, for every a and b in its domain, if $f(a) = f(b)$, then $a = b$. We can determine if a function is one-to-one by looking at its graph.

Do you remember the vertical line test to determine whether a graph represents a function? If there are no vertical lines that intersect the graph at more than one point, then we said the graph is a function.

Similarly, a function is one-to-one if it also passes the **horizontal line test**. If there are no horizontal lines that intersect the graph of a function at more than one point, then it is a one-to-one function. Only one-to-one functions have inverse functions.

Examples: The quadratic function $f(x)$ represented by the parabola below is *not* one-to-one, but the cubic function $g(x)$ is. Therefore, $f(x)$ does not have an inverse function, but $g(x)$ does. Note, however, that if we restrict the domain of $f(x) = x^2$ to $x \geq 0$, then this new function would be one-to-one, and its inverse function would be $f^{-1}(x) = \sqrt{x}$.

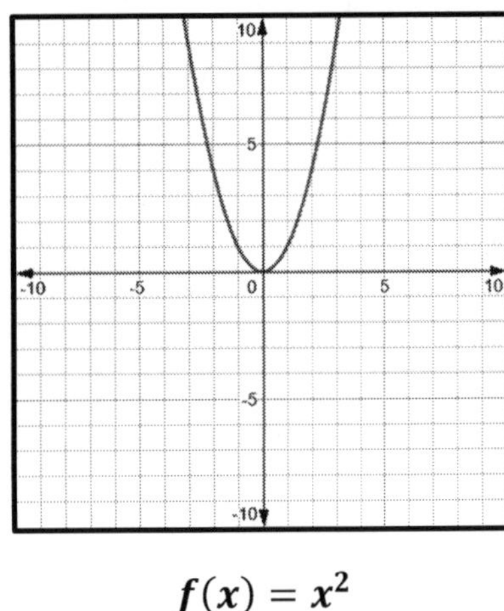

$f(x) = x^2$

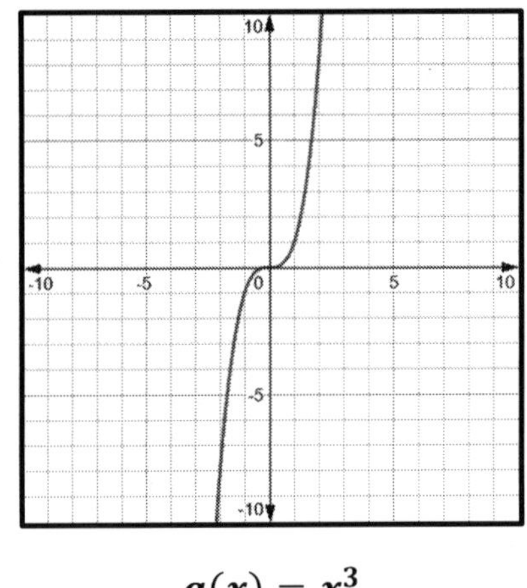

$g(x) = x^3$

If function f has an inverse f^{-1}, then the domain of f is the range of f^{-1} and the range of f is the domain of f^{-1}.

To find the equation of an inverse function: substitute y for x and x for y, and solve for y.

Example: To find the inverse of $f(x) = 3x + 4$, start with $y = 3x + 4$. Then, substitute y for x and x for y (in other words, switch the variables), giving you $x = 3y + 4$. Then solve for y:

$$x = 3y + 4$$
$$x - 4 = 3y$$
$$y = \frac{x-4}{3}$$

Therefore, $f^{-1}(x) = \frac{x-4}{3}$

As was mentioned earlier, $y = \log_b x$ and $y = b^x$ are inverse functions. This means that if we swap the x and y for one of the functions, the two equations are equivalent. That is, swapping the x and y in the second equation gives us $x = b^y$, and as we saw, $y = \log_b x$ and $x = b^y$ are equivalent.

A constant function, such as $f(x) = 2$, is graphed as a horizontal line. It does not pass the horizontal line test; therefore, constant functions do not have inverse functions.

The graphs of a function and its inverse are symmetric with respect to the line $y = x$.

Example: To graph $y = \log_2 x$, we can reflect the graph of $y = 2^x$ over the line $y = x$.

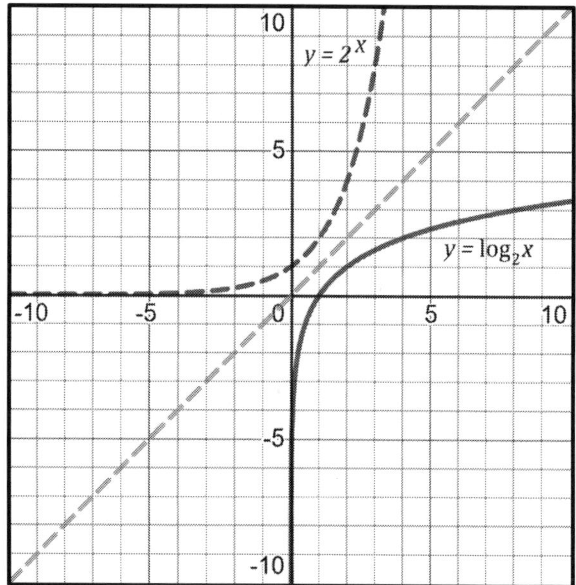

The fact that the graph of the inverse is a reflection over $y = x$ shouldn't be surprising. Remember, to find the inverse function algebraically, we switch the variables x and y. A reflection over $y = x$ does the same thing: $(x, y) \rightarrow (y, x)$.

Below are graphs of two more pairs of inverse functions: exponential and logarithmic.

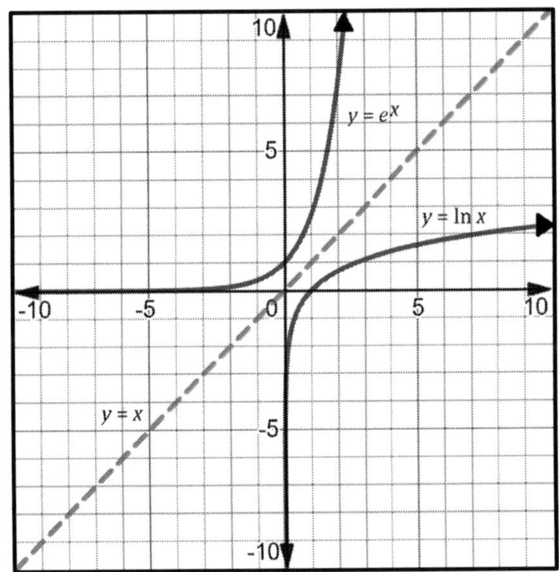

inverse functions $y = e^x$ *and* $y = \ln x$

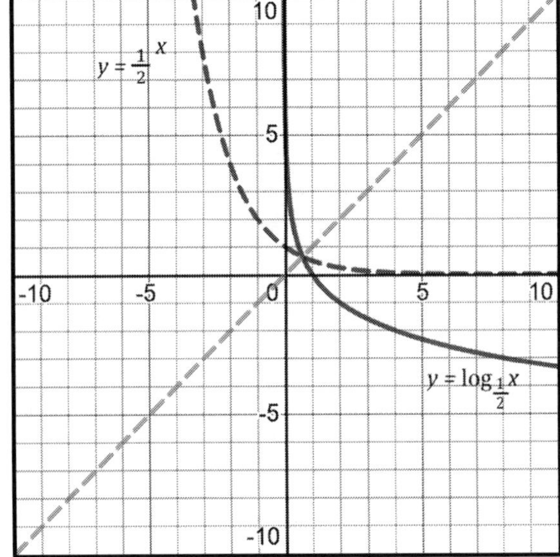

inverse functions $y = \frac{1}{2}^x$ *and* $y = log_{\frac{1}{2}} x$

MODEL PROBLEM

To the right is a graph of $f(x) = \frac{1}{8}x^3 + 2$.

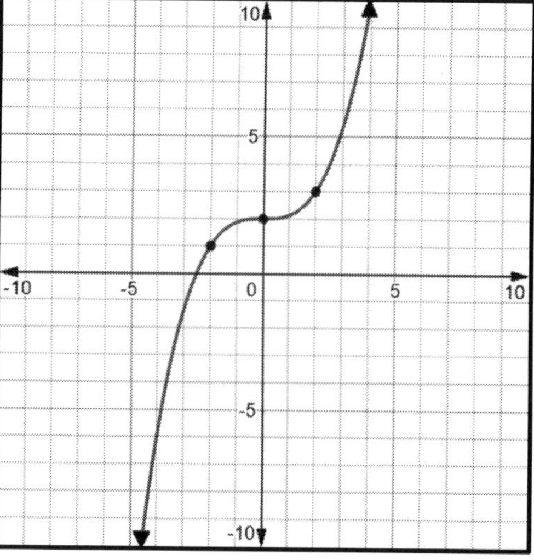

Write a definition of the inverse function, $f^{-1}(x)$.

Sketch a graph of the $f^{-1}(x)$ on the same grid.

Solution:

(A) $x = \frac{1}{8}y^3 + 2$

(B) $x - 2 = \frac{1}{8}y^3$

$8x - 16 = y^3$

$\sqrt[3]{8x - 16} = y$

(C) $f^{-1}(x) = \sqrt[3]{8x - 16}$

(D)

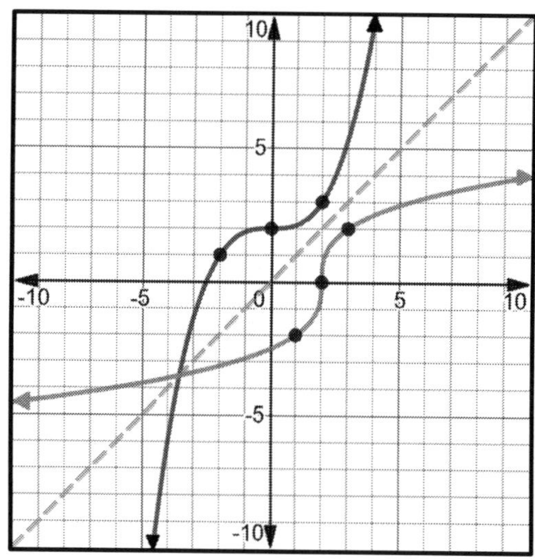

Explanation of steps:

(A) Substitute y for x and x for $f(x)$.

(B) Solve for y in terms of x.

(C) Write the inverse function.

(D) The graph of the inverse function is a reflection of the function over $y = x$. So, graph $y = x$ and sketch the image. It may be helpful to graph the image of a few points first (especially those with integer coordinates).

[$(3,2) \to (2,3)$,
$(0,2) \to (2,0)$,
$(-2,1) \to (1,-2)$]

PRACTICE PROBLEMS

1. Which graph represents a one-to-one function?

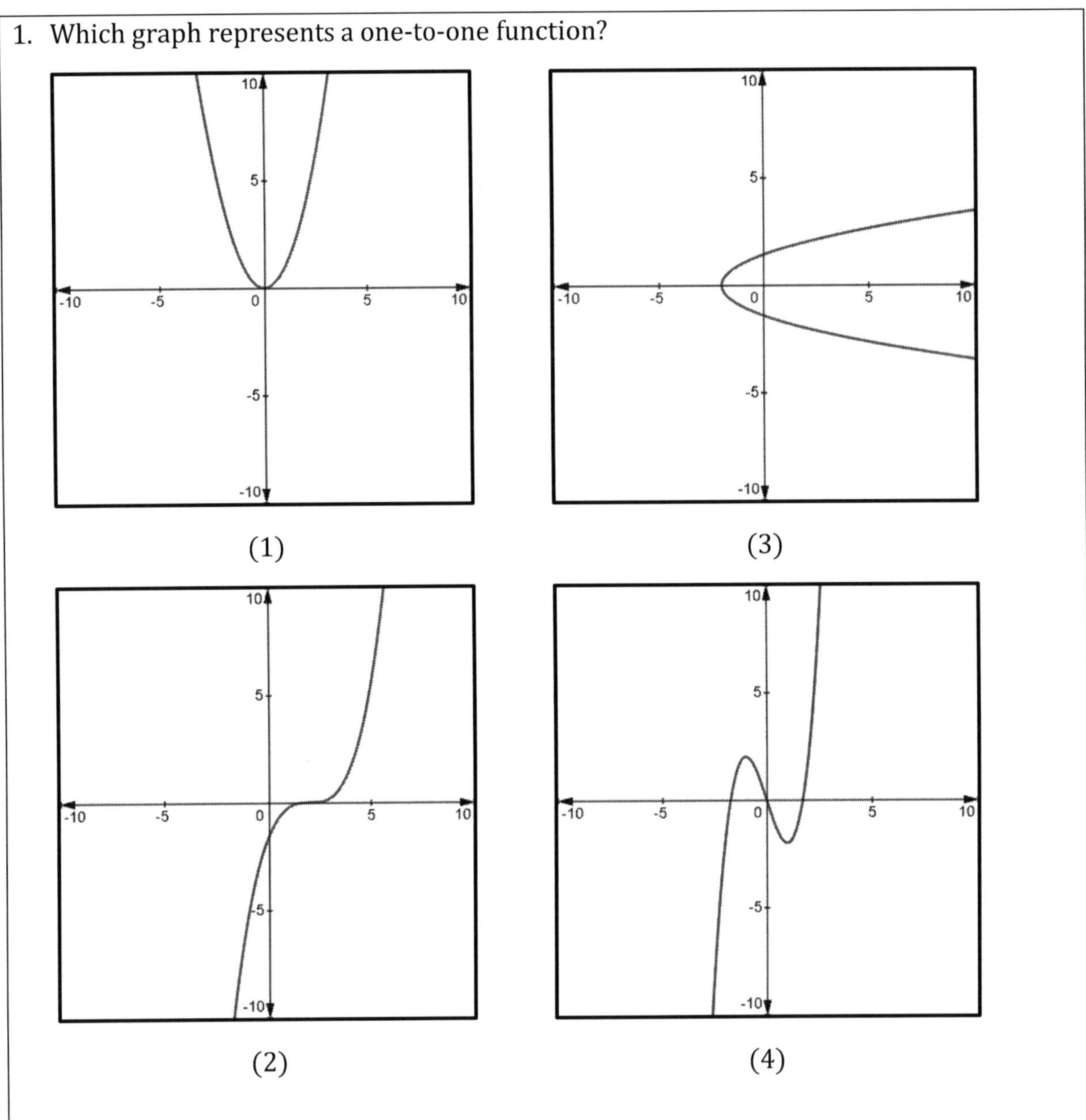

(1) (3)

(2) (4)

2. If $f(x) = 5x + 2$, what is $f^{-1}(x)$?	3. If $f(x) = \frac{2x + 5}{3}$, what is $f^{-1}(x)$?
4. If $f(x) = -\frac{2}{3}x$, what is $f^{-1}(x)$?	5. If $f(x) = \frac{1}{3}x + 2$, what is $f^{-1}(x)$?
6. If $f(x) = x^2 - 5$, what is $f^{-1}(x)$?	7. If $f(x) = (x - 2)^3$, what is $f^{-1}(x)$?

8. If $f(x) = 3^x + 1$, what is $f^{-1}(x)$?	9. If $f(x) = \log(x - 2) + 3$, what is $f^{-1}(x)$?
10. If $f(x) = \frac{x}{x+2}$, what is $f^{-1}(x)$?	11. Given $p(x) = \sin x - 1$ for $-\frac{\pi}{2} < x < \frac{\pi}{2}$, what is $p^{-1}(x)$?

12. Below is a graph of $f(x) = x^2 - 2$ for $x \geq 0$.

a) Find the positive root of $f(x)$.

b) What is $f^{-1}(x)$?

c) On the same grid, sketch a graph of $f^{-1}(x)$ for $x \geq -2$.

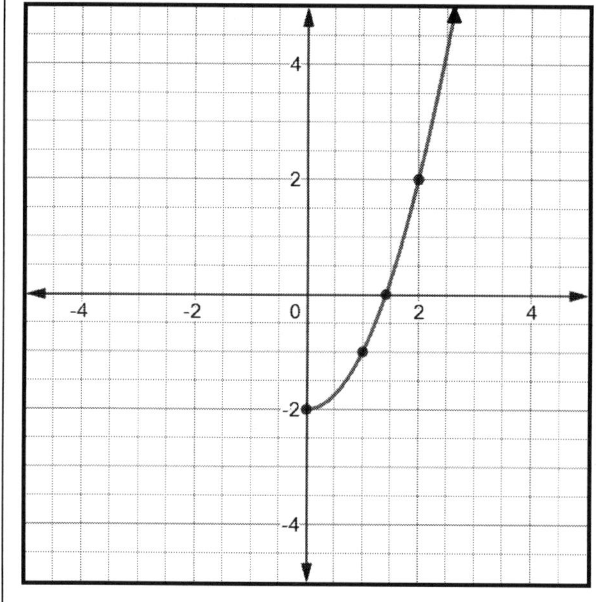

13. Below is a graph of the function $f(x) = 2\sqrt{x+1} - 1$ for $x \geq -1$.

a) Find the root of $f(x)$.

b) What is $f^{-1}(x)$?

c) On the same grid, sketch a graph of $f^{-1}(x)$ for $x \geq -1$.

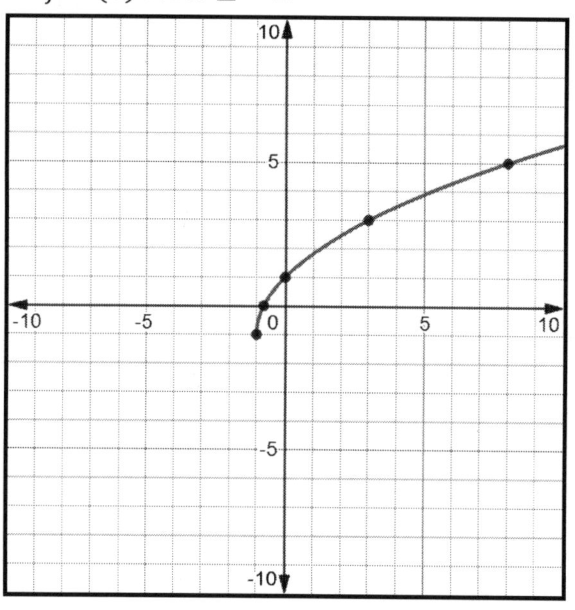

13.4 Average Rate of Change

Key Terms and Concepts

The **average rate of change** between any two points on a curve is the slope of the line through those two points, which is called the **secant line**.

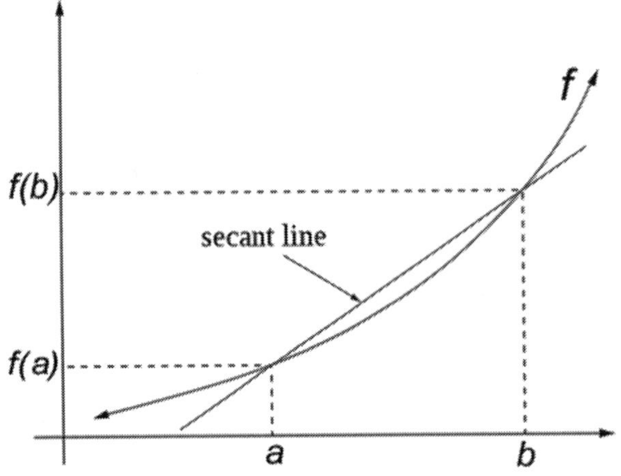

As we know, the formula for the slope of a line through points (x_1, y_1) and (x_2, y_2) is $m = \frac{y_2 - y_1}{x_2 - x_1}$. So, the average rate of change R over the interval $a \leq x \leq b$ is the slope of the secant line through points $(a, f(a))$ and $(b, f(b))$, which is $R = \frac{f(b) - f(a)}{b - a}$. We will use this formula to calculate average rate of change.

For example, suppose you take a car trip and record the distance that you travel every hour. The graph below shows the distance s (in miles) as a function of time t (in hours).

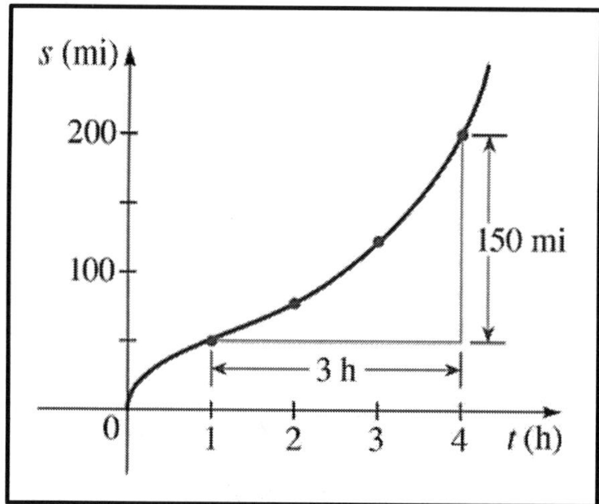

If we want to calculate the average rate of change over the interval $1 \leq t \leq 4$, we simply find the slope between the points (1,50) and (4,200). So, the average rate of change (or average speed in this case) is $R = \dfrac{s(b) - s(a)}{b - a} = \dfrac{200 - 50}{4 - 1} = \dfrac{150}{3} = 50$ mph.

For non-linear functions, the **average rate of change may vary** for different intervals.

Example: For $s(t)$ above, the average rate of change over the interval $2 \leq t \leq 4$ is $\dfrac{200 - 75}{4 - 2} = \dfrac{125}{2} = 62.5$ mph, represented by the slope of the secant line in the graph below.

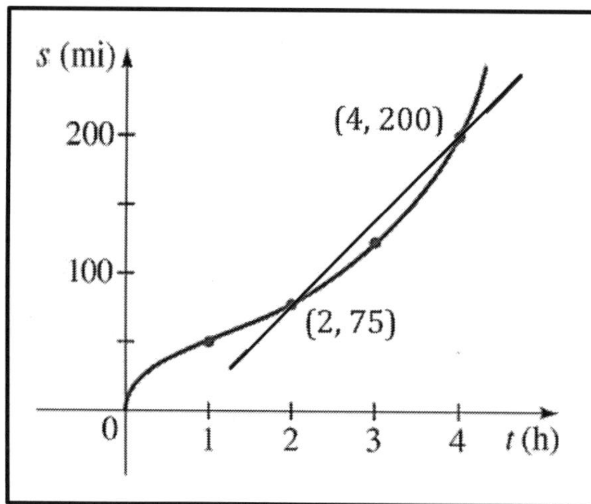

CALCULATOR TIP

To find the average rate of change on your calculator for an interval of a function:

1. Press [Y=] and enter the function into Y1. Then press [2nd] [QUIT].
2. For the interval $a \le x \le b$, where a and b are given, enter $\dfrac{Y_1(b) - Y_1(a)}{b - a}$.

   Be sure to use parentheses around the numerator as well as around the denominator. On the TI-84, you can enter each Y1 by pressing [ALPHA][F4][ENTER].

   [On the TI-83, you'll need to enter each Y1 by pressing [VARS] [Y-VARS] [1][1].*]*

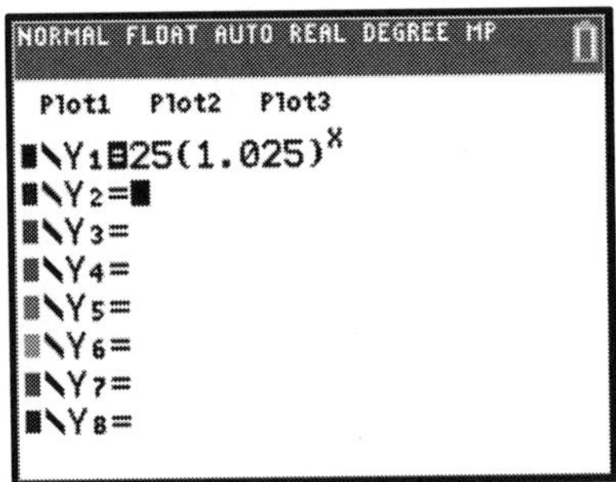

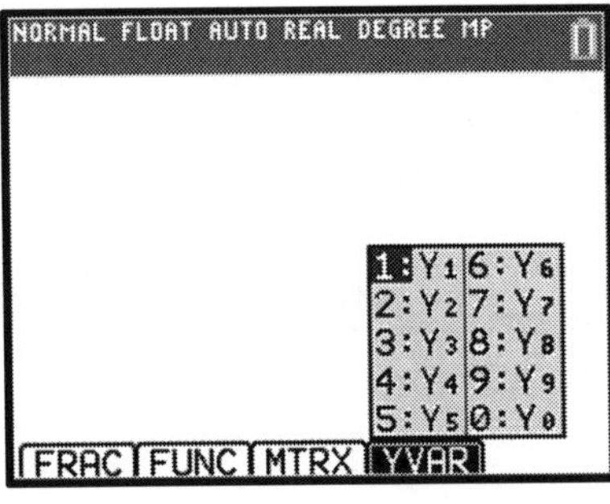

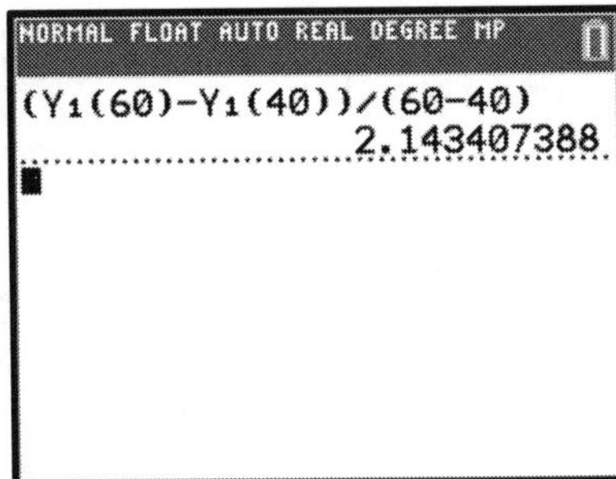

Given $f(x) = 25(1.025)^x$ graphed to the right, we can find the average rate of change over the interval $40 \le x \le 60$ by entering the function into the calculator as Y1, then calculating $\dfrac{f(60) - f(40)}{60 - 40}$ by using Y1 in place of the function name, *f*, as shown in the calculator screenshots above.

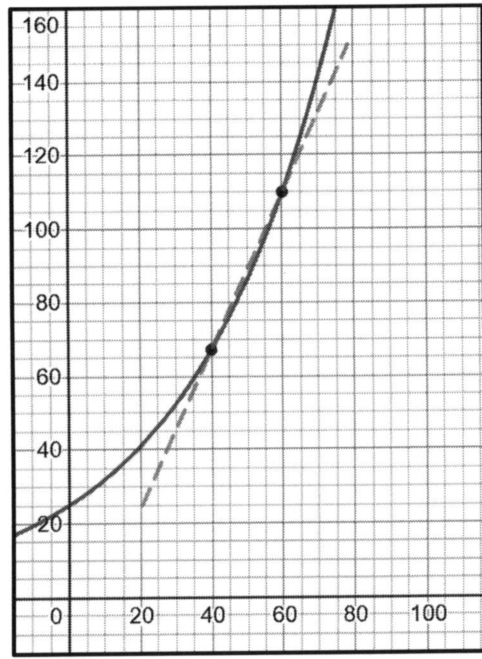

The average rate of change for $f(x)$ over the interval $40 \le x \le 60$ is approximately 2.14, which is the slope of the secant line shown as a dashed line through the two points on the graph.

Model Problem

If an object is dropped from a tall building, then the distance it has fallen after t seconds is given by the function $d(t) = 16t^2$. Find its average speed (average rate of change) over the interval between 1 second and 5 seconds.

Solution:

(A) At 1 second, $d(1) = 16(1)^2 = 16$. At 5 seconds, $d(5) = 16(5)^2 = 16 \cdot 25 = 400$.

(B) Slope of the line through points (1,16) and (5,400) is $\dfrac{400-16}{5-1} = \dfrac{384}{4} = 96$ ft/sec.

Explanation of steps:

(A) Find the points at the start and end of the given interval.
[Think of each point as the ordered pair of t and $d(t)$. At $t = 1$, $d(t) = 16$, and at $t = 5$, $d(t) = 400$, so the points are (1,16) and (5,400)]

(B) Find the slope of the line between the two points.

Practice Problems

1. The graph shown below includes points $A(60, 79)$, $B(80, 139)$, and $C(90, 49)$.

Find the average rate of change between

a) points A and B

b) points B and C

c) points A and C

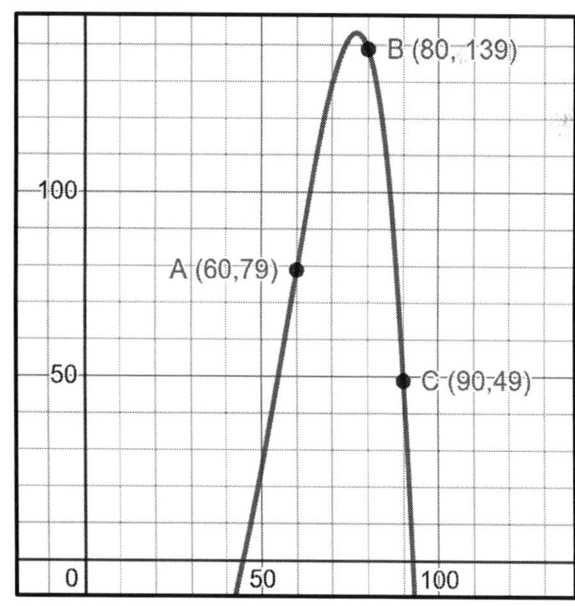

2. The following table shows the average prices of movie tickets over a period of years.

Year	1987	1991	1995	1999	2003	2007	2009
Price ($)	3.91	4.21	4.35	5.06	6.03	6.88	7.50

a) Find the average rate of change for the interval 1987 – 1999.

b) Find the average rate of change for the interval 1999 – 2009.

3. Calculate the average rate of change of a function, $f(x) = x^2 + 2$ as x changes from 5 to 15?

4. Find the average rate of change for the function $f(x) = x^2 + 10x + 16$ over the interval $[-3,3]$.

5. Find the average rate of change for the function $f(x) = x^4 + 2x^3$ over the interval $[-1,1]$.

6. Find the average rate of change for the function $f(x) = 2 \sin x$ over the interval $\left[\frac{\pi}{2}, \frac{3\pi}{2}\right]$.

13.5 Equation of Two Functions

Key Terms and Concepts

If we are given two functions, $f(x)$ and $g(x)$, we may want to know the values of x for which $f(x) = g(x)$.

If $f(x)$ and $g(x)$ are both polynomial or both exponential, we have learned how to solve the system of functions algebraically. However, for systems of other types of functions, or of two different types, it may be difficult to solve algebraically. Nevertheless, we can solve most systems graphically, using the calculator, by seeing where the two graphs intersect. The solutions to the equation $f(x) = g(x)$ would be the x-coordinates of the points of intersection of the graphs of the two functions.

When sketching a copy of the graph, be sure to label the two functions with their equations. Also, for accuracy, plot a few points having integer values, if possible.

CALCULATOR TIP

To find where two graphs intersect on the calculator:

1. Press [Y=] and enter both equations.
2. Press [GRAPH] to graph them.
3. Press [2nd][CALC][5] for intersect.
4. Move the blinking cursor onto each curve and press [ENTER] for the "First curve?" and "Second curve?" prompts.
5. For the "Guess?" prompt, use the arrow keys to move the cursor near one of the points of intersection. Then press [ENTER].
6. The coordinates of the closest point of intersection will be shown.
7. If there appear to be additional points of intersection, repeat steps 2 to 5 but move the cursor closer to the next point in response to the "Guess?" prompt.

Example: If $f(x) = \ln x$ and $g(x) = \frac{1}{x}$, to find the solutions to $f(x) = g(x)$, enter both equations into the calculator and use the intersect function to determine the points of intersection. Looking at these graphs, below, there is only one point of intersection at $(1.763, 0.567)$, so the solution is approximately 1.763.

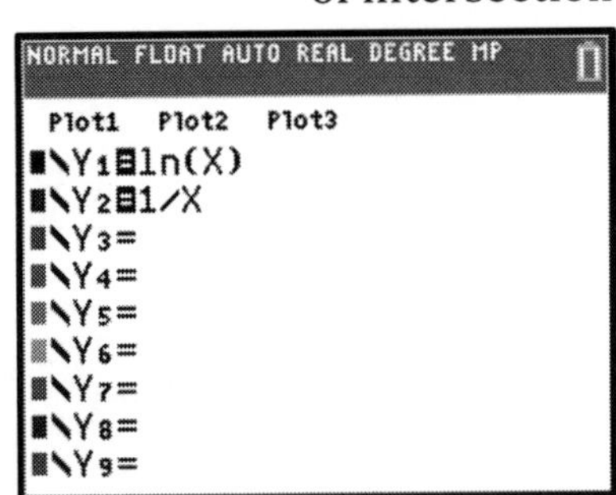

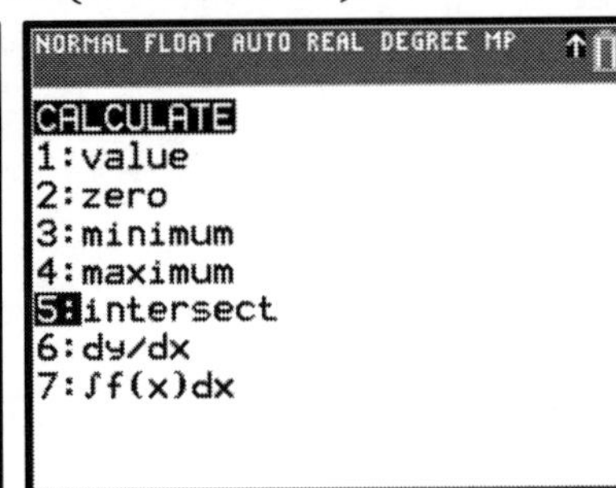

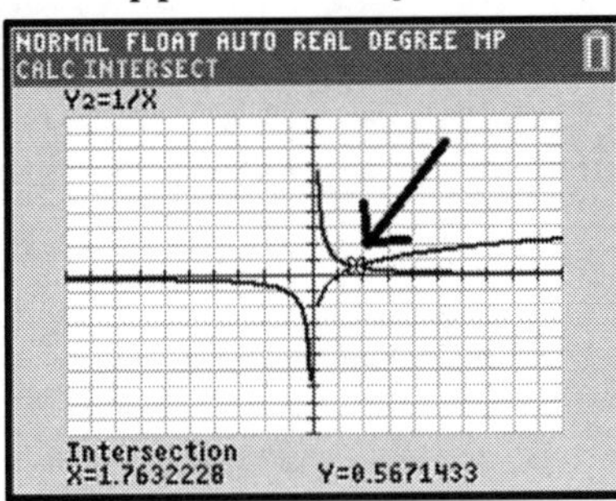

MODEL PROBLEM

Given $f(x) = e^x$ and $g(x) = \sqrt{3x+4}$, graphically find the solutions of $f(x) = g(x)$, to the *nearest hundredths.*

Solution:

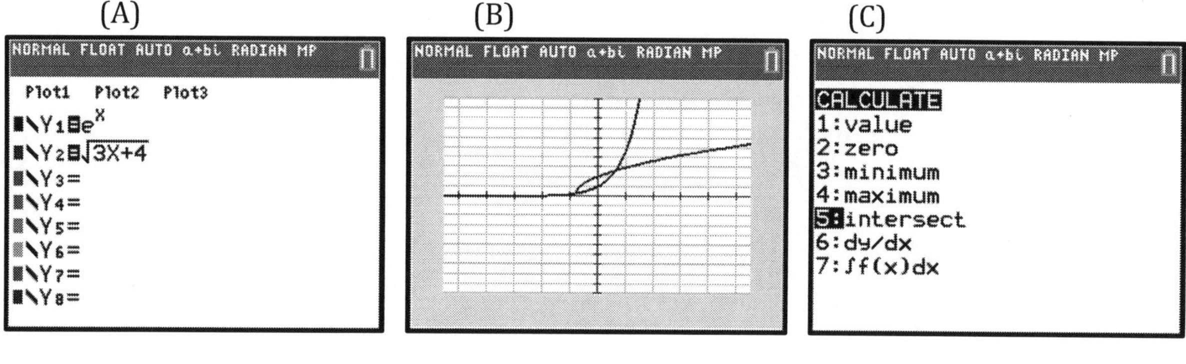

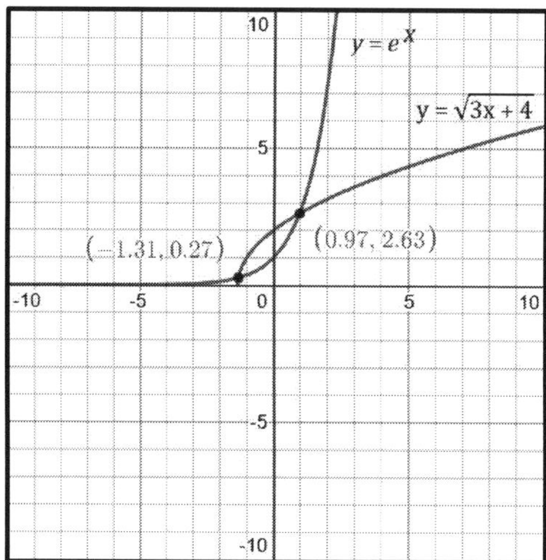

(D) Graphs intersect at $(-1.31, 0.27)$ and $(0.97, 2.63)$, so the solutions are $\{-1.31, 0.97\}$.

Explanation of steps:

(A) On the calculator, enter the two equations.

(B) Press [GRAPH] to graph them.

(C) Press [2nd][CALC][5] to find the points of intersection. Press [ENTER] twice, then, at the "Guess?" prompt, use the arrow keys to move close to the first point of intersection and press [ENTER] again. The coordinates of this point will be shown. Repeat for each additional point of intersection.

(D) The solutions are the set of x-coordinates of the points of intersection.

PRACTICE PROBLEMS

1. Solve graphically, and sketch the graphs on the grid below. Express solutions to the *nearest hundredth*. $y = 2^x - 3$ $y = 2x^2 - 3$	2. Solve graphically, and sketch the graphs on the grid below. Express solutions to the *nearest hundredth*. $y = 2^x$ $y = \log(x + 5) + 1$
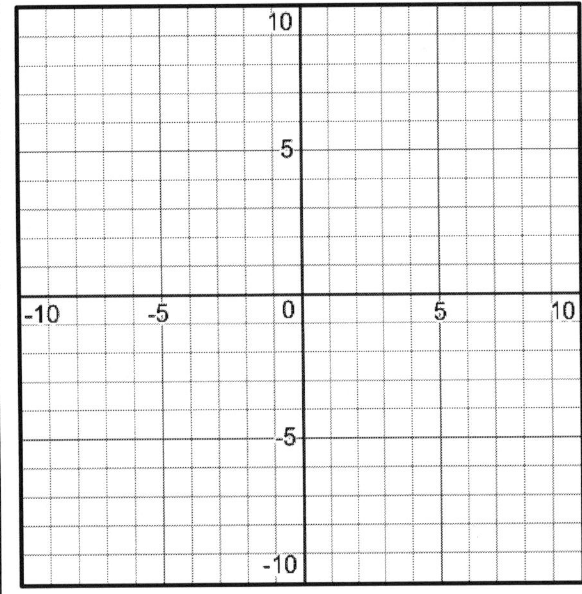	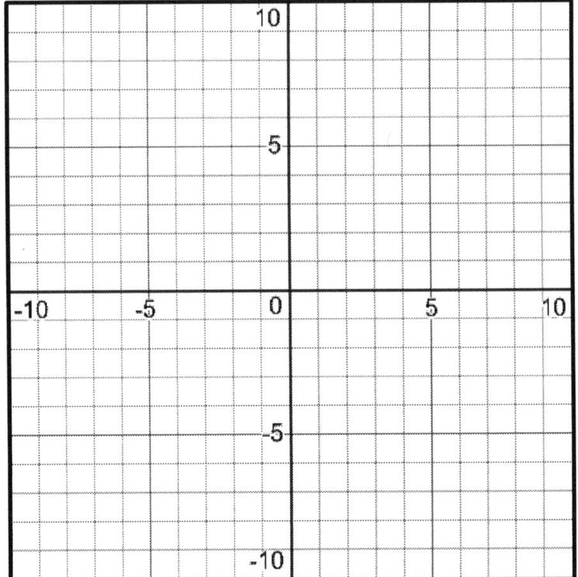

3. Solve graphically, and sketch the graphs on the grid below. Express solutions to the *nearest hundredth*.

$$y = |x^3 + x|$$
$$y = \sqrt{x+3}$$

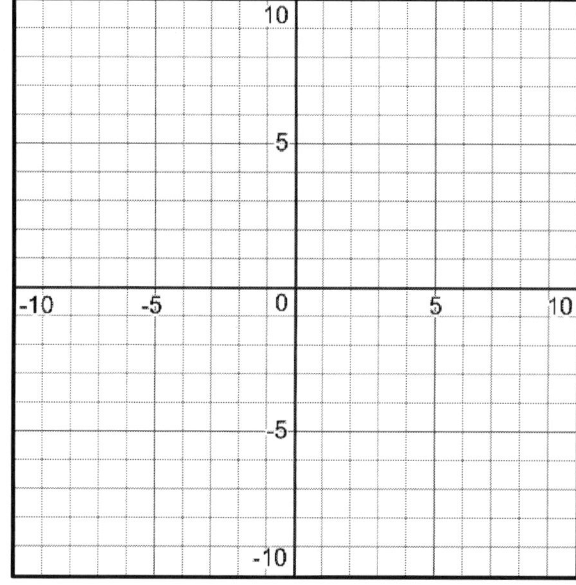

4. Solve graphically, and sketch the graphs on the grid below. Express solutions to the *nearest hundredth*.

$$y = -\frac{x^5}{6} + x^3 \qquad y = \frac{x^3}{6} + \frac{x^2}{2} - 2$$

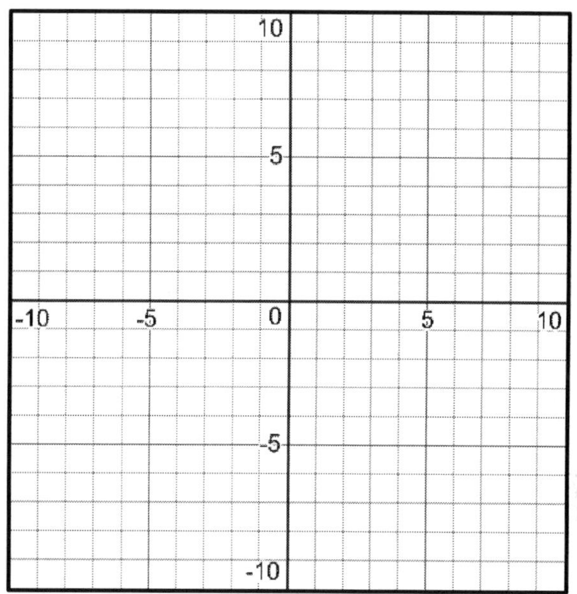

13.6 Inequality of Two Functions [NG]

Key Terms and Concepts

In the previous section, we saw how to find the solutions to $f(x) = g(x)$ graphically. We may also want to know for which values of x one function is less than (or less than or equal to, or greater than, or greater than or equal to) the other function.

The solutions to the inequality $f(x) < g(x)$ would be the set of x-coordinates where the graph of $f(x)$ is below the graph of $g(x)$. For the solutions to $f(x) \leq g(x)$, include the solutions to both $f(x) < g(x)$ and $f(x) = g(x)$.

To determine the solution set, graph the two functions on the calculator and find the points of intersection. Then state the interval(s) of the domain for which the graph of $f(x)$ is below the graph of $g(x)$, usually as compound inequalities or in interval notation.

Model Problem

Given $f(x) = x^2$ and $g(x) = |x + 6|$, graphically find the solutions of $f(x) < g(x)$.

Solution:

(A)

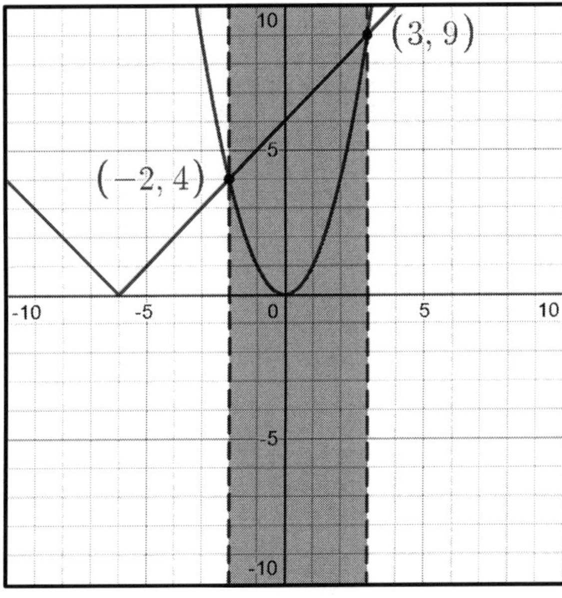

(B) $-2 < x < 3$, or the interval $(-2, 3)$

Explanation of steps:

(A) Graph the two functions on the calculator and find the points of intersection. [(−2,4) and (3,9)].

(B) State the interval(s) of the domain for which the graph of $f(x)$ is below the graph of $g(x)$, usually as compound inequalities or in interval notation.

PRACTICE PROBLEMS

1. Graphically find the solutions of $f(x) \leq g(x)$.

$$f(x) = x - 1$$
$$g(x) = -2x^2 + 2x + 5$$

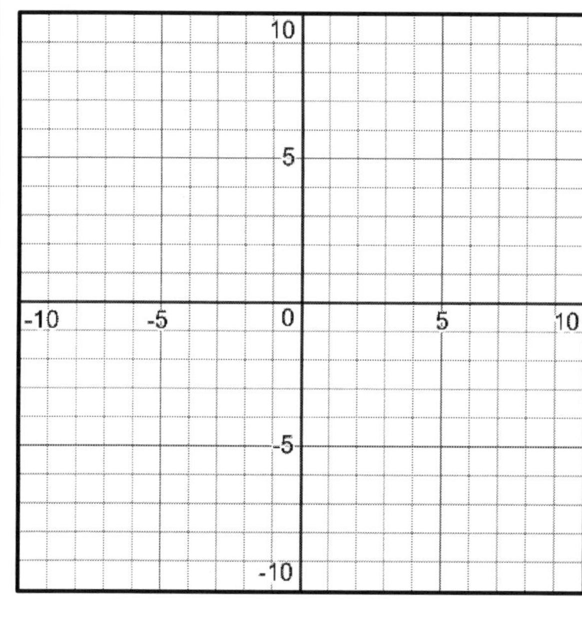

2. Graphically find the solutions of $f(x) < g(x)$.

$$f(x) = x^3$$
$$g(x) = -x^2 + 2$$

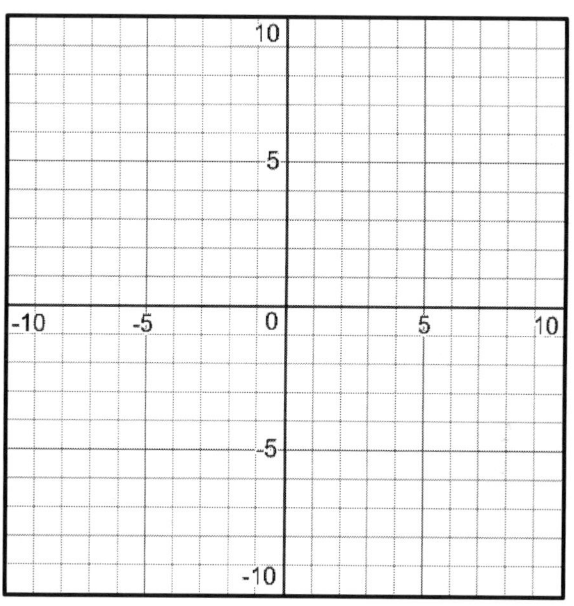

3. Graphically find the solutions of $f(x) \geq g(x)$.

$$f(x) = x^2$$
$$g(x) = -x^2 - 2x$$

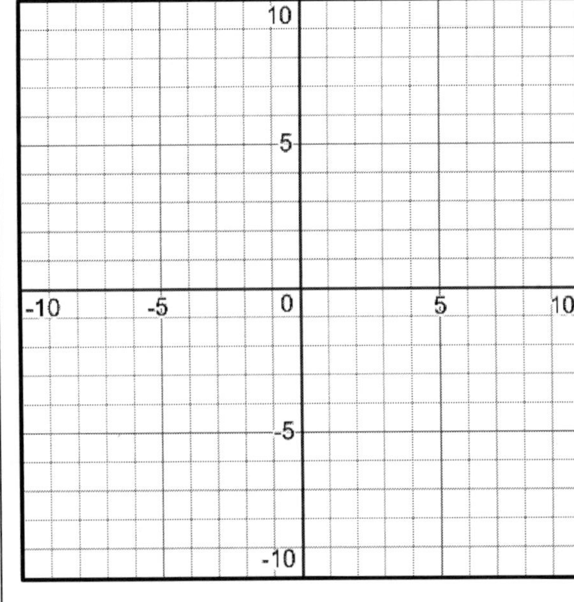

4. Graphically find the solutions of $f(x) > g(x)$.

$$f(x) = x^3 + 2x^2 + x + 1$$
$$g(x) = 2x + 3$$

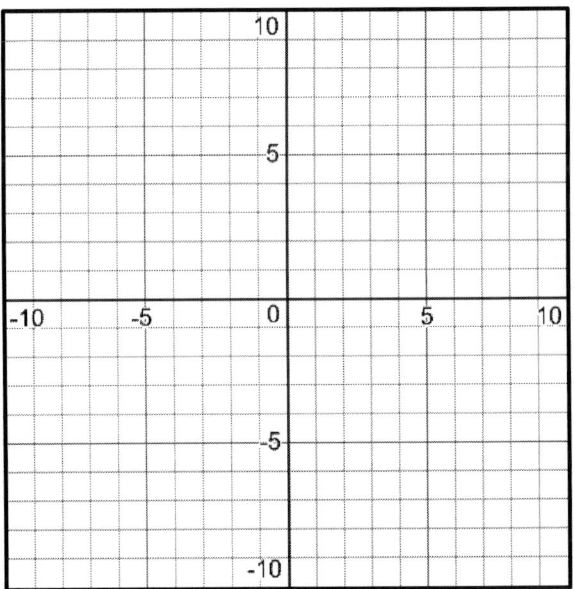

Chapter 14. Sequences and Series

14.1 Arithmetic Sequences

KEY TERMS AND CONCEPTS

A **sequence** is an ordered list of numbers, called terms. If we let the integer n represent the term number in the sequence, then a_n represents the nth term in the sequence. Unless otherwise specified, you can assume that the sequence is an **infinite sequence**, meaning that the values of n are the set of all counting numbers (or sometimes, whole numbers).

Examples: $9, 11, 13, 15, \dots$

$\uparrow \; \uparrow \; \uparrow \; \uparrow$

$a_1, a_2, a_3, a_4, \dots$

An **explicit formula** describes how to calculate each term (a_n) based on its term number (n). The formula allows us to determine the value of a specified term. By substituting the term number for n, we can evaluate the formula just as we would for any function.

Examples: The formula for the sequence above is $a_n = 2n + 7$. We can calculate the value of any term using this formula by substituting for n; for example, to calculate what the fourth term is, $a_4 = 2(4) + 7 = 15$.

An **arithmetic sequence** is one in which each term is obtained by *adding* the same number (called the **common difference**, represented by d) to the preceding term.

Example: $9, 11, 13, 15, \dots$

$+2 \; +2 \; +2$ so, $d = 2$

We can find the common difference of an arithmetic sequence by subtracting consecutive terms, such as $a_2 - a_1$ or $a_3 - a_2$. In general, the common difference $d = a_n - a_{n-1}$.

We can define an arithmetic sequence using the formula $a_n = a_1 + (n-1)d$ where a_1 is the first term and *d* is the common difference.

Example: The sequence $9, 11, 13, 15, \ldots$ can be written as $a_n = 9 + (n-1) \cdot 2$. By distributing the *d* and combining terms, we can simplify the formula:

$$a_n = 9 + (n-1) \cdot 2$$
$$a_n = 9 + 2n - 2$$
$$a_n = 2n + 7$$

We can write sequence definitions in function notation. An *arithmetic* sequence is always a *linear* function and its common difference, *d*, is its slope.

Example: $a_n = 2n + 7$ can be written in function notation, such as $a(n) = 2n + 7$.

To find a specific term of an arithmetic sequence:

1. Determine a_1 (the first term) and *d* (the common difference) from the given terms.
2. In the general formula for arithmetic sequences, substitute for *n*, a_1, and *d*, and evaluate.

Example: To find the 20th term of the arithmetic sequence $32, 37, 42, 47, \ldots$
The general formula is $a_n = a_1 + (n-1)d$ and in this case, $a_1 = 32, d = 5$, and $n = 20$, so $a_{20} = 32 + (20-1)(5) = 127$.

CALCULATOR TIP

If the formula for an arithmetic sequence is known, the calculator can also be used to find the value of any terms.

1. Press [2nd][LIST][5] for the seq function.
2. On the TI-84 models with Stat Wizards turned on (in the [MODE] screen), you'll be prompted to enter the formula, the variable used in the formula, the starting and ending term numbers to display, and a step of 1 (see center screenshot below).

   Then press [ENTER] twice to Paste and display the result.

   [On the TI-83, this screen is skipped, so you will need to type these directly within the parentheses of the seq function, separated by commas, and then press [ENTER].]

Example: To display the 25th term of the sequence $a_n = 2n + 7$, follow the above steps, as shown below. The variable X is used for convenience, but N could be used instead. Just be sure the variable used in the expression (Expr) is the same variable named on the next line. The display shows that 57 is the 25th term.

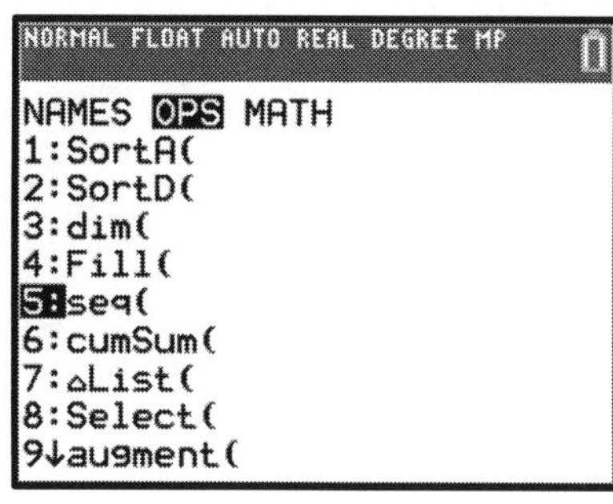

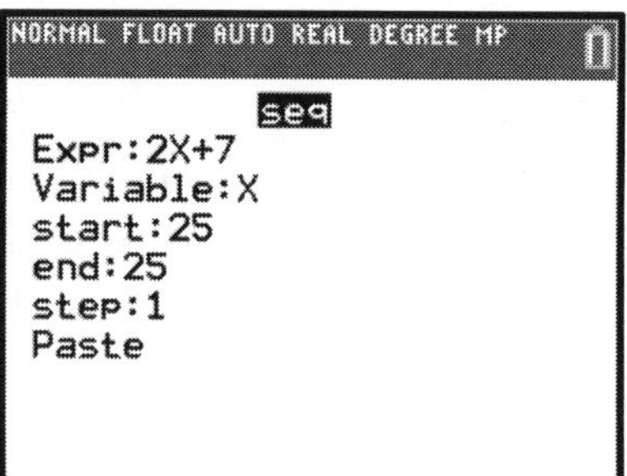

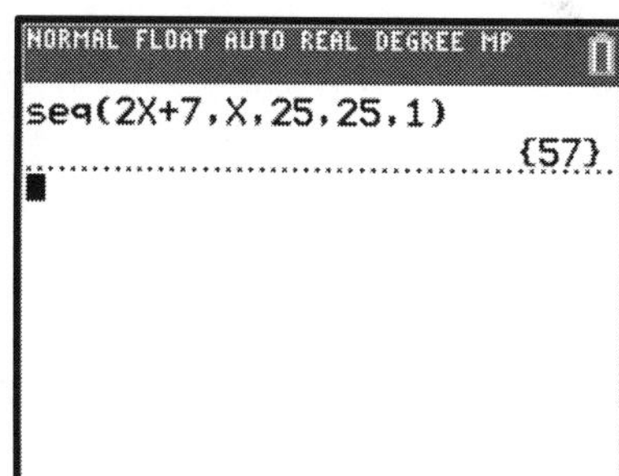

Because the common difference, d, is also the slope of the linear function, we can use the slope formula to calculate d given any two terms.

To write the formula for an arithmetic sequence given two terms:

1. Express the two terms as points. Find d, which is the slope of the line through these points, using the slope formula.
2. Substitute d, as well as n and a_n from one of the terms, into the general formula for an arithmetic sequence. Solve for a_1.
3. Write the general formula, substituting known values for a_1 and d, and simplify.

Example: To write a formula for the arithmetic sequence where $a_5 = 63$ and $a_8 = 99$, find d, which is the slope of the line through the points (5,63) and (8,99):

$d = \frac{99-63}{8-5} = \frac{36}{3} = 12$. Then, substitute $d = 12$ and, using the first point, $n = 5$ and $a_n = 63$ into the general formula $a_n = a_1 + (n-1)d$ to find a_1:

$63 = a_1 + (5-1)(12)$

$63 = a_1 + 48$

$a_1 = 15$

Now that we know $a_1 = 15$ and $d = 12$, we can write the formula:

$a_n = a_1 + (n-1)d$

$a_n = 15 + (n-1)(12)$

$a_n = 12n + 3$

MODEL PROBLEM 1: *FIND THE NTH TERM*

Find the twelfth term of the arithmetic sequence: 18, 22, 26, 30, ...

Solution:

(A) $a_1 = 18$

$d = 22 - 18 = 4$

(B) $a_n = a_1 + (n-1)d$

$a_{12} = 18 + (12-1)(4)$

$a_{12} = 62$

Explanation of steps:

(A) Determine a_1 (the first term) and d (the common difference) from the given terms.

(B) In the general formula for arithmetic sequences, substitute for n, a_1, and d, and evaluate *[n is the term number, 12].*

PRACTICE PROBLEMS

1. Write an explicit formula for the arithmetic sequence $32, 37, 42, 47, \ldots$	2. Write an explicit formula for the arithmetic sequence $24, 17, 10, 3, \ldots$
3. Write an explicit formula for the arithmetic sequence $-1, -\frac{1}{2}, 0, \frac{1}{2}, \ldots$	4. Find the eighth term of the arithmetic sequence for which $a_1 = 21$ and $d = 9$.
5. Find the twelfth term of the arithmetic sequence: $16, 27, 38, \ldots$	6. Find the ninth term of the arithmetic sequence: $35, 30, 25, \ldots$

7. Find the 27th term of the arithmetic sequence, $5, 8, 11, 14, \ldots$	8. Find the 20th term of the arithmetic sequence, $-8, -2, 4, \ldots$

MODEL PROBLEM 2: ***WRITE A FORMULA GIVEN TWO TERMS***

Two terms of an arithmetic sequence are $a_8 = 21$ and $a_{27} = 97$. Write a formula for the nth term.

Solution:

(A) $(8,21)$ and $(27,97)$

$$d = \frac{97 - 21}{27 - 8} = \frac{76}{19} = 4$$

(B) $a_n = a_1 + (n-1)d$

$21 = a_1 + (8-1) \cdot 4$

$21 = a_1 + 32 - 4$

$-7 = a_1$

(C) $a_n = a_1 + (n-1)d$

$a_n = -7 + (n-1) \cdot 4$

$a_n = -7 + 4n - 4$

$a_n = 4n - 11$

Explanation of steps:

(A) Express the two terms as points. Find d, which is the slope of the line through these points, using the slope formula.

(B) Substitute d, as well as n and a_n from one of the terms [$a_8 = 21$, *so use n = 8 and* $a_n = 21$], into the general formula for an arithmetic sequence. Solve for a_1.

(C) Write the general formula, substituting known values for a_1 and d, and simplify.

PRACTICE PROBLEMS

9. Two terms of an arithmetic sequence are $a_6 = 10$ and $a_{21} = 55$. Write a formula for the nth term.

10. Two terms of an arithmetic sequence are $a_4 = -23$ and $a_{22} = 49$. Write a formula for the nth term.

11. In a concert hall, there are 29 seats in the third row and 53 seats in the fifteenth row. Suppose the number of seats in the nth row can be determined by an arithmetic sequence. Determine a formula for a_n, the nth term of this sequence.

    Use the formula to determine how many seats there are in the tenth row.

14.2 Geometric Sequences

KEY TERMS AND CONCEPTS

A **geometric sequence** is one in which each term is obtained by *multiplying* the same number (called the **common ratio**, represented by *r*) to the preceding term.

Example: $5, 10, 20, 40, \ldots$

$\times 2 \quad \times 2 \quad \times 2$ so, $r = 2$

We can find the common ratio by dividing consecutive terms, such as $\frac{a_2}{a_1}$ or $\frac{a_3}{a_2}$. In general, the common ratio $r = \frac{a_n}{a_n - 1}$.

We can define a geometric sequence using the formula $a_n = a_1 r^{n-1}$ where a_1 is the first term and *r* is the common ratio ($r \neq 0$).

Example: The formula for the above sequence can be written as $a_n = 5 \cdot 2^{n-1}$.

Since $r^{n-1} = \frac{r^n}{r}$, an alternate way of writing the geometric sequence formula is $a_n = \frac{a_1 r^n}{r}$.

This will allow us to simplify our formulas so that they don't include binomial exponents.

Example: $a_n = 5 \cdot 2^{n-1}$ can be rewritten as $a_n = \frac{5 \cdot 2^n}{2} = \frac{5}{2}(2)^n$ or $a_n = 2.5(2)^n$.

A *geometric* sequence is always an *exponential* function. By simplifying the formula as described above, we can express the formula as an exponential function in standard form.

Example: $a_n = 2.5(2)^n$ can be rewritten as $a(n) = 2.5(2)^n$.

To find a specific term of a geometric sequence:

1. Determine a_1 (the first term) and *r* (the common ratio) from the given terms.
2. In the general formula for geometric sequences, substitute for *n*, a_1, and *r*, and evaluate.

Example: To find the sixth term of the geometric sequence $8, 32, 128, \ldots$, we know $a_1 = 8, r = 4$, and $n = 6$, so substitute into the general formula $a_n = a_1 r^{n-1}$ to get $a_6 = 8(4)^{6-1} = 8(4)^5 = 8{,}192$.

CALCULATOR TIP

If the formula for a geometric sequence is known, the calculator can also be used to find the value of any terms, just like we saw with arithmetic sequences.

3. Press [2nd][LIST][5] for the seq function.
4. On the TI-84 models, enter the formula, the variable used in the formula, the starting and ending term numbers to display, and a step of 1 next to the prompts on the next screen. Then press [ENTER] twice to Paste and display the result.
   [On the TI-83, this screen is skipped, so you will need to type these directly within the parentheses of the seq function, separated by commas, and then press [ENTER].]

Example: To display the 6th through 8th terms of the sequence $a_n = 5(3)^n$, follow the above steps, as shown below. The terms are 3645, 10935, and 32805.

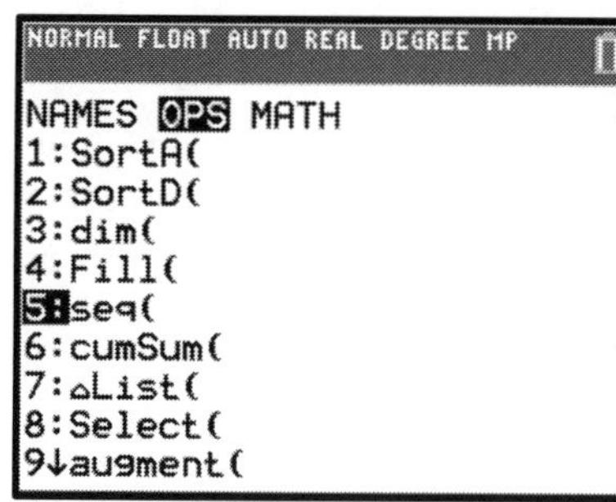

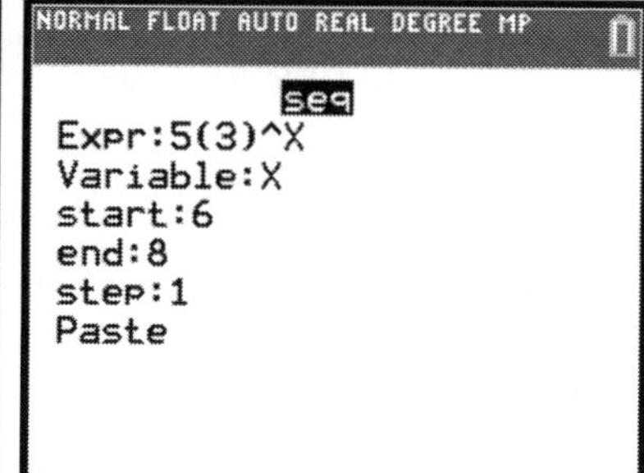

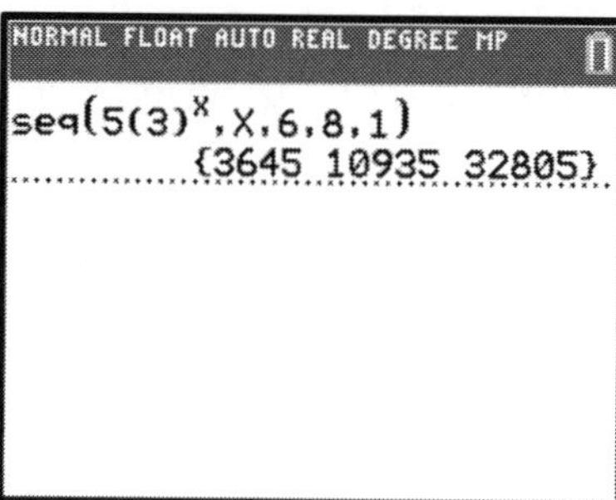

In a geometric sequence, since we multiply each term by *r* to get to the next term, then given *r* and the value of a term, we can determine the value of any later term.

Example: Suppose we know $a_4 = 1{,}875$ and $r = 5$. How can we find a_7?

$$a_4, a_5, a_6, a_7$$

$$\times r \quad \times r \quad \times r$$

Since a_4 and a_7 are three terms apart, we can multiply a_4 by r^3 to get a_7.
So, $a_7 = a_4 r^3 = 1{,}875(5)^3 = 234{,}375$

For arithmetic sequences, we were able to calculate the common difference *d* given two terms by using the slope formula. Although the slope formula won't help us with geometric sequences, there is a formula we can use to find the common ratio *r* given two terms. We can derive this formula using an example.

Suppose we are given two terms of a geometric sequence, $a_3 = 40$ and $a_6 = 320$. As we did with arithmetic sequences, we can express these terms as ordered pairs (n, a_n): that is, $(3,40)$ and $(6,320)$. We know that $40r^3 = 320$, or more generally, $y_1 r^{x_2 - x_1} = y_2$. Dividing both sides of this equation by y_1 gives us a formula we can use to find *r*:

$$r^{x_2 - x_1} = \frac{y_2}{y_1}$$

In this example, $r^{6-3} = \frac{320}{40}$, so $r^3 = 8$. Therefore, $r = \sqrt[3]{8} = 2$.

It is important to note that, given two terms, it is possible that two different values of *r* could yield the same two terms. For example, suppose our formula for *r* resulted in $r^4 = 16$. This would mean $r = \pm\sqrt[4]{16} = \pm 2$, and so a value of $r = 2$ or $r = -2$ could yield the same pair of terms. In other words, there are two different geometric sequences that would yield those same two terms, one in which the signs remain the same throughout the sequence and one in which the signs alternate between terms.

To write a formula for a geometric sequence given two terms:

1. Express the two terms as points, (n, a_n). Calculate the value(s) of *r* using the formula $r^{x_2 - x_1} = \frac{y_2}{y_1}$. If there are two distinct values of *r*, then there are two different geometric sequences that would yield those same two terms.
2. Substitute (each) *r*, as well as *n* and a_n from one of the terms, into the general formula for a geometric sequence. Solve for a_1.
3. Write the general formula, substituting known values for (each) a_1 and *r*, and simplify.

Example: For the preceding example where $a_3 = 40$ and $a_6 = 320$, we found $r^{6-3} = \frac{320}{40}$, or $r^3 = 8$. So, $r = \sqrt[3]{8} = 2$. We can write a formula for this sequence by substituting $r = 2$ and, from the term $a_3 = 40$, $n = 3$ and $a_n = 40$ into the general formula $a_n = a_1 r^{n-1}$. This gives us $40 = a_1 2^{3-1}$. Solving for a_1, we get $a_1 = 10$. Now that we know $a_1 = 10$ and $r = 2$, we can write the formula $a_n = 10(2)^{n-1}$. This could be simplified to $a_n = 5(2)^n$.

MODEL PROBLEM 1: *FIND THE NTH TERM*

Find the seventh term of the geometric sequence: $3, -6, 12, -24, \ldots$

Solution:

(A) $a_1 = 3; r = -\frac{6}{3} = -2$

(B) $a_n = a_1 r^{n-1}$

$a_7 = 3(-2)^{7-1} = 192$

Explanation of steps:

(A) Determine a_1 (the first term) and r (the common ratio) from the given terms.

(B) In the general formula for geometric sequences, substitute for n, a_1, and r, and evaluate *[n is the term number, 7].*

PRACTICE PROBLEMS

1. Write a formula for the nth term of the geometric sequence, $6, 18, 54, 162, \ldots$	2. Write a formula for the nth term of the geometric sequence, $12, 6, 3, 1.5, \ldots$
3. Write a formula for the nth term of the geometric sequence, $-2, 8, -32, 128, \ldots$	4. Write a formula for the nth term of the geometric sequence, $-1, 2, -4, 8, \ldots$

5. Write a formula for the nth term of the geometric sequence, $4, 10, 25, 62.5, \ldots$	6. Find the ninth term of the geometric sequence: $16, 32, 64, \ldots$
7. Find the fifth term of the geometric sequence: $12, -36, 108\ \ldots$	8. Find the seventh term of the geometric sequence: $8, 12, 18, \ldots$
9. What is the fifteenth term of the geometric sequence $5, -10, 20, -40, \ldots$?	10. Find the seventh term of the geometric sequence for which $a_1 = 6$ and $r = -\frac{1}{2}$.

MODEL PROBLEM 2: *WRITE A FORMULA GIVEN TWO TERMS*

Two terms of a geometric sequence are $a_6 = 192$ and $a_{10} = 3{,}072$. Write the two formulas for a_n that could yield these terms.

Solution:

(A) (6,192) and (10,3072)

$$r^{10-6} = \frac{3072}{192}$$

$$r^4 = 16$$

$$r = \pm 2$$

(B) $$a_n = a_1 r^{n-1}$$

$192 = a_1(2)^5$ or $192 = a_1(-2)^5$

$a_1 = \frac{192}{32} = 6$ $\quad a_1 = \frac{192}{-32} = -6$

$r = 2$ and $a_1 = 6$ $\quad r = -2$ and $a_1 = -6$

(C) $a_n = 6(2)^{n-1} = \frac{6(2)^n}{2}$ $\quad a_n = -6(-2)^{n-1} = \frac{-6(-2)^n}{-2}$

$a_n = 3(2)^n$ or $a_n = 3(-2)^n$

Explanation of steps:

(A) Express the two terms as points, (n, a_n). Calculate the value(s) of *r* using the formula $r^{x_2 - x_1} = \frac{y_2}{y_1}$. If there are two distinct values of *r*, then there are two different geometric sequences that would yield those same two terms.

(B) Substitute (each) *r*, as well as *n* and a_n from one of the terms *[we'll use (6,192)]*, into the general formula for a geometric sequence. Solve for a_1. *[The left side calculates a_1 if $r = 2$ and the right side calculates a_1 if $r = -2$.]*

(C) Write the general formula, substituting known values for (each) a_1 and *r*, and simplify. *[Both geometric sequences, $a_n = 3(2)^n$ and $a_n = 3(-2)^n$, would yield the terms $a_6 = 192$ and $a_{10} = 3{,}072$. The first formula represents the sequence, 6, 12, 24, 48, ... and the second formula represents the sequence, −6, 12, −24, 48, ...]*

PRACTICE PROBLEMS

11. Given the terms $a_3 = 135$ and $a_6 = 3{,}645$ of a geometric sequence, a) find a_{10} b) write a general formula for a_n	12. The daily number of new viewers of an online video can be represented by a geometric sequence. On Day 5, there were 3,520 new viewers, and on Day 10, there were 112,640 new viewers. a) Determine how many new viewers there were on Day 8. b) Write a general formula for the number of new viewers on the nth day.

14.3 Recursively Defined Sequences

Key Terms and Concepts

Sequences can often be defined using a **recursive formula**. A recursive formula first defines the starting term (called the *seed term*), and then defines the next terms based on values of previous terms.

Examples: (a) The arithmetic sequence $9, 11, 13, 15, \ldots$, can be defined as:

$$a_1 = 9$$
$$a_n = a_{n-1} + 2$$

common difference $d = 2$

This means the first term a_1 is 9 and the nth term a_n is defined as the value of the previous term a_{n-1} plus the common difference, 2.

(b) Similarly, the geometric sequence $5, 10, 20, 40, \ldots$, can be defined as:

$$a_1 = 5$$
$$a_n = 2a_{n-1}$$

common ratio $r = 2$

This means the first term a_1 is 5 and the nth term a_n is defined as 2 times the value of the previous term a_{n-1}.

Not all sequences are arithmetic or geometric. For example, this sequence has neither a common difference nor a common ratio. Each term is 2 times the previous term, plus 3.

$$10, 23, 49, 101, \ldots$$

However, it can still be defined using a recursive formula:

$$a_1 = 10$$
$$a_n = 2a_{n-1} + 3$$

A famous recursively defined sequence is the **Fibonacci sequence**, in which each term (after the first two seed terms) equals the sum of the two previous terms:

$$1, 1, 2, 3, 5, 8, 13, 21, 34, \ldots$$

We can define this as:

$$a_1 = 1$$
$$a_2 = 1$$
$$a_n = a_{n-1} + a_{n-2}$$

CALCULATOR TIP

The calculator can be used to generate the terms of a recursive sequence. The calculator names its sequences starting with *u* instead of *a*.

1. Press [MODE] and change from FUNCTION to SEQ mode.
2. Press [Y=]. Since we are in SEQ mode, this screen will look very different.
3. *n*Min is the number of the first term, which is usually 1.
4. Next to u(n), enter the recursive formula for u_n by pressing [2nd][7] for u and [X,T,Θ,*n*] for *n*. When entering the formula, place subscripts in parentheses (using function notation).
5. Next to u(nMin) *[or* u(1) *on some models]*, enter the value of the seed term, u_1. *[Some models allow a second seed term to be entered as* u(2) *if needed.]*
6. Press [2nd][TABLE] to view the terms in a table.
7. When finished with sequences, remember to switch back to FUNCTION mode.

Example: The screens below show the calculator generating terms of the sequence,

$$u_1 = 10$$
$$u_n = 2u_{n-1} + 3$$

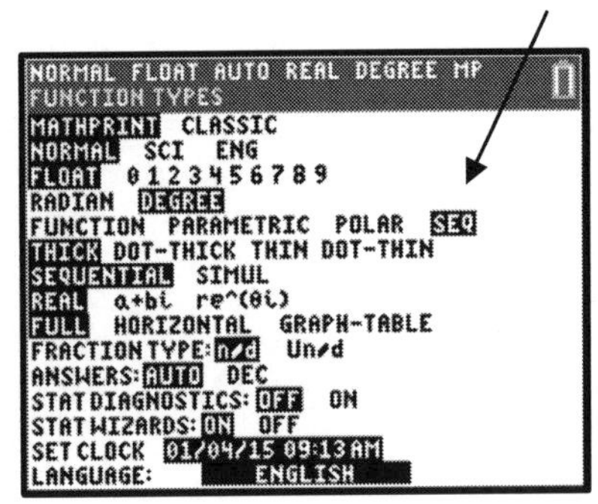

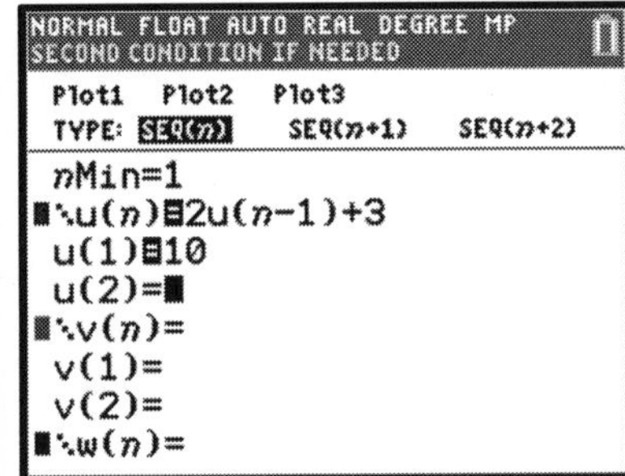

NORMAL FLOAT AUTO REAL DEGREE MP
PRESS + FOR △Tbl

n	u(*n*)
1	10
2	23
3	49
4	101
5	205
6	413
7	829
8	1661
9	3325
10	6653
11	13309

n=1

Model Problem

The first 5 terms of an arithmetic sequence are $12, 7, 2, -3, -8$.

1. Write an explicit formula for this sequence.
2. Write a recursive formula for this sequence.

Solution:

1.

(A) $a_n = a_1 + (n-1)d$

(B) $a_n = 12 + (n-1)(-5)$

(C) $a_n = 17 - 5n$

2.

(D) $a_1 = 12$

(E) $a_n = a_{n-1} - 5$

Explanation of steps:

(A) Start with the template for explicit formulas of arithmetic sequences.

(B) Substitute the first term *[12]* for a_1 and the common difference *[−5]* for *d*.

(C) Simplify.

(D) State the value of the first term.

(E) State the value of each new term based on the previous term. *[Subtract 5 from the previous term.]*

Practice Problems

1. Write a recursive formula for the arithmetic sequence, $6, 10, 14, 18, \ldots$	2. Write a recursive formula for the geometric sequence, $8, 24, 72, 216, \ldots$
3. Write a recursive formula for the geometric sequence, $16, -8, 4, -2, \ldots$	4. Write the first 4 terms of the sequence, $a_1 = 3$ $a_n = 2a_{n-1} + 1$

5. Write the first 4 terms of the sequence, $a_1 = 3$ $a_n = 2a_{n-1} - 4$	6. A sequence is recursively defined as: $a_1 = 3$ and $a_n = a_{n-1} + n$ a) Write the first four terms of this sequence. b) Is this sequence arithmetic, geometric, or neither? Justify your answer.
7. Write the first 4 terms of the sequence, $a_1 = 2$ $a_n = 3a_{n-1} + n$	8. Write the first 4 terms of the sequence, $a_1 = -3$ $a_n = 2a_{n-1} - n$

9. Write a formula for the sequence

$$2, 3, 6, 18, 108, 1944, \ldots$$

such that $a_1 = 2$, $a_2 = 3$, and a_n is defined recursively for $n > 2$.

10. The first four terms in a sequence are

$$40, 8, 24, 16, \ldots$$

Each term after the first two terms is found by taking the mean (average) of the two preceding terms.

a) Write a recursive formula for this sequence.

b) Which term is the first odd number in this sequence?

14.4 Sigma Notation

Key Terms and Concepts

The Greek capital letter, Σ (**sigma**), is used to represent **summation.**

Example: Let's look at $\sum_{k=1}^{5} k^2$. The notation above and below the sigma symbol means that k starts at 1 and increments up to 5. For each value of k, we add the value of k^2 to the sum. In other words,

$$\sum_{k=1}^{5} k^2 = 1^2 + 2^2 + 3^2 + 4^2 + 5^2 = 55.$$

The variable below the sigma symbol is called the **index** or **summation variable**. Any letter may be used. The index is generally restricted to the set of integers. The starting value is called the **lower bound** (or *lower limit*) and the ending value is called the **upper bound** (or *upper limit*). The index starts at the lower bound and increments in steps of 1 up to the upper bound. With each iteration, the expression to the right of the sigma is evaluated and its value is added to the running sum.

To evaluate a summation by hand, it is helpful to create a table.

Example: To evaluate $\sum_{k=0}^{4} (2k-1)$, we can create the table below.

k	$2k - 1$
0	$2(0) - 1 = -1$
1	$2(1) - 1 = 1$
2	$2(2) - 1 = 3$
3	$2(3) - 1 = 5$
4	$2(4) - 1 = 7$

The sum is $-1 + 1 + 3 + 5 + 7 = 15$.

When we use an expression of more than one term in a summation, parentheses are required.

Example: We used parentheses to write $\sum_{k=0}^{4}(2k-1)$. Note how this is different from $\sum_{k=0}^{4}2k-1$, which means find $\sum_{k=0}^{4}2k$ and then subtract 1 from the sum.

CALCULATOR TIP

We can also use the calculator to evaluate a summation.

The TI-84 has a summation function available. To enter the summation:

1. Press [MATH][0] or [ALPHA][F2][2]. This selects the summation function.
2. Type the index variable K, the lower and upper bounds, and the expression in the appropriate places on the screen. To type the variable in the expression, press [ALPHA][K].
3. Press [ENTER].

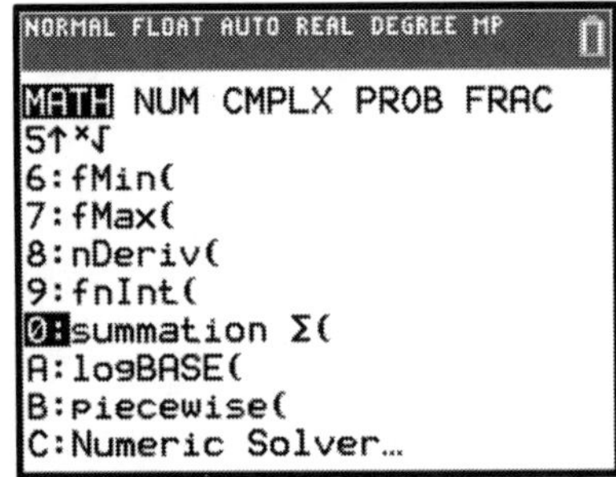

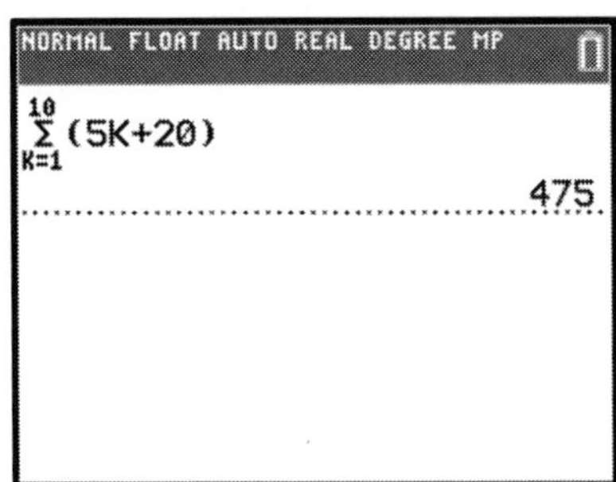

[Note to teachers: If you set calculators to Test Mode by holding down [◄][►][ON] keys simultaneously, make sure that logBASE and summation Σ are not disabled.]

Unfortunately, the TI-83 calculator does not have the summation function. However, we can still calculate a summation by using the sum and sequence functions.

1. Press [2nd][STAT] MATH [5] to select the sum function and press [ENTER].
2. Press [2nd][STAT] OPS [5] to select the seq function and press [ENTER].
3. Type the expression. For the variable, press [ALPHA] followed by the letter.
4. Press [,], then type the index variable (using the [ALPHA] button again).
5. Press [,], then type the lower and upper bounds, separated by [,].
6. Press [)][)] and [ENTER].

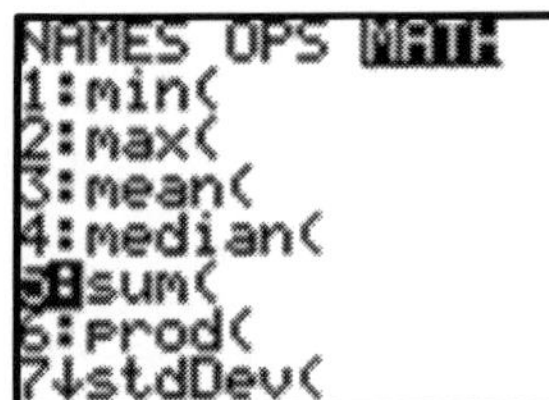

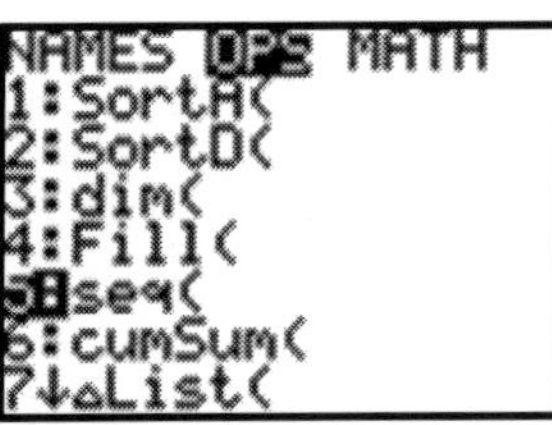

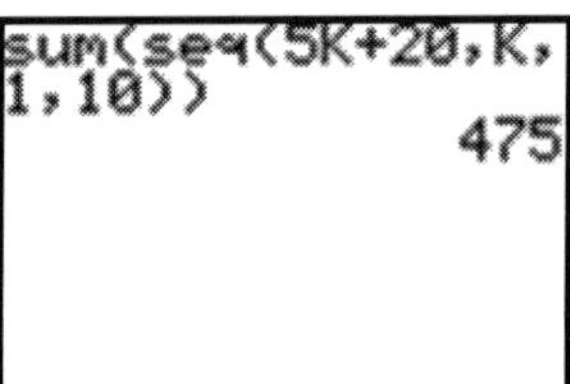

MODEL PROBLEM

Evaluate $\sum_{k=1}^{3} 2^{k-1}$.

Solution:

(A)

k	2^{k-1}
1	$2^{1-1} = 2^0 = 1$
2	$2^{2-1} = 2^1 = 2$
3	$2^{3-1} = 2^2 = 4$

(B)

(C) The sum is 7.

Explanation of steps:

(A) Create a table with the index *[k]* in column 1 and the expression *[*2^{k-1}*]* in column 2.

(B) For each value of the index, evaluate the expression

[$\sum_{k=1}^{3}$ *means k goes from 1 to 3].*

(C) Add the results to find the sum *[*$1 + 2 + 4 = 7$*].*

PRACTICE PROBLEMS

1. Evaluate $\sum_{x=2}^{5} 3x$ by hand. Then, check your answer using the calculator.	2. Evaluate $\sum_{m=2}^{5} (m^2 - 1)$ by hand. Then, check your answer using the calculator.
3. Evaluate $\sum_{n=1}^{5} (-2n + 100)$ by hand. Then, check your answer using the calculator.	4. Evaluate $\sum_{r=3}^{5} (2^r + 2)$ by hand. Then, check your answer using the calculator.

5. Evaluate $\sum_{n=1}^{5}(n^2+n)$ by hand. Then, check your answer using the calculator.	6. Evaluate $\sum_{k=1}^{3}(2-k)^2$ by hand. Then, check your answer using the calculator.
7. Evaluate $\sum_{k=0}^{2}3(2)^k$ by hand. Then, check your answer using the calculator.	8. Evaluate $\sum_{r=3}^{5}(-r^2+r)$ by hand. Then, check your answer using the calculator.
9. Evaluate $\sum_{n=1}^{3}(-n^4-n)$ by hand. Then, check your answer using the calculator.	10. Evaluate $\sum_{m=1}^{3}(2m+1)^{m-1}$ by hand. Then, check your answer using the calculator.

11. Which expression is equivalent to the sum of the terms 2, 8, 32, 128?

(1) $\sum_{n=1}^{4} 2(n-1)^4$ (3) $\sum_{n=1}^{4} 2(4)^{n-1}$

(2) $\sum_{n=1}^{4} 4(2)^{n-1}$ (4) $\sum_{n=1}^{4} 4(n-1)^2$

12. Which expression is equivalent to the sum of the terms 6, 12, 20, 30?

(1) $\sum_{n=4}^{7} 2^n - 10$ (3) $\sum_{n=2}^{5} 5n-4$

(2) $\sum_{n=3}^{6} \frac{2n^2}{3}$ (4) $\sum_{n=2}^{5} n^2+n$

13. Which of the following does *not* express the sum $\frac{2}{3}+\frac{3}{4}+\frac{4}{5}+\frac{5}{6}+\frac{6}{7}$ using sigma notation?

(1) $\sum_{k=3}^{7} \frac{k-1}{k}$ (3) $\sum_{k=1}^{5} \frac{k+1}{k+2}$

(2) $\sum_{k=1}^{5} \frac{k}{k+1}$ (4) $\sum_{k=2}^{6} \frac{k}{k+1}$

14. The expression $1+\sqrt{2}+\sqrt[3]{3}$ is equivalent to

(1) $\sum_{n=1}^{3} \sqrt{n}$ (3) $\sum_{n=1}^{3} n^{-n}$

(2) $\sum_{n=0}^{3} n^n$ (4) $\sum_{n=1}^{3} n^{\frac{1}{n}}$

15. The projected total profits, in dollars, for Kite Shoes from 2020 to 2022 can be estimated using the model below, where n is the year and $n = 0$ represents 2020.

$$\sum_{n=0}^{2}(13{,}567n+294)$$

Use this model to find the company's projected total profits for the period 2020 to 2022.

14.5 Arithmetic Series

Key Terms and Concepts

A **series** is the sum of the terms of a sequence. A **partial sum** (or **finite series**) is the sum of a finite number of consecutive terms of a sequence beginning with the first term.

In the equation below, a_k represents the kth term of the sequence a. With each iteration, we add the value of a_k to the running sum.

$$\sum_{k=1}^{n} a_k = a_1 + a_2 + a_3 + \cdots + a_n$$

Note that when we work with series, the last term of the partial sum is often called the nth term. In the chapters on sequences, we often used n to represent the index of each term of the sequence, whereas in the summation above, we are using k for this purpose.

An **arithmetic series** is the sum of an arithmetic sequence.

We can find the partial sum of the first n terms of an **arithmetic series** using the formula $S_n = \dfrac{n(a_1 + a_n)}{2}$.

Examples: (a) The sum of $17 + 25 + 33 + 41 + 49$, which is the sum of the first 5 terms of $a_n = 8n + 9$, can be calculated as $S_5 = \dfrac{5(17 + 49)}{2} = 165$.

(b) The sum of the first n positive integers is $\dfrac{n(n+1)}{2}$.

If we don't know the value of a_n we can substitute the expression $a_1 + (n-1)d$ for a_n.

Example: To find the sum of the first 10 terms of the arithmetic sequence $25, 30, 35, 40, \ldots$, we can start with $S_n = \dfrac{n(a_1 + a_n)}{2}$.

However, since a_n is unknown, we can replace it with $a_1 + (n-1)d$, giving us $S_n = \dfrac{n(a_1 + a_1 + (n-1)d)}{2}$. We know $a_1 = 25$, $d = 5$, and $n = 10$.

Therefore, $S_{10} = \dfrac{10(25 + 25 + (10-1)\cdot 5)}{2} = 475$.

Another method for finding the partial sum of an arithmetic series is by writing the series in sigma notation using $\sum_{k=1}^{n}\left[a_1+(k-1)d\right]$ and then entering the summation into the calculator. Note that in the arithmetic sequence expression, we replaced *n* with *k*.

Example: We can write the sum of the first 10 terms of the arithmetic sequence $25, 30, 35, 40, \ldots$ in sigma notation. Again, we know that $a_1 = 25$ and $d = 5$, so we can substitute these values into the expression $a_1 + (k-1)d$ and simplify to get, $5k + 20$. So, the sum of the first 10 terms can be written as $\sum_{k=1}^{10}(5k+20)$. Entering the summation into the calculator, we get 475.

If we don't know how many terms (n) are included, we can set the expression for *n* equal to a_n and solve for *n*.

Example: To express $8 + 11 + 14 + 17 + \cdots + 50$ in sigma notation, we can start with $\sum_{k=1}^{n}(3k+5)$. Since the last term $a_n = 50$, we have $3n + 5 = 50$, so $n = 15$.

Therefore, the summation can be rewritten as $\sum_{k=1}^{15}(3k+5)$. Now, we can calculate the sum either by using the summation function of the calculator or by the partial sum formula: $S_{15} = \dfrac{15(8+50)}{2} = 435$.

MODEL PROBLEM

Find the sum of the first 10 terms of the arithmetic sequence 20, 26, 32, ...

Solution:

(A) $a_1 = 20, d = 6, n = 10$

(B) Method 1

$$\sum_{k=1}^{10}(6k+14) = 470$$

Method 2

$$S_{10} = \frac{10(20 + 20 + (10 - 1)\cdot 6)}{2} = 470$$

Explanation of steps:

(A) Determine the first term, a_1, the common difference, d, and the number of terms, n.

(B) Choose to use either summation (Method 1) or the partial sum formula (Method 2).

- For Method 1, write an expression for the arithmetic sequence *[*$a_1 + (k-1)d = 20 + (k-1)\cdot 6 = 6k + 14$*]*. Then use the summation feature of the calculator to calculate the sum of the terms *[from term* $k = 1$ *to term 10]*.
- For Method 2, substitute $a_1 + (n-1)d$ for a_n in the partial sum formula, $S_n = \frac{n(a_1 + a_n)}{2}$, to get $S_n = \frac{n(a_1 + a_1 + (n-1)d)}{2}$. Then, substitute for a_1, d, and n, and evaluate.

PRACTICE PROBLEMS

1. Write $3 + 5 + 7 + 9 + 11$ using sigma notation. Use the partial sum formula to calculate the sum.	2. Write $7 + 11 + 15 + 19 + 23 + 27 + 31$ using sigma notation. Use the partial sum formula to calculate the sum.

3. Write $25 + 29 + 33 + 37 + 41 + 45$ using sigma notation. Use the partial sum formula to calculate the sum.	4. Write $5 + 7 + 9 + 11 + \cdots + 43$ using sigma notation
5. Write $1 + 3 + 5 + 7 + 9 + \cdots + 39$ using sigma notation.	6. Write $11 + 23 + 35 + \cdots + 119$ using sigma notation. Use your calculator to find the sum.

7. Write $7 + 14 + 21 + 28 + \cdots + 105$ using sigma notation. Use your calculator to find the sum.

8. Write $18 + 26 + 34 + 42 + \cdots + 122$ using sigma notation. Use your calculator to find the sum.

9. Use the partial sum formula to find the sum of the first 19 terms of the sequence $3, 10, 17, 24, 31, \ldots$

10. Use the partial sum formula to find the sum of the first twenty terms of the sequence whose first five terms are $5, 14, 23, 32, 41$.

11. In a theater with 30 rows, the number of seats in a row increases by two with each successive row. The front row has 15 seats. Use the partial sum formula to find the total seating capacity of the theater.	12. An auditorium has 21 rows of seats. The first row has 18 seats, and each succeeding row has two more seats than the previous row. Use the partial sum formula to determine how many seats there are in the auditorium.

14.6 Geometric Series

Key Terms and Concepts

A **geometric series** is the sum of the terms of a geometric sequence.

We can find the partial sum of the first n terms of a **geometric series** with a common ratio of r using the formula $S_n = \dfrac{a_1 - a_1 r^n}{1 - r}$.

Example: To find the sum of the first 10 terms of the sequence 8, 24, 72, 216, ..., we can calculate $S_{10} = \dfrac{8 - 8(3)^{10}}{1 - 3} = 236{,}192$.

Note that some texts will factor out a_1 from the numerator and write this formula as $S_n = a_1\left(\dfrac{1 - r^n}{1 - r}\right)$.

Another way to find the partial sum of a geometric series is by writing the series in sigma notation using $\sum_{k=1}^{n} a_1 r^{k-1}$. The summation contains the same expression we used for geometric sequences except that we now are using k instead of n. Once written in sigma notation, we can enter it into the calculator to find the sum.

Example: For the sum of the first 10 terms of the geometric sequence 8, 24, 72, 216, ..., $a_1 = 8$ and $r = 3$, so $a_1 r^{k-1} = 8(3)^{k-1}$. Therefore, the sum of the first 10 terms can be written as $\sum_{k=1}^{10} 8(3)^{k-1}$. Entering this into the calculator will also yield 236,192.

As with arithmetic series, if we don't know how many terms (n) are included, we can set the expression for n equal to a_n and solve for n. In this case, since n is a term of an exponent, we'll solve for n by using logarithms.

Example: To express $5 + 20 + 80 + 320 + \cdots + 327{,}680$, we can start with $\sum_{k=1}^{n} 5(4)^{k-1}$.

Since the last term $a_n = 327{,}680$, we can set $5(4)^{n-1} = 327{,}680$.
We can now use logarithms to solve for n:

$$5(4)^{n-1} = 327{,}680$$
$$4^{n-1} = 65{,}536$$
$$\log 4^{n-1} = \log 65{,}536$$
$$(n-1)\log 4 = \log 65{,}536$$
$$n - 1 = \frac{\log 65{,}536}{\log 4}$$
$$n - 1 = 8$$
$$n = 9$$

Therefore, 327,680 is the ninth term, and our sum is $\sum_{k=1}^{9} 5(4)^{k-1} = 436{,}905$.

Model Problem

Find the sum of the first 10 terms of the geometric sequence 20, 120, 720, ….

Solution:

(A) $a_1 = 20, r = 6, n = 10$

(B) Method 1

$$\sum_{k=1}^{10}\left(20 \cdot 6^{k-1}\right) = 241{,}864{,}700$$

Method 2

$$S_{10} = \frac{20 - 20 \cdot 6^{10}}{1 - 6} = 241{,}864{,}700$$

Explanation of steps:

(A) Determine the first term, a_1, the common ratio, r, and the number of terms, n.

(B) Choose to use either summation (Method 1) or the partial sum formula (Method 2).

- For Method 1, write an expression for the geometric sequence *[*$a_1 r^{k-1} = 20 \cdot 6^{k-1}$*]*. Then use the summation feature of the calculator to calculate the sum of the terms *[from term $k = 1$ to term 10]*.
- For Method 2, use the partial sum formula, $S_n = \dfrac{a_1 - a_1 r^n}{1 - r}$.

  Then, substitute for a_1, r, and n, and evaluate.

PRACTICE PROBLEMS

1. Write $7 + 21 + 63 + 189 + 567$ in sigma notation.	2. Write $1 - 2 + 4 - 8 + 16 - 32$ in sigma notation.
3. Write $200 + 100 + 50 + 25$ in sigma notation.	4. Use the partial sum formula to find the sum of the first 7 terms of the series, $9 + 36 + 144 + 576 + \cdots$ Write the series in sigma notation and use the calculator to check that it calculates the same summation.

5. Use the partial sum formula to find the sum of the first eight terms of the series $3 - 12 + 48 - 192 + \ldots$	6. Write $10 + 20 + 40 + 80 + \cdots + 5120$ in sigma notation.
7. Write $\frac{1}{2} + 1 + 2 + 4 + 8 + \cdots + 1024$ in sigma notation.	8. Write $6 + 30 + 150 + \cdots + 93{,}750$ in sigma notation. Use your calculator to find the sum.

9. a) Find $\sum_{n=1}^{7} 9(-3)^{n-1}$ on the calculator.

b) Use the partial sum formula to calculate the same sum.

10. A jogger ran $\frac{1}{3}$ mile on day 1, and $\frac{2}{3}$ mile on day 2, and $1\frac{1}{3}$ miles on day 3, and $2\frac{2}{3}$ miles on day 4, and this pattern continued for 3 more days. Which expression represents the total distance the jogger ran?

(1) $\sum_{d=1}^{7} \frac{1}{3}(2)^{d-1}$

(2) $\sum_{d=1}^{7} \frac{1}{3}(2)^{d}$

(3) $\sum_{d=1}^{7} 2\left(\frac{1}{3}\right)^{d-1}$

(4) $\sum_{d=1}^{7} 2\left(\frac{1}{3}\right)^{d}$

Chapter 15. Probability

15.1 Theoretical and Empirical Probability

Key Terms and Concepts

Probability is the likelihood that an event will occur, expressed as a number between 0 and 1 inclusive, where 0 means impossibility and 1 means certainty. We will use the notation $P(E)$ to mean the probability that event E will occur. So, $0 \leq P(E) \leq 1$.

To express the **probability of an event** as a fraction, use

$$P(E) = \frac{successful\ outcomes}{all\ outcomes}$$

where the numerator is the number of possible outcomes that would meet the criteria for success and the denominator is the total number of equally likely outcomes.

Example: The probability of rolling an odd number on a standard 6-sided die is $\frac{3}{6} = \frac{1}{2}$ because, of all 6 possible outcomes, 3 outcomes $\{1,3,5\}$ would be successful.

The set of all possible outcomes of an event is called a **sample space**. An **event** is a set of outcomes (a subspace of the sample space) to which a probability is assigned.

Examples: The sample space for tossing a coin is $\{heads, tails\}$. The sample space for rolling a six-sided die is $\{1,2,3,4,5,6\}$.

If the probability of an event is *x*, the probability of its **complement** (i.e., the probability that the event does *not* occur) is $1 - x$.

Example: Since the probability of rolling a 2 on a 6-sided die is $\frac{1}{6}$, the probability of *not* rolling a 2 is $1 - \frac{1}{6} = \frac{5}{6}$.

In other words, if E is an event and $\bar{E}$ is its complement, $P(E) + P(\bar{E}) = 1$.

It is often useful to compare the probabilities of two or more events to determine which of them is most likely to occur. If we calculate each event's probability as a value between 0 and 1, inclusive, the probability with the greater value represents the event that is more likely to occur.

Example: If $P(A) = \frac{3}{5}$ and $P(B) = \frac{4}{5}$, then the event B is more likely to occur.

In **theoretical probability**, a theoretical knowledge or analysis of probabilities is applied. For example, when rolling a fair 6-sided die, each outcome (1 through 6) is equally likely to occur, so each outcome theoretically has a $\frac{1}{6}$ probability.

With **empirical probability** (or *experimental probability*), only **past results** are considered. For example, if an experiment was performed in which a 6-sided die was tossed 100 times and a 4 was rolled 20 of those times, the empirical probability of rolling a 4 on the next toss is $\frac{20}{100}$ or $\frac{1}{5}$.

In this experiment, there were 100 **trials**, of which the desired outcome had a **frequency** of 20. The formula for empirical probability is $\frac{frequency}{trials}$. The more trials we run, the more likely it will be that the empirical probability will approach the theoretical probability of $\frac{1}{6}$.

A **simulation** is a model of real events, usually performed by a computer. A simulation may use random number generators to simulate the outcome of the events.

Examples: (a) A random integer between 1 and 6 can represent the roll of a die.

(b) If an event has a 37% probability, a random number between 0 and 99 can be generated, with the values 0 to 36 representing that the event occurred.

MODEL PROBLEM

A whole number between 1 and 100, inclusive, is selected at random. What is the probability that the selected number is a perfect square?

Solution:

$$\frac{10}{100} = \frac{1}{10}$$

Explanation of steps:

Express as a fraction, $\frac{successful}{all}$, and reduce if possible.

[10 out of the 100 possible outcomes are perfect squares: 1, 4, 9, 16, ... 100.]

Practice Problems

1. A spinner has 4 equal sectors colored yellow, blue, green and red. What is the probability of landing on blue after spinning the spinner?	2. A box contains six black balls and four white balls. What is the probability of selecting a black ball at random from the box?
3. A glass jar contains 6 red, 5 green, 8 blue and 3 yellow marbles. If a single marble is chosen at random from the jar, what is the probability of choosing a red marble?	4. Mary chooses an integer at random from 1 to 6. What is the probability that the integer she chooses is a prime number?
5. If the probability that it will rain on Thursday is $\frac{5}{6}$, what is the probability that it will *not* rain on Thursday?	6. A box contains 6 dimes, 8 nickels, 12 pennies, and 3 quarters. What is the probability that a coin drawn at random is *not* a dime?

7. A dartboard consists of 20 equal sectors numbered 1 through 20. If a dart thrown is equally likely to land on any point on the dartboard, what is the probability that it lands in a sector numbered 15 or greater?	8. A deck of cards contains 52 cards, 13 from each suit. If a random card is flipped over, what is the probability that it is a spade?
9. The spinner below is divided into eight equal regions and is spun once. What is the probability of *not* getting red? 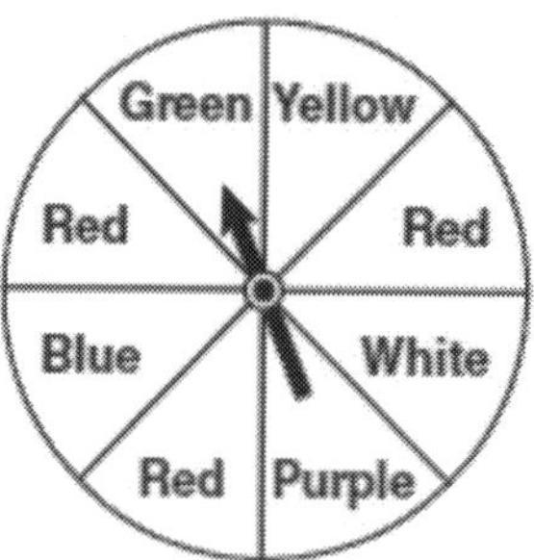 	10. Three boxes contain colored blocks. Box 1 contains 15 red and 14 blue blocks. Box 2 contains 16 white and 15 blue blocks. Box 3 contains 15 red and 15 white blocks. All of the blocks from the three boxes are placed into one storage bin. If one block is randomly selected from the bin, which color block would most likely be picked?

11. A factory recorded that over the past year, it shipped 78,000 good widgets and discarded 2,000 faulty widgets. What is the empirical probability that a widget produced by the factory will be faulty?	12. Selena plays on a softball team. She has 8 hits in 20 times at bat. Based on past performance, what is the probability that she will get a hit in her next time at bat?
13. Once a month, a light bulb factory collects 100 random products to determine how many of the bulbs are faulty. In the last 10 months, a total of 20 faulty bulbs were found. What is the empirical probability that a light bulb that the factory produces next month will be faulty?	14. A random number generator outputs a value from 0 to 9. If an event has a 20% probability of occurring, how can I use the random number generator in a simulation to determine whether the event occurred?

15.2 Probability Involving And or Or

Key Terms and Concepts

Let's consider situations in which two criteria for a successful outcome need to be considered. For example, when selecting a random card from a deck, one event could be "the card is red **and** it is a king." Another event could be "the card is red **or** it is a king."

In either case, to determine the probability, we could

a) look at the list of all possible outcomes and **count the outcomes** that meet both criteria (for "and") or either criteria (for "or"), or
b) we could consider two separate events (two sets of outcomes) – for example, A as the event that the card is red and B as the event that the card is a king – and then find the probability of the **intersection** ($A \cap B$ for "and") or the **union** ($A \cup B$ for "or") of these two events.

$P(A\ and\ B)$

When the successful outcome of an event must satisfy two criteria, joined by the word "and," we can determine the probability either by checking each element in the sample space or by using set intersection of the two separate events.

Example: A 6-sided die is tossed. Let's find the probability that the result is an even number ***and*** a number greater than 3.

1) *Check Each Outcome*: From the list of all possible outcomes, count the outcomes that meet both conditions, as shown in boxes here: $\{1, 2, 3, \boxed{4}, 5, \boxed{6}\}$. Only two out of the six outcomes are both even and greater than 3, so the probability is $\frac{2}{6} = \frac{1}{3}$.

2) *Set Intersection*: The set of successful outcomes would equal the **intersection** of the sets of successful outcomes from each condition. Let A, the event that an even number is rolled, equal $\{2, 4, 6\}$, and let B, the event that a number greater than 3 is rolled, equal $\{4, 5, 6\}$. So, the probability that an even number ***and*** a number greater than 3 is rolled is:

$P(A\ and\ B) = P(A \cap B) =$

$P(\{2, 4, 6\} \cap \{4, 5, 6\}) =$

$P(\{4, 6\}) = \frac{2}{6} = \frac{1}{3}$.

We can model this on a Venn Diagram:

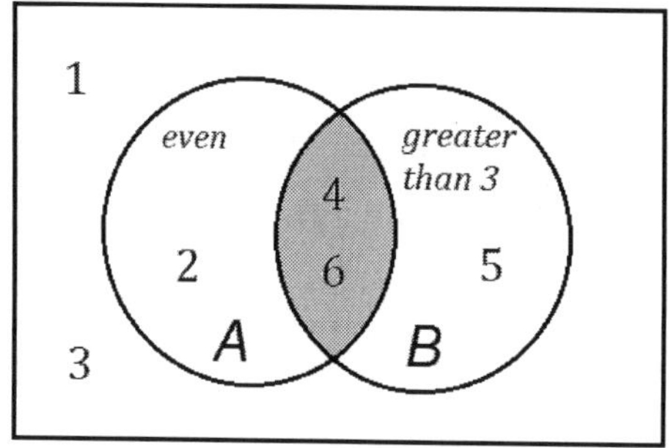

$P(A \text{ or } B)$

Similarly, when the successful outcome of an event may satisfy either of two criteria, joined by the word "or," we can determine the probability either by checking each element in the sample space or by using set union of the two separate events.

Example: A 6-sided die is tossed. Let's find, by each of the three methods, the probability that an even number ***or*** a number greater than 3 is rolled.

1) *Check Each Outcome*: From the list of all possible outcomes, count the outcomes that meet either or both conditions, as shown in boxes here: $\{1, \boxed{2}, 3, \boxed{4}, \boxed{5}, \boxed{6}\}$. Four out of the six outcomes are even or greater than 3 (or both), so the probability is $\frac{4}{6} = \frac{2}{3}$.

2) *Set Union*: The set of successful outcomes would equal the **union** of the sets of successful outcomes from each condition. Let A, the event that an even number is rolled, equal $\{2, 4, 6\}$, and let B, the event that a number greater than 3 is rolled, equal $\{4, 5, 6\}$. So the probability that an even number ***or*** a number greater than 3 is rolled is:

$P(A \text{ or } B) = P(A \cup B) =$

$P(\{2, 4, 6\} \cup \{4, 5, 6\}) =$

$P(\{2, 4, 5, 6\}) = \frac{4}{6} = \frac{2}{3}$.

We can model this on a Venn Diagram:

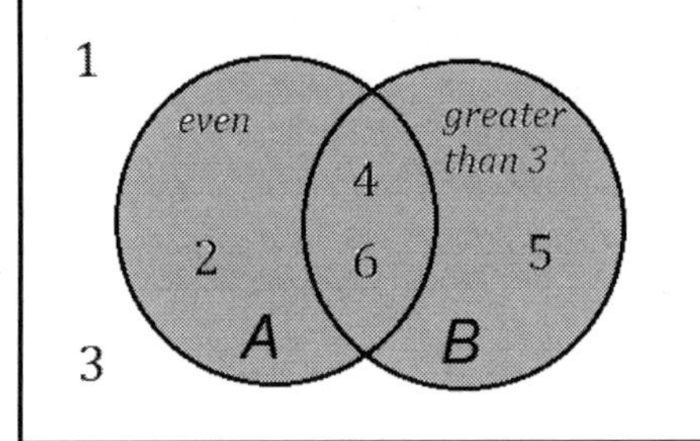

A third way of finding $P(A \text{ or } B)$ is by using the Addition Rule, which we can derive by looking at the Venn Diagram above. Suppose we want to know the number of elements in the union of the two sets, $|A \cup B|$, where $||$ represents the **cardinality** (or number of elements). If we add the number of elements in set A (there are 3 even outcomes) to the number of elements in set B (there are 3 outcomes that are greater than 3), we see that we are counting the elements in the intersection region twice. So, to find the number of elements in the union of these two sets, we need to subtract the number of elements in the intersection region: $|A \cup B| = |A| + |B| - |A \cap B|$. This gives us $3 + 3 - 2 = 4$.

Since the probability of an event also depends on the number of successful outcomes (i.e., cardinality of an event set), we can apply this concept to probability.

The **Addition Rule** states that $\boldsymbol{P(A\ or\ B) = P(A) + P(B) - P(A\ and\ B)}$.

Example: Using the example problem above, if A is the set of even outcomes and B is the set of outcomes greater than 3, then

$$P(A\ or\ B) = P(A) + P(B) - P(A\ and\ B) =$$
$$P(\{2,4,6\}) + P(\{4,5,6\}) - P(\{4,6\}) =$$
$$\frac{3}{6} + \frac{3}{6} - \frac{2}{6} = \frac{4}{6} = \frac{2}{3}$$

Using set notation, the Addition Rule may also be written as:

$$\boldsymbol{P(A \cup B) = P(A) + P(B) - P(A \cap B)}$$

For some problems, it may be helpful to create Venn Diagrams. Usually, it is best to work from the center out; that is, start with the intersection of the sets.

Example: Among the students in Mr. Krane's home room, 16 students are taking physics, 7 students are taking calculus, 3 students are taking both classes, and 9 students are taking neither class. To determine how many students are in Mr. Krane's home room, we can set up a Venn Diagram.

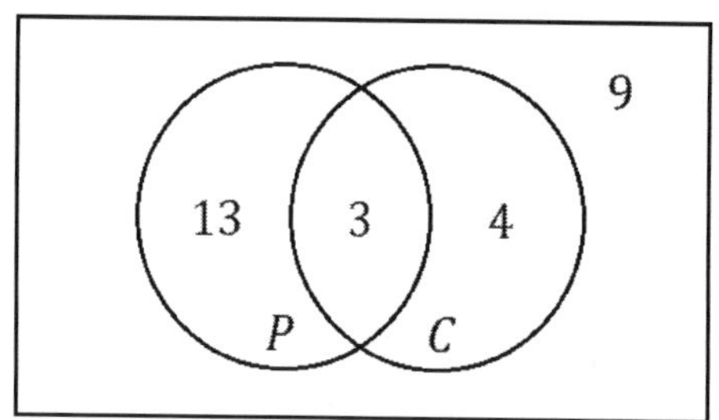

Draw two overlapping circles for the set P of students taking physics and the set C of students taking calculus. Then start by writing a 3 in the intersection for the number of students taking both. Now, since 16 are taking physics, including the 3 who are already counted, write the difference 13 in the part of P that is outside the intersection. Similarly, since 7 are taking calculus and 3 are already counted, write the difference 4 in the part of C that is outside the intersection. For the 9 students who are taking neither class, write the 9 outside the two circles. Now, to find the number of students in the home room, add the numbers in all the parts: $13 + 3 + 4 + 9 = 29$.

In the Venn diagrams above, the circles representing the sets were overlapping, representing the fact that they had one or more elements in common; that is, their intersection was not empty. Events that have one or more outcomes in common are called **overlapping** events.

Sometimes, however, the criteria for two events cannot both be met at the same time; in such cases, the events are called **mutually exclusive** (or *disjoint*). The intersection of mutually exclusive sets is an empty set; that is, they have no elements in common.

Example: When rolling a die, A = "getting an even number" and B = "getting a 5" are mutually exclusive events.

When A and B are mutually exclusive, then $P(A\ and\ B) = 0$.
For mutually exclusive events only, the Addition Rule reduces to $P(A\ or\ B) = P(A) + P(B)$.
In a Venn Diagram, mutually exclusive events may be drawn as below, with no intersection.

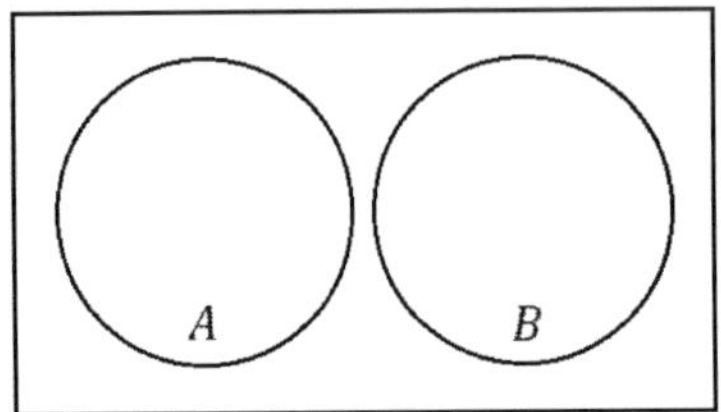

Model Problem

In a standard deck of 52 cards, there are four kings. Half of the 52 cards are red, including two of the kings; the rest of the cards are black.

What is the probability that a card drawn at random is *red or a king*?

Solution:

$$P(R\ or\ K) = P(R) + P(K) - P(R\ and\ K) =$$

$$\frac{26}{52} + \frac{4}{52} - \frac{2}{52} = \frac{28}{52} = \frac{7}{13}$$

Explanation of steps:

(A) We can use the Addition Rule.

[There are 26 red cards out of 52, so $P(R) = \frac{26}{52}$*. There are 4 kings out of 52 cards, so* $P(K) = \frac{4}{52}$*. There are 2 red kings, so* $P(R\ and\ K) = \frac{2}{52}$*. Apply the Addition Rule by calculating* $P(R) + P(K) - P(R\ and\ K)$*.]*

(B) Reduce the fraction.

Practice Problems

1. Shannon's pencil box contains four pens, two markers, and five pencils. Find the probability that an item chosen at random is a pen or a marker.	2. What is the probability that a letter selected randomly from the word APPLE is a vowel or a P?
3. The probability that a red block is selected from a bucket is $\frac{3}{8}$, and the probability that a blue block is selected is $\frac{2}{8}$. What is the probability that a red block or a blue block is selected?	4. A case contains 4 red pens, 2 blue pens, 5 red pencils, and 3 black markers. Find the probability that an item chosen at random is a pen or red.

5. A number between 1 and 10, inclusive, is randomly selected. The set of outcomes that satisfy condition A is $\{2, 4, 5, 7, 8\}$ and the set of outcomes that satisfy condition B is $\{3, 5, 8, 9\}$.

   a) Find $P(A \text{ and } B)$.

   b) Find $P(A \text{ or } B)$.

6. From the Addition Rule, develop a formula for $P(A \text{ and } B)$.

7. In a kindergarten class, there are 9 boys and 11 girls. Of the students in the class, 4 boys and 5 girls have perfect attendance. If a student is selected at random, what is the probability of selecting a girl *or* a student with perfect attendance?

8. In a class of 500 students, 350 are taking a mathematics course and 300 are taking a science course. If 200 of these students are taking both courses, what is the probability that a student selected at random is *not* taking either of these courses?

15.3 Two-Way Frequency Tables

Key Terms and Concepts

A **two-way table** (sometimes called a *pivot table* or *contingency table*) shows frequencies for bivariate data belonging to two different categories. One category of data, with its possible values, is represented by rows and the other category, with its possible values, is represented by columns.

Example: A bakery holds a taste test in which participants select their favorite cup cake icing flavor. Each data item consists of two categorical values, gender and flavor preference. The two-way table below shows the results for 50 adults - 20 women and 30 men. In this survey, only 2 out of 20 women preferred vanilla, but 16 out of the 30 men chose vanilla.

	Vanilla	Chocolate	Strawberry	Total
Women	2	10	8	20
Men	16	6	8	30
Total	18	16	16	50

Entries in the body of the table are called **joint frequencies**. Entries in the "Total" row and "Total" column are called **marginal frequencies.** The marginal frequencies are the sums of the joint frequencies on that row or column of the table. The grand total in the lower right hand cell is the total number of data points. It should equal the sum of each set of marginal frequencies.

Two-way tables help us to find conditional probabilities. A **conditional probability** is the probability of an event *given that* a certain condition is true. $P(B|A)$ represents the conditional probability of B given that A is true, read as "the probability of B given A." With a table, we set the given condition to be true by isolating a row or column.

Example: (a) Using the frequency table above, we could say that among *women*, 2 out of 20 prefer *vanilla*. The given condition, "among women," restricts us to the top row. So, $P(vanilla|woman)$, the conditional probability that a randomly selected person prefers *vanilla*, given that she is a *woman*, is $\frac{2}{20}$ or $\frac{1}{10}$.

(b) Among those who prefer *chocolate*, 6 out of 16 are *men*. In this case, the given condition restricts us to the column labelled "Chocolate." Therefore, $P(man|chocolate)$, the conditional probability that a randomly selected person is a *man*, given that the person prefers *chocolate*, is $\frac{6}{16}$ or $\frac{3}{8}$.

We can also use the two-way table to decide whether events appear to be independent. Two events, A and B, are **independent** if $P(A) = P(A|B)$; otherwise, they are **dependent**.

Example: Using the same table below, the participants' gender and their choice of icing flavor do not appear to be independent. For example, we can consider the event that the participant is a man, M, and the event that the participant prefers vanilla, V. If these were independent events then the probability of selecting a man, $P(M) = \frac{30}{50} = \frac{3}{5}$, should be equal to the probability of selecting a man given a preference for vanilla, $P(M|V) = \frac{16}{18} = \frac{8}{9}$. Since these probabilities are not equal, they are dependent events. The participants' gender and their choice of icing flavor are dependent as long as *any pair* of categorical events are dependent.

	Vanilla	Chocolate	Strawberry	Total
Women	2	10	8	20
Men	16	6	8	30
Total	18	16	16	50

MODEL PROBLEM

A public opinion survey explored the relationship between age and support for increasing the minimum wage. The results are summarized in the two-way table below.

	For	Against	No opinion	Total
21 – 40	25	20	5	50
41 – 60	20	35	20	75
Over 60	55	15	5	75
Total	100	70	30	200

Find the exact probability that a participant selected at random supports an increase in the minimum wage, given that the participant is between 21 and 40 years old.

Solution: $P(For|21\ to\ 40) = \frac{25}{50} = 0.5$

Explanation:

A total of 50 people in the 21 to 40 age group were surveyed (isolate the first row). Of those, 25 were "for" increasing the minimum wage. Therefore, 50% ($25 \div 50$) of the respondents in this age group supported the increase.

PRACTICE PROBLEMS

1. In a survey of eighth and ninth grade students, participants were asked what grade they were in and whether they planned to watch the Super Bowl. Results are shown in the table below. Round your answers to the *nearest tenth of a percent.*

	Watching	Not Watching	Undecided	Total
8th Grade	25	20	8	53
9th Grade	31	22	7	60
Total	56	42	15	113

a) What percent of the students are undecided?

b) What percent of the ninth graders are watching?

2. You go to a dance and help clean up afterwards. To help, you collect the soda cans, Coca-Cola and Sprite, and organize them. Some cans were on the table and some were in the garbage. 72 total cans were found. 42 total cans were found in the garbage and 50 total cans were Coca-Cola. 14 Sprite cans were found on the table. From the given information, complete the two-way joint frequency table below.

	Coca-Cola	Sprite	Total
Table			
Garbage			
Total			

3. The table below contains data on a random sample of cats and dogs as they were brought to an animal shelter and checked for fleas.

	Fleas	No Fleas	Total
Cats	24	56	80
Dogs	48	112	160
Total	72	168	240

a) Calculate the probability that an animal has fleas, given that it is a cat.

b) Determine whether having fleas and being a cat are independent events.

4. A group of fifth graders were asked about their preferred pet. The survey responses, broken down by gender and preference, is represented by the table below.

	Dogs	Cats	Rabbits
Girls	53	72	25
Boys	62	28	40

a) What is the probability that a randomly selected student is a girl, given that the student prefers rabbits?

b) What is the probability that a randomly selected girl prefers rabbits?

c) What is the probability that a randomly selected student is a boy, given that the student prefers dogs or cats?

15.4 Series of Events [CC]

Key Terms and Concepts

We will now look at how to calculate the probability for a **series of events**. A series of events could be the repetition of an event, such as tossing a coin several times, or could be separate events performed in sequence, such as tossing a coin and then rolling a die.

Two events are **independent** if the outcome of the first does not affect the second. For example, if a die is rolled twice, the outcome of the first roll does not affect the outcome of the second roll in any way. When two events are independent, the probability that **both** occur is the **product** of the two probabilities: $\boldsymbol{P(A \text{ and } B) = P(A) \cdot P(B)}$.

Example: The probability of rolling a die twice and getting a 2 both times is $\frac{1}{6} \times \frac{1}{6} = \frac{1}{36}$.

Note that rolling a die two times is essentially the same as rolling a pair of dice together. Therefore, the probability of rolling a pair of dice and getting two 2s is also $\frac{1}{36}$.

We can extend this to any number of independent events.

Example: If we roll a die ten times, the probability of getting 2s on all ten rolls is
$\frac{1}{6} \cdot \frac{1}{6} \cdot \frac{1}{6} \cdot \frac{1}{6} \cdot \frac{1}{6} \cdot \frac{1}{6} \cdot \frac{1}{6} \cdot \frac{1}{6} \cdot \frac{1}{6} \cdot \frac{1}{6} = \left(\frac{1}{6}\right)^{10}$
This is the same probability as rolling ten dice all at once and getting all 2s.

A common example of a series of independent events is a random drawing **with replacement**.

Example: A bag contains 4 red and 6 blue cubes. A cube is randomly drawn and placed back into the bag (*replaced*). Then a cube is randomly drawn again.
The probability that both are red is $\frac{4}{10} \times \frac{4}{10} = \frac{16}{100} = \frac{4}{25}$.

We can use a tree diagram to represent the sample space for multiple events. Counting the leaves at the end of the branches will tell us the size of the sample space.

Example: This tree gives us all the outcomes from tossing a coin and then rolling a die.

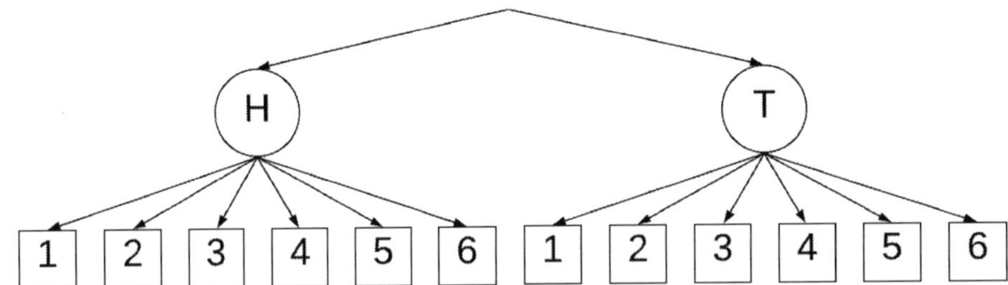

The sample space S = {H1, H2, H3, H4, H5, H6, T1, T2, T3, T4, T5, T6}.

To check whether two events are independent, we can test $\boldsymbol{P(A\ and\ B) = P(A) \cdot P(B)}$. If this equation does not hold true, then the events are dependent.

Two events are **dependent** if the outcome of the first affects the outcome of the second. For a series of two dependent events, take the probability of the first event times the probability of the second event, **given that the first event has already occurred**. This happens, for example, when items are randomly drawn **without replacement**.

Example: From the bag containing 4 red and 6 blue cubes, a cube is randomly drawn but *not replaced*. Then a second cube is randomly drawn.

The probability that the first cube is red is $\frac{4}{10}$. However, given that a red cube has already been drawn and not replaced and now only 3 of the remaining 9 cubes are red, the probability that the second cube is red is $\frac{3}{9}$.

So, the probability that both are red is $\frac{4}{10} \times \frac{3}{9} = \frac{12}{90} = \frac{2}{15}$.

For the two dependent events, we can say that $\boldsymbol{P(A\ and\ B) = P(A) \cdot P(B|A)}$, where $P(B|A)$ represents the **conditional probability** of B given that the event A has already occurred, usually read as "the probability of B given A."

Example: In the above example, the probability of picking two reds, $P(R1\ and\ R2)$, is equal to the probability of picking red on the first drawing, $P(R1)$, times the probability of picking a red on the second drawing given that a red was already picked on the first drawing, $P(R2|R1)$. That is,

$$P(R1\ and\ R2) = P(R1) \cdot P(R2|R1) = \frac{4}{10} \cdot \frac{3}{9} = \frac{2}{15}.$$

If we need to calculate $P(B|A)$, we can use the formula, $\boldsymbol{P(B|A) = \frac{P(A \text{ and } B)}{P(A)}}$, $P(A) \neq 0$.

Example: Suppose a board game player rolls a die and tells you that the roll was greater than 3 (G). You can calculate the probability that the roll was even (E) given that it was greater than 3 by:

$$P(E|G) = \frac{P(G \text{ and } E)}{P(G)} = \frac{\frac{2}{6}}{\frac{1}{2}} = \frac{4}{6} = \frac{2}{3}.$$

Note that $P(G \text{ and } E) = P(G) \cdot P(E|G) = \frac{1}{2} \times \frac{2}{3} = \frac{2}{6}$.

(2 of the 6 outcomes are greater than 3 and even: outcomes 4 and 6.)

If two events, A and B, are **independent**, then $\boldsymbol{P(B|A) = P(B)}$. This can be shown algebraically, since we know that for independent events, $P(A \text{ and } B) = P(A) \cdot P(B)$.

$$P(B|A) = \frac{P(A \text{ and } B)}{P(A)} = \frac{\cancel{P(A)} \cdot P(B)}{\cancel{P(A)}} = P(B)$$

Also, if two events, A and B, are *mutually exclusive*, then they must be dependent, since $P(B|A) = 0$. If A occurs, then B cannot, so the occurrence of A affects the probability of B.

Model Problem

In a standard deck of 52 cards, four of the cards are aces. If a player is dealt two cards in a game of Texas Hold'em Poker, what is the probability that they are a pair of aces?

Solution:

(A) (B) (C)

$$\frac{4}{52} \cdot \frac{3}{51} = \frac{12}{2652} = \frac{1}{221}$$

Explanation of Steps:

(A) Calculate the probability of the first event *[4 out of 52 cards are aces]*.

(B) When drawing without replacement *[as we do when dealing cards]*, we need to consider that the first event has taken place *[an ace was already selected and removed]* before determining the second probability *[3 out of the remaining 51 cards are aces]*.

(C) Multiply the probabilities of the two events. *[You are likely to be dealt a pair of aces one time out of every 221 hands.]*

PRACTICE PROBLEMS

1. The probability that the Cubs win their first game is $\frac{1}{3}$. The probability that the Cubs win their second game is $\frac{3}{7}$. What is the probability that the Cubs win both games?

2. A coin is flipped five times. What is the probability of getting five tails?

3. A board game uses the two spinners shown below. A player spins the arrow on each spinner once. The first spinner determines how many spaces to move. The second spinner determines whether the move from the first spinner will be forward or backward. Find the probability that a player will move *fewer than* four spaces and *backward.*

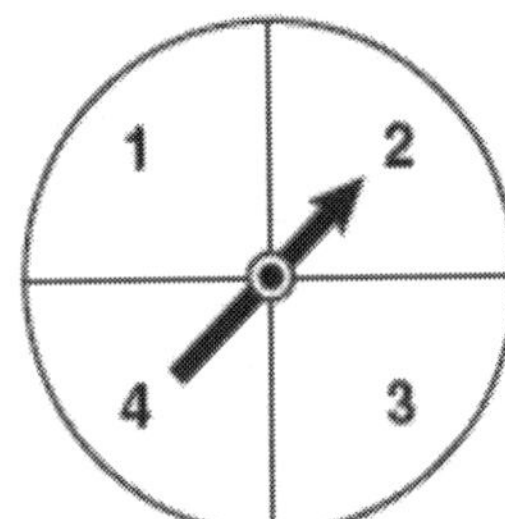

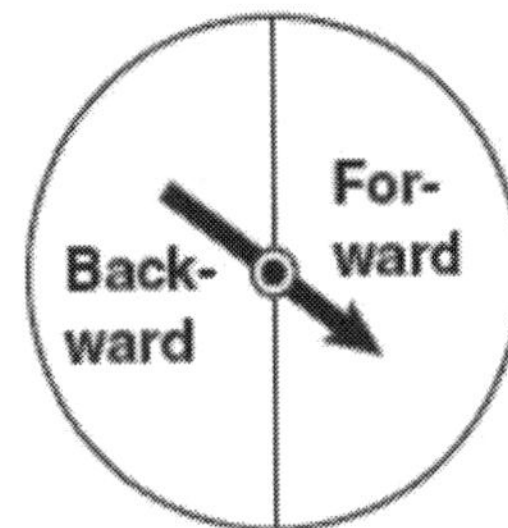

4. Barbie is going on a trip. She will be taking a taxi, a flight, and then a train. She chose the following three companies based on their claims.
   Bart's Taxi Service claims that it is on time 95 percent of the time.
   New Heights Airlines claims that it is on time 93 percent of the time.
   Trax Railways claims that it is on time 98 percent of the time.

   Based on the three companies' claims, what is the probability, rounded to the *nearest percent*, that all three parts of Barbie's trip will be on time, assuming that all three events are independent?

5. Assume that a child is equally likely to be a girl or a boy, and that each child is an independent event. For a family of 3 children, draw a tree diagram to show a sample space of all possible combinations.

   a) What is the probability that a family of 3 children will have 2 girls and 1 boy?

   b) What is the probability that a family of 3 children will have at least one girl?

6. Out of 18 people in a survey, 11 have dogs and 7 have cats. Four people have both cats and dogs. Draw a Venn Diagram to represent this situation.

   a) How many people have only cats?

   b) How many have neither?

7. A bag contains a red, a white, and a blue marble. Jack picks a marble at random five times with replacement. What is the probability of getting at least one blue marble?

8. Bob and Laquisha have volunteered to serve on the Junior Prom Committee. The names of twenty volunteers, including Bob and Laquisha, are put into a bowl. If two names are randomly drawn from the bowl without replacement, what is the probability that Bob's name will be drawn first and Laquisha's name will be drawn second?

9. A student council has seven officers, of which five are girls and two are boys. If two officers are chosen at random to attend a meeting with the principal, what is the probability that the first officer chosen is a girl and the second is a boy?	10. Liang places three playing cards face down on a table. The cards are two aces and a king. What is the probability, if two cards are selected at random without replacement, that they are both aces?
11. Ling places four playing cards face down on a table: three aces and a king. What is the probability, if two cards are selected at random without replacement, that they are both aces?	12. The probability that it will snow on Sunday is $\frac{3}{5}$. The probability that it will snow on both Sunday and Monday is $\frac{3}{10}$. What is the probability that it will snow on Monday, if it snowed on Sunday?

13. In a standard deck of 52 cards, 13 of the cards are hearts. If a player is dealt two cards, what is the probability that they are both hearts?

14. In a standard deck of 52 cards, there are 13 of each of the four suits (hearts, diamonds, clubs, and spades). If a player is dealt two cards, what is the probability that they are the same suit? *[Hint: use your answer to #12.]*

15. A bag contains 10 red and 15 blue marbles.

a) What is the probability of selecting 5 marbles *with replacement* and getting 5 reds?

b) What is the probability of selecting 5 marbles *without replacement* and getting 5 reds?

Chapter 16. Statistics

16.1 Observational Studies and Experiments

Key Terms and Concepts

One of the purposes of collecting statistical data is to find patterns or relationships among the data. There are several different ways to collect statistical data.

In the field of statistics, the **population** is the universal set. Depending on what is being studied, the population may be living beings, such as all residents of a country, or all students of a school, or all deer in a state. Or, the population may be inanimate objects, such as all cars in a city, or all products of a certain manufacturer.

It is often difficult to gather information about an entire population. Usually, the subjects of the collection method are selected as a random sample of a larger population with the intent of learning more about the general characteristics of the population. The **sample** is a subset of the population.

Example: A company that makes thousands of lightbulbs each day may pull aside a random sample of 100 bulbs for testing. If they find that 3% of the bulbs in the sample are faulty, they can infer that approximately 3% of all of the bulbs that they produce, known as the population, may be faulty.

A **statistic** is a characteristic of a sample, such as a person's height, weight, or test score in a sample set of people, or in the example above, whether a bulb is faulty. A **parameter** is a corresponding characteristic of the population.

Selecting the subjects for a study or survey could involve **systematic sampling**, in which typically every *n*th element in a list is selected.

Example: In the example above in which the company produces lightbulbs, it can select every 10th bulb it produces until 100 bulbs are selected and then conduct the study on this sample.

A **survey** is a way to collect information by asking participants to answer questions. A **census** is used to collect data from the entire population, whereas a **sample survey** is used to collect data from a sample, or subset, of the population.

Examples: Types of surveys include public opinion polls and market-research surveys.

In an **observational study**, data is collected by simply observing and/or testing the subjects. For example, instead of asking the subjects about their shopping habits at the mall, we can observe their actions at the mall without asking questions or interfering. This type of study measures variables without controlling the subjects or their environment.

Example: In the example above, the company that tests for faulty lightbulbs is conducting an observational study.

In an **experimental study**, the subjects are randomly divided into groups: one or more groups are selected to receive some type of treatment and one group will not receive any treatment. Data is collected to examine the possible effects of the treatment. This method could be used to find relationships involving cause and effect.

If all the factors in an experiment are kept constant except for the one variable (that is, the amount of treatment that is received), then the study is called a **controlled experiment**. In a controlled experiment, the groups that receive the treatment are called the **experimental groups** and the group that does not receive the treatment is called the **control group**. Often, there will be only one experimental group and one control group.

In a controlled experiment, the factor that is different between the experimental groups and the control group is the **independent variable**. The response that is measured to see if the treatment had an effect is the **dependent variable**.

Example: If an experiment is used to determine the effect of a drug on a person's blood pressure, the independent variable is the dosage of the drug (which is zero for the control group), and the dependent variable is the blood pressure.

MODEL PROBLEM

Suppose you want to compare the average class sizes at your school. What is the most appropriate method for gathering the needed data?

(1) census
(2) sample survey
(3) observational study
(4) experiment

Solution: (1)

Explanation of steps:

A census is appropriate when counting the entire population. *[Since the sizes of all the classes in a single school is data that can easily be gathered, it makes sense to collect the data from the entire population.]*

PRACTICE PROBLEMS

1. A doctor wants to test the effectiveness of a new drug on her patients. She separates her sample of patients into two groups and administers the drug to only one of these groups. She then compares the results. Which type of study best describes this situation?
(1) census
(2) sample survey
(3) observational study
(4) controlled experiment

2. A school cafeteria has five different lunch periods. The cafeteria staff wants to find out which items on the menu are most popular, so they give every student in the first lunch period a list of questions to answer in order to collect data to represent the school. Which type of study does this represent?
(1) census
(2) sample survey
(3) observational study
(4) controlled experiment

3. Which task is *not* a component of an observational study?
(1) The researcher decides who will make up the sample.
(2) The researcher divides the sample into two groups, with one group acting as a control group.
(3) The researcher gathers data from the sample.
(4) The researcher analyzes the data received from the sample.

4. Scientists want to run a controlled experiment to determine the effect of elevated atmospheric levels of carbon dioxide (CO_2) on the photosynthesis rates of certain plants. The normal level of CO_2 in the air (pre-1900) is about 300 ppm. The scientists want to test the effects of elevated amounts of CO_2, both at levels of 400 ppm (the present atmospheric level) and 500 ppm.

   a) How should the plants be divided into groups?

   b) What are the independent and dependent variables?

16.2 Statistical Bias

Key Terms and Concepts

In a good survey, the **sample** (a subgroup of the population that is surveyed) should accurately represent the **population**. When it does not, the cause may be statistical bias.

Bias is a type of systemic error that can cause statistical data to be incorrect. These errors can occur in the planning or data collection stages of the process. Examples of bias include the faulty design of a questionnaire, such as misleading or ambiguous questions, or a non-random selection of a sample set from the population. Bias results in a misrepresentation of the population.

To avoid bias, it is important to survey as large a sample as possible, preferably selected at random, and to not limit the sample to a specific subgroup of the population that may have a reason to share a similar opinion or response.

Example: To determine the most popular current movie among neighborhood residents, a survey of 20 teenagers shopping at a comic book store would be more biased than a survey of 100 randomly selected shoppers at the mall.

It is also important to avoid the self-selection of the sample. For example, by inviting people to respond to a survey about sports could lead to bias, especially because sports fans are more likely to choose to respond.

Model Problem

To survey students at a particular college about their health and fitness routines, which survey method would likely produce more bias?

a) Survey all 12 members of the men's basketball team

b) Survey every fifth student to enter the campus until 100 students have been measured?

Solution:

Choice (a).

Explanation of steps:

If a sample does not accurately represent the overall population, it is more likely biased. *[The fitness routines of basketball team members will probably not be representative of the whole school, since fitness is essential to success in the sport. Also, sampling 12 men is not as large and representative a sample as 100 randomly selected students, including men and women.]*

Practice Problems

1. The principal of a high school would like to determine why there has been a large decline during the year in the number of students who buy food in the school's cafeteria. To do this, 25 students from the school will be surveyed. Which method would be the most appropriate for selecting the 25 students to participate in the survey?

   (1) Randomly select 25 students from the senior class.

   (2) Randomly select 25 students from those taking physics.

   (3) Randomly select 25 students from a list of all students at the school.

   (4) Randomly select 25 students from a list of students who eat in the cafeteria.

2. Which method of collecting data would most likely result in an unbiased random sample?
   (1) placing a survey in a local newspaper to determine how people voted in the last presidential election
   (2) surveying honor students taking calculus to determine the average amount of time students in a school spend doing homework each night
   (3) selecting every third teenager leaving a movie theater to answer a survey about entertainment
   (4) selecting students by the last digit of their school ID number to participate in a survey about cafeteria food

3. A survey is conducted about the proposed increase in the sports budget in a town's school district. Which survey method would likely contain the most bias?
   (1) Ask every third person entering the town's grocery store.
   (2) Ask every third person leaving the town's shopping mall this weekend.
   (3) Ask every fifth student entering town's high school on Monday morning.
   (4) Ask every fifth person leaving Saturday's town's high school football game.

4. Four hundred licensed drivers participated in the math club's survey on driving habits. The table below shows the number of drivers surveyed in each age group.

Ages of People in Survey on Driving Habits

Age Group	Number of Drivers
16–25	150
26–35	129
36–45	33
46–55	57
56–65	31

Which statement best describes a conclusion based on the data in the table?

(1) It may be biased because no one younger than 16 was surveyed.

(2) It may be biased because the majority of drivers surveyed were in the younger age intervals.

(3) It would be fair because many different age groups were surveyed.

(4) It would be fair because the survey was conducted by the math club students.

5. A large company wants to find out which medical plan its employees would prefer. Which procedure would be most likely to obtain a statistically unbiased sample?

(1) survey a random sample of employees from a list of all employees

(2) invite all employees to indicate their choices by e-mail

(3) place suggestion boxes throughout the company's workplace

(4) select one member from each department and record their preferences

16.3 Normal Distribution

Key Terms and Concepts

Statistical data is said to have a **normal distribution** when its graph approximates a symmetrical bell-shaped curve.

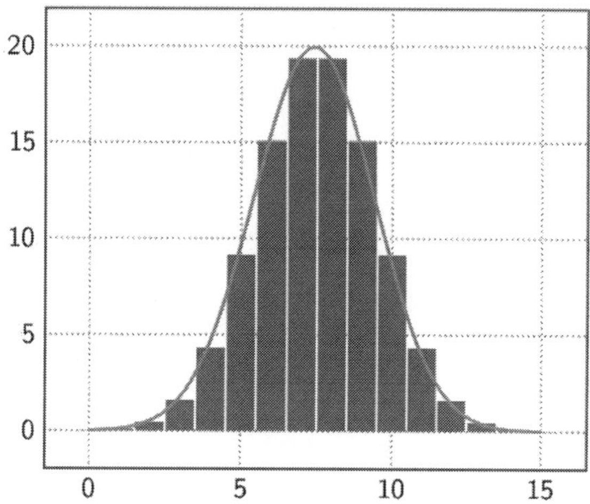

Standard deviation is a measure of how spread out the data values are. A low standard deviation indicates that the values tend to be close to the mean, while a high standard deviation indicates that the values are spread out over a wider range.

The letter ***s*** is used for the sample standard deviation and the Greek letter **σ** (sigma) is used for the population standard deviation.

In a normal distribution, the standard deviation is useful in describing general characteristics of the data, including the percentage of data that fall within specific ranges.

A data value's **z-score** (also known as its **standard score**) represents the number of standard deviations the value is above or below the mean.

Example: If a set of normally distributed data has a mean of 10 and a standard deviation of 2, a data value of 8 can be described as lying one standard deviation below the mean, with a standard score of -1, and a value of 16 as three standard deviations above the mean, with a standard score of $+3$.

We can calculate standard score as $z = \dfrac{value - mean}{SD} = \dfrac{x - \mu}{\sigma}$.

The **Empirical Rule** states that, for a normal distribution with enough data points,

- about 68% of the values are within 1 standard deviation of the mean
- about 95% of the values are within 2 standard deviations of the mean

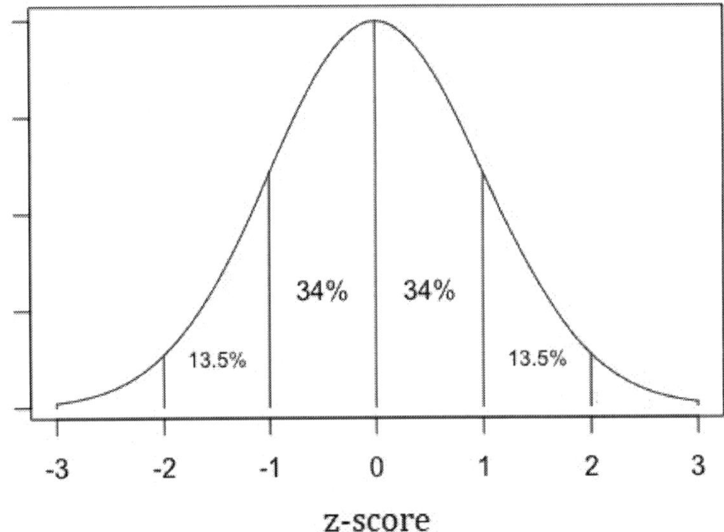

The percentage of data between two values is called a **population percentage**.

Example: If a set of normally distributed data has a mean of 10 and a standard deviation of 2, we can infer that about 68% of the data are between 8 and 12 and about 95% of the data are between 6 and 14.

The Empirical Rule can be used to find the percentage of numbers within one or two standard deviations of the mean, but sometimes we will want to know the population percentage between *any* two values. For this, we can use the calculator's **normal cumulative density function** (normalcdf) function.

CALCULATOR TIP

To calculate the percentage of numbers between two values in a normal distribution, we'll need the mean and standard deviation as well as the lower and upper boundary values.

To find the percentage of numbers between two values in a normal distribution on your calculator:

1) Press [2nd][DISTR][2] to select normalcdf.
2) Type the lower boundary, upper boundary, mean, and standard deviation. *[On the TI-83, enter these four values in parentheses, separated by commas.]*
3) Press [ENTER]. The answer is given as a decimal.

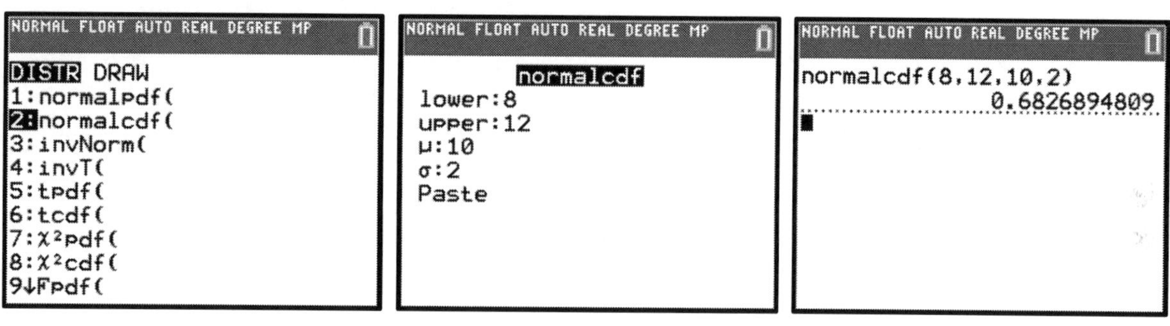

Note that this result also represents the *probability* that a randomly selected data value lies between the two given values.

MODEL PROBLEM

In a normally distributed set of test scores with a mean of 75 and a standard deviation of 8, find the probability, to the *nearest percent*, that a student earned between a 65 and 75.

Solution:

39%

Explanation of steps:

Use the calculator to enter the lower and upper values, the mean, and the standard deviation into the normalcdf function *[*normalcdf(65,75,75,8) $\approx$ 0.39435*]*.

Practice Problems

1. On a standardized test, a student had a score of 74, which was exactly 1 standard deviation below the mean. If the standard deviation for the test is 6, what is the mean score for the test?

2. On a standardized test, a student had a score of 85, which was exactly 2 standard deviations above the mean. If the standard deviation for the test is 4, what is the mean score for this test?

3. In the diagram below, about 68% of the scores fall within the shaded area, which is symmetric about the mean. The distribution is normal and the scores in the shaded area range from 50 to 80. What is the standard deviation of the scores in this distribution?

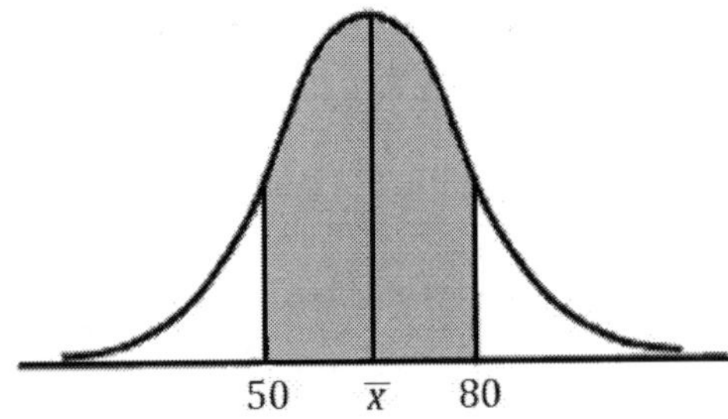

4. In the diagram below, the shaded area represents approximately 95% of the scores on a standardized test. The distribution is normal and the scores in the shaded area range from 78 to 92. What is the standard deviation of the scores in this distribution?

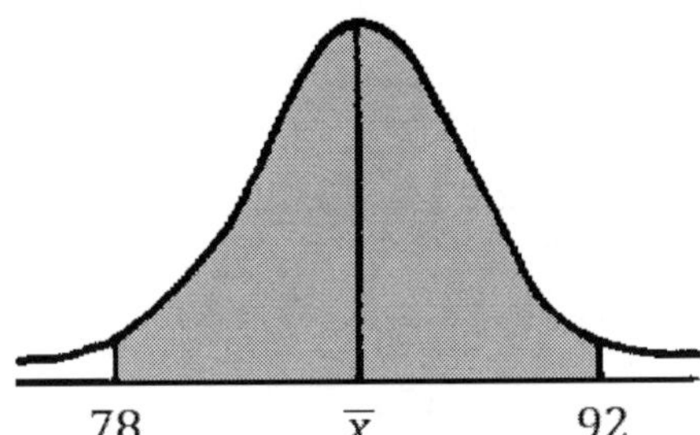

5. The heights of a group of girls are normally distributed with a mean of 66 inches. If 95% of the heights of these girls are between 63 and 69 inches, what is the standard deviation for this group?

6. On a test that has a normal distribution of scores, a score of 57 falls one standard deviation below the mean, and a score of 81 is two standard deviations above the mean. Determine the mean score of this test.

7. The mean of a normally distributed set of data is 56, and the standard deviation is 5. In which interval do approximately 95% of all the data lie?

8. The sale prices of 75 homes in a certain neighborhood approximated a normal distribution with a mean of $150,000 and a standard deviation of $25,000. Approximately how many of the homes were priced between $125,000 and $175,000?

9. If the mean on a standardized test with a normal distribution is 54.3 and the standard deviation is 4.6, what percent of the scores, to the *nearest percent,* fall between 54.3 and 63.5?	10. The scores on a 100-point test approximate a normal distribution with a mean score of 72 and a standard deviation of 9. To the *nearest percent,* what percent of the students taking the test received a score greater than 80?
11. A set of normally distributed student test scores has a mean of 80 and a standard deviation of 4. Determine the probability, to the *nearest hundredth,* that a randomly selected score will be between 74 and 82.	12. A shoe manufacturer collected data regarding men's shoe sizes and found that the distribution of sizes fits the normal curve. If the mean shoe size is 11 and the standard deviation is 1.5, find the probability, to the *nearest hundredth,* that a man's shoe size is greater than or equal to 12.5.

16.4 Confidence Interval

KEY TERMS AND CONCEPTS

Often in statistics, we want to determine whether an outcome is plausible. For example, suppose we toss a coin 10 times and the coin lands on heads 3 out of 10 times. This may not seem like a fair coin. To decide, we could run computer simulations. A **simulation** is a way to model random events, usually by a computer, such that the simulated outcomes closely match real-world outcomes.

We can have a computer simulate the tossing of a fair coin 10 times as one sample, and record the number (or proportion) of heads. Then, we can repeat this simulation 50 times (50 samples) and report the results of these simulations in a dot plot, as shown below, with each dot representing the result of one sample.

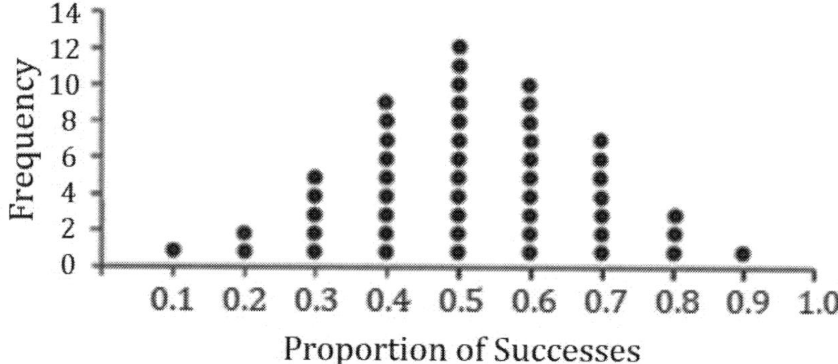

In this dot plot, the horizontal axis is labeled with the **proportion** of heads in each sample. For example, the column labeled 0.3 represents how many times 0.3 or 30% (3 out of 10) of the flips were heads. Note that in this case, there are 5 dots in the column for 0.3. This means, out of our 50 samples, there were 5 samples in which the simulator generated 3 heads out of 10 flips. The dots line up so neatly into columns in this dot plot because the possible outcomes for each sample are discrete; the integers from 0 heads to 10 heads correspond to the proportions from 0.0 to 1.0.

For an outcome to be plausible, it should fall within the **margin of error** from the mean. The margin of error is considered to be two standard deviations, $\boldsymbol{ME = 2\sigma}$, since the data values within this margin of error from the mean represents 95% of the outcomes.

An outcome that falls within the margin of error from the mean is considered **plausible**. An outcome is *not* plausible (also known as **statistically significant**) if it falls outside of this interval. A statistically significant outcome is one that is expected to occur less than 5% of the time.

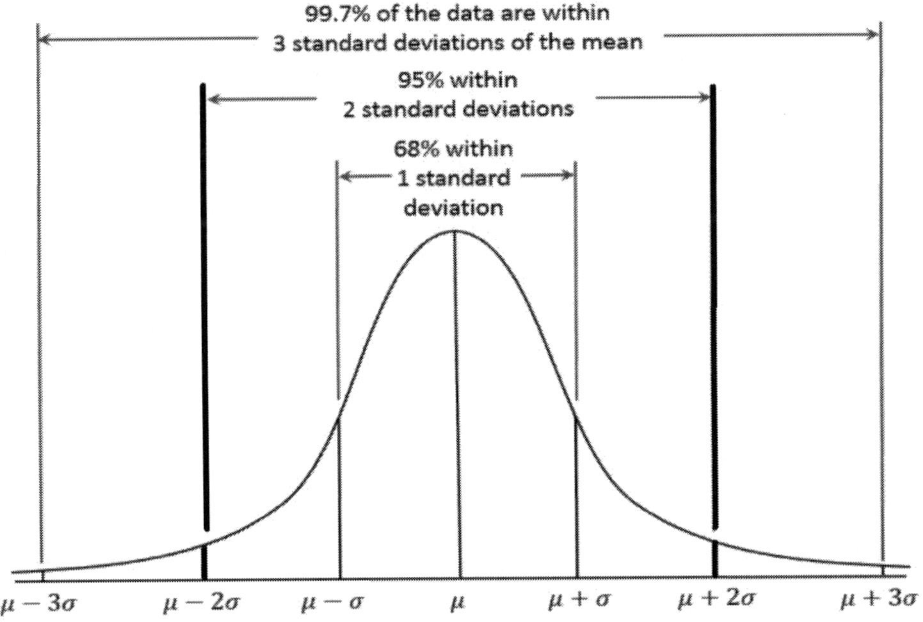

If we calculate the mean and standard deviation of the results from the simulations, we can use these values to determine the confidence interval. The **confidence interval** is the range of plausible values, centered at the mean and within the margin of error. About 95% of the data should fall within the confidence interval. Since the margin of error is two standard deviations (2σ), the confidence interval is $[\boldsymbol{\mu - 2\sigma, \mu + 2\sigma}]$, where μ (the Greek letter, mu) is the population mean and σ (sigma) is the population standard deviation. We can also write this as $\boldsymbol{\mu \pm 2\sigma}$.

Example: In the simulations of 50 samples of 10 coin flips described above, the mean is 5.16 and the standard deviation is approximately 1.71. Therefore, the confidence interval $\mu \pm 2\sigma$ is from $5.16 - 2(1.71)$ to $5.16 + 2(1.71)$, which is the interval from 1.74 to 8.58. According to these results, any outcome between 2 to 8 heads would be plausible, and any outcome outside of this range ($h \leq 1$ or $h \geq 9$) would be statistically significant (*not* plausible). In our example, a result of 3 heads is plausible. Therefore, we shouldn't conclude that the coin is unfair based on these results.

MODEL PROBLEM

Samir rolls a pair of dice 10 times and notices that, in 8 of the 10 rolls, the sum was greater than 7. He's not sure if the dice are fair, so he runs a computer simulation to generate 1,000 samples of rolling dice 10 times, and records the number of rolls that are greater than 7 for each sample. The histogram below shows the results of the simulations.

a) Samir calculates the mean and standard deviation for the simulated samples and finds that the mean is 4.135 and the standard deviation is approximately 1.516. State the 95% confidence interval to the *nearest hundredth*, and explain what it means in this context.
b) Based on these results, state whether an outcome of rolling higher than 7 in 8 out of 10 rolls is plausible.

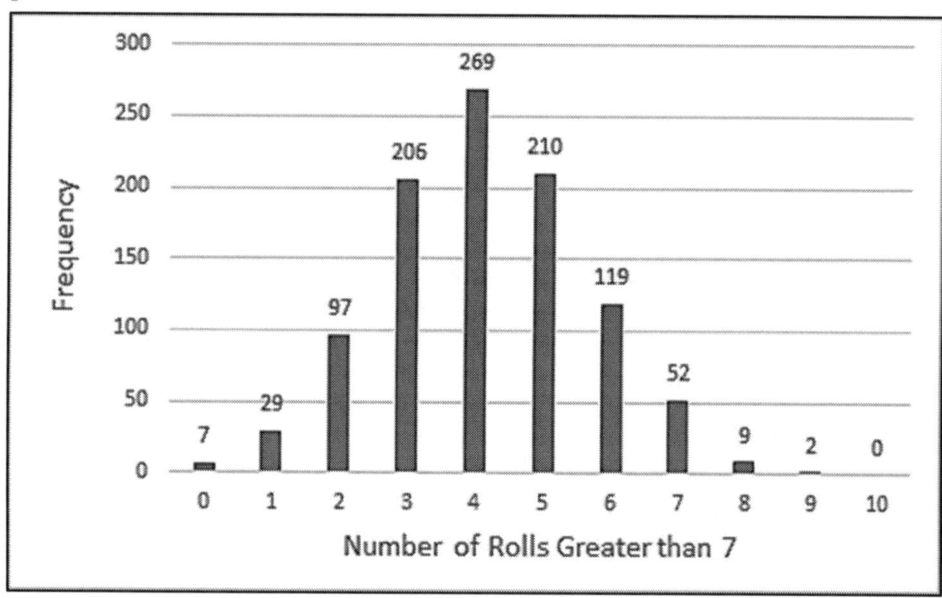

Solution:

(A) The 95% confidence interval is between 1.10 and 7.17, which means that 95% of the simulations resulted in a number of rolls greater than 7 that was within this interval. Any number below 1.10 or above 7.17 would be statistically significant.

(B) Getting 8 out of 10 is *not* plausible.

Explanation of steps:

(A) The interval is calculated as between 2 standard deviations below the mean *[*$4.135 - 2(1.516) \approx 1.10$*]* and 2 standard deviations above the mean *[*$4.135 + 2(1.516) \approx 7.17$*]*.

(B) The result of 8 rolls higher than 7 is outside of the confidence interval *[1.10 to 7.17]* and is therefore *not* plausible.

PRACTICE PROBLEMS

1. For a set of data, the mean is 5.6 and the standard deviation is 0.2. Determine the 95% confidence interval and state whether a score of 5.9 is within the margin of error.

2. A variable is normally distributed with a mean of 16 and a standard deviation of 6. Find the percent of the data set

   a) that is greater than 16

   b) that falls between 10 and 22

   c) that is greater than 28

3. 24 juniors recently took the SAT exam. On the math portion of the exam, the scores were normally distributed with a mean score of 540 and a standard deviation of 40.

   a) What percent of the math scores fell between 500 and 620?

   b) What percent of the math scores were higher than 660?

   c) Approximately how many scores were higher than 660?

   d) To the nearest whole number, how many scores fell between 500 and 660?

4. A chemist wants to determine the reaction time for a newly developed antacid. The pharmaceutical company claims that the reaction time is 10 minutes. The chemist runs a computer simulation to generate the mean reaction times for 1,000 samples. The graph below shows a sampling distribution of the means. The overall mean of all the samples is 11.3095 and the standard deviation of 0.7625.

Based on the simulation results, is the claim of a 10-minute reaction time for the antacid plausible? Explain your response including an interval containing the middle 95% of the data, rounded to the *nearest hundredth*.

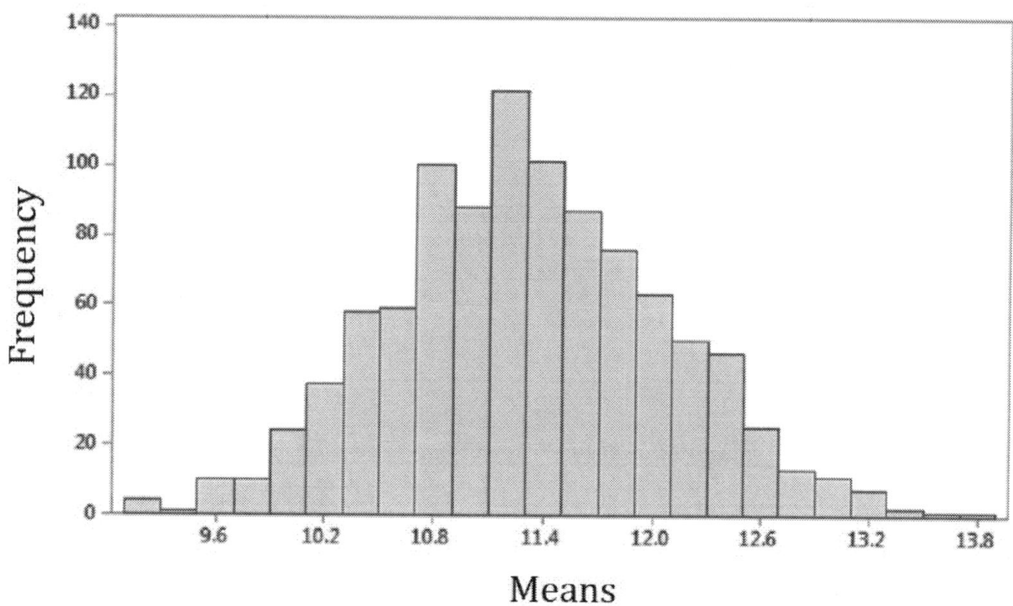

16.5 Estimate Population Parameters

Key Terms and Concepts

Recall that we used the value 2 in the margin of error formula, $ME = 2\sigma$. This is called the **critical value**. However, when estimating population parameters, statisticians will use slightly different critical values, depending on the sample size. For a 95% confidence interval, the critical value may be greater than 2 for very small sample sizes, but can be as low as 1.96 for large sample sizes. To keep our formulas simple in this section, we will use a constant critical value of 1.96. Bear in mind, however, that this may cause our results to vary slightly from the calculator's results.

 CALCULATOR TIP

To find the critical value that the calculator uses for a specific sample size:

1. Press [STAT] [TESTS] [8] for TInterval.
2. For Inpt, select Stats. Enter zero for the sample mean, $\bar{x}$, enter the square root of the sample size for the standard deviation, *Sx*, and enter the sample size as *n*. Use a C-Level of 0.95 and press [ENTER] to calculate.
3. The upper bound of the interval that shows on the next screen is the critical value used for this sample size of data.

Example: For a sample size of 400, the screenshots below show that a critical value of 1.9659 is used, which is pretty close to 1.96, the value we will use in our formulas.

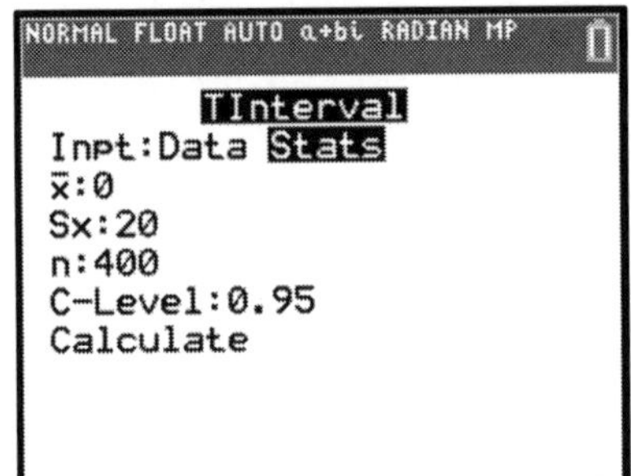

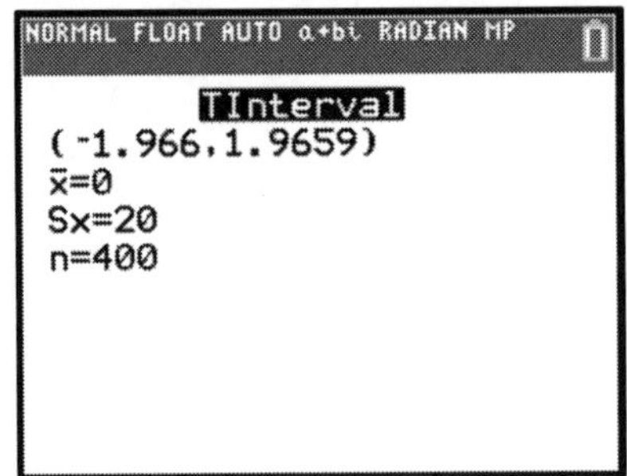

Sometimes, we would like to draw conclusions about parameters of the population based on sample statistics. This is called **inferential statistics**. For example, if the population represents test scores of thousands of test takers, we could select a random sample of test takers and use the **sample mean** of their scores to estimate the **population mean**.

To estimate the population mean from a sample:

1. Calculate the mean and standard deviation of the sample, which are represented by $\bar{x}$ and s, respectively.
2. Calculate the margin of error using the formula $ME = 1.96 \cdot \left(\frac{s}{\sqrt{n}}\right)$, where s is the sample standard deviation and n is the sample size.
3. The population mean, represented by μ, can be estimated to fall within the interval: $\bar{x} - ME \leq \mu \leq \bar{x} + ME$, where $\bar{x}$ is the sample mean and μ is the population mean.

Example: If a sample of 400 items from a normally distributed population has a sample mean $\bar{x} = 22.1$ and a standard deviation $s = 12.8$, then the margin of error $ME = 1.96 \cdot \left(\frac{s}{\sqrt{n}}\right) = 1.96 \cdot \left(\frac{12.8}{\sqrt{400}}\right) \approx 1.3$. Therefore, the population mean should fall within the interval $20.8 \leq \mu \leq 23.4$.

 CALCULATOR TIP

To have the calculator find the confidence interval when estimating a population mean from a sample mean:

1. Press [STAT] [TESTS] [8] for TInterval.
2. For Inpt, select Stats. Enter the sample mean as $\bar{x}$, the standard deviation as Sx, and the sample size as n. Use a C-Level of 0.95 (for 95%) and press [ENTER] to calculate.
3. The next screen shows the confidence interval.

Example: The screenshots below use the data from the previous example, above.

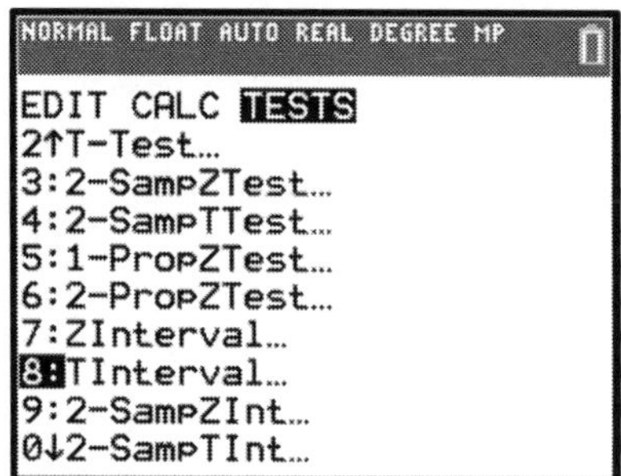

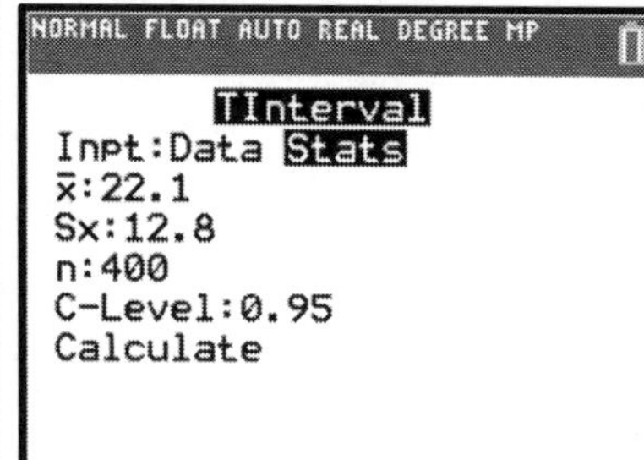

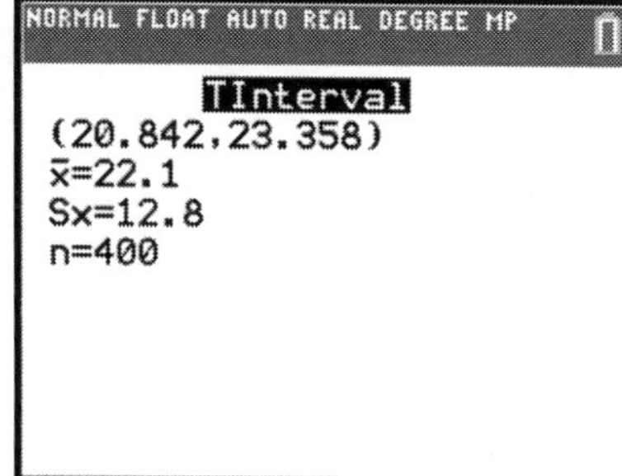

To find a more accurate approximation of the population mean, we can take multiple samples or run computer simulations to generate multiple samples. We can then plot the **sampling distribution** of all of the sample means on a dot plot or histogram and determine the overall mean and standard deviation of all the samples. The population mean would very likely fall within the interval between 2 standard deviations below the mean and 2 standard deviations above the mean (using the overall mean of all the samples and the standard deviation of these means).

We can also use a sample proportion to estimate the population proportion. For example, suppose we want to find out what proportion of a city's voters plans to choose a certain candidate in an upcoming election. We can randomly select a sample, such as by polling 1,000 registered voters, and record their responses. Let's say 460 responders, or 46% of the sample, say that they will vote for the candidate. We would say that the **sample proportion**, represented by the symbol $\hat{p}$, is 0.46. The sample proportion is called a **point estimate** because it is used to estimate a population parameter, in this case the population proportion.

From this information, we can estimate a *range of values* that would almost certainly include the **population proportion**, the proportion of city voters who will vote for the candidate. The population proportion is represented by the letter, p.

We can calculate the *margin of error* by using a formula that estimates the standard deviation:

$ME = 1.96 \cdot \sqrt{\frac{(\hat{p})(1-\hat{p})}{n}}$, where $\hat{p}$ is the sample proportion and n is the sample size.

In our example, $ME = 1.96 \cdot \sqrt{\frac{(\hat{p})(1-\hat{p})}{n}} = 1.96 \cdot \sqrt{\frac{(0.46)(1-0.46)}{1000}} \approx 0.03.$

We can approximate that the population proportion will almost certainly fall within the margin of error from the sample proportion: $\hat{p} - ME \leq p \leq \hat{p} + ME$.

In our example, this would be the interval between $0.46 - 0.03$ and $0.46 + 0.03$, which is from 0.43 to 0.49. From this, we can confidently predict that between 43% to 49% of the voters will choose the candidate.

 CALCULATOR TIP

To have the calculator find the confidence interval when estimating a population proportion from a sample proportion:

4. Press [STAT] TESTS [ALPHA][A] for 1-PropZInt.
5. Enter the number of successes as the x value and the sample size as the value of n. Use a C-Level of 0.95 (for 95%) and press [ENTER] to calculate.
6. The next screen shows the confidence interval, as well as the sample proportion $\left(\frac{x}{n}\right)$ as $\hat{p}$, and the sample size, n, that was entered.

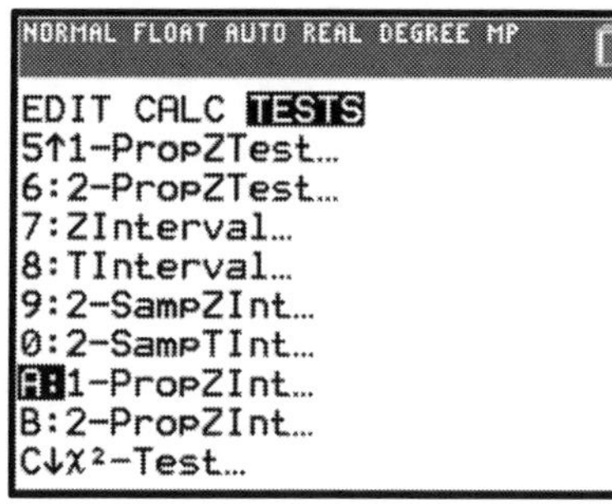

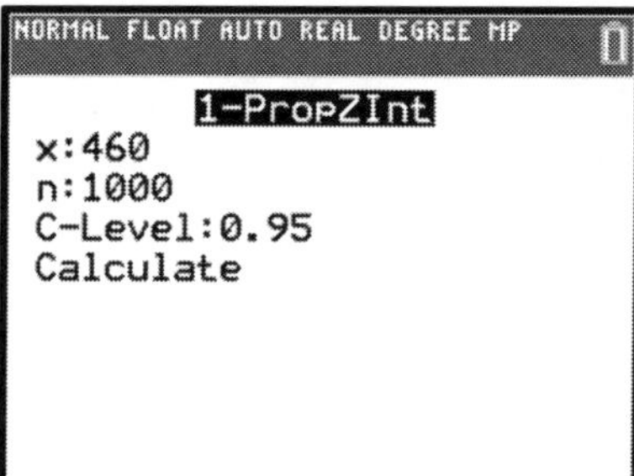

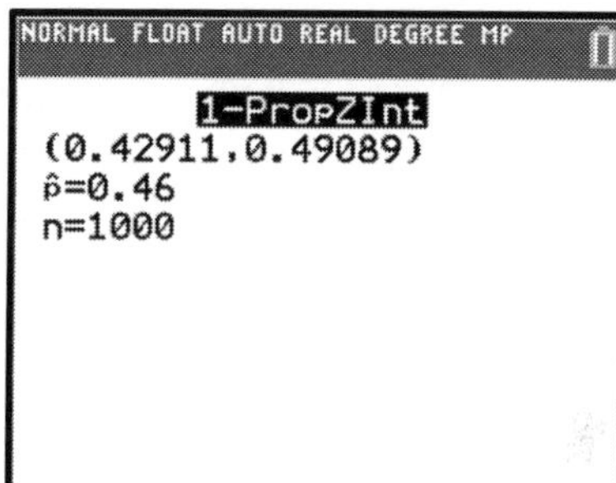

Once again, the calculator's result may differ slightly from our manual formula calculation due to the fact that it may use a critical value that is slightly larger than 1.96.

Model Problem

To estimate the mean heart rate in young adults, a random sample of 100 young adults are selected and their heart rates are recorded. For the sample, the mean heart rate $\bar{x} = 74.16$ beats per minute and the standard deviation $s = 5.375$. Calculate the confidence interval for the population mean, μ.

Solution:

(A) $ME = 1.96 \cdot \left(\frac{s}{\sqrt{n}}\right) = 1.96 \cdot \left(\frac{5.375}{\sqrt{100}}\right) = 1.0535$

(B) $74.16 - 1.0535 < \mu < 74.16 + 1.0535$

(C) $73.11 < \mu < 75.21$

Explanation of steps:

(A) Calculate the margin of error, $ME = 1.96 \cdot \left(\frac{s}{\sqrt{n}}\right)$, where s is the sample standard deviation and n is the sample size.

(B) The lower bound of the interval is the sample mean, $\bar{x}$, minus the margin of error and the upper bound is the sample mean plus the margin of error.

(C) Write the confidence interval as a compound inequality or in interval notation.

Alternatively, the calculator could have been used to find the interval. Again, bear in mind that the results on the calculator (below) differ slightly from our manual formula's estimation, but either may be used as your answer.

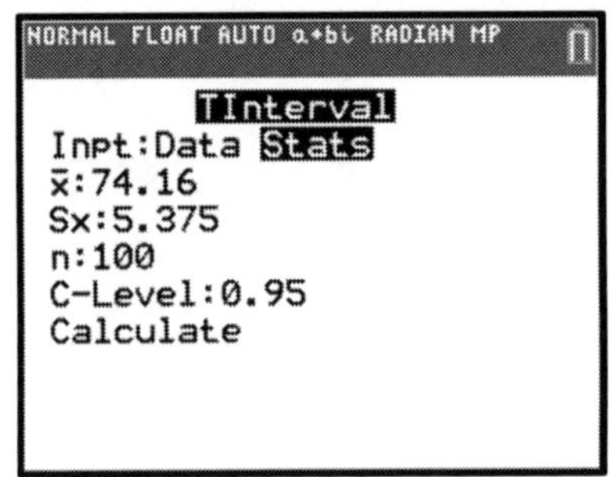

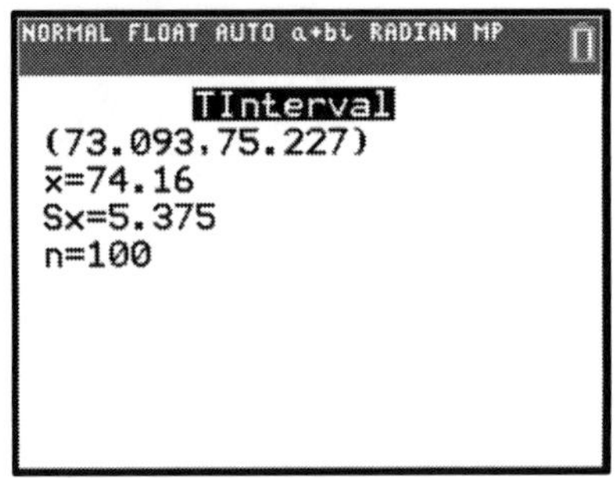

PRACTICE PROBLEMS

1. Use the calculator's TInterval function to determine the critical value that the calculator uses for a sample size of

   a) 9

   b) 900

   b) 90,000

2. Given $s = 12$, and $n = 200$ for a set of sample data, use a critical value of 1.96 to calculate the margin of error, to the *nearest hundredth*, for estimating the population mean.

3. To estimate the average age of its patrons, a theater selects a random sample of 100 patrons and finds that the average age for the sample is 44.25 with a standard deviation of 10.2. How can this information be used to estimate the average age of all its patrons?

4. To estimate the proportion of its customers who prefer vanilla over chocolate, an ice cream company surveyed a sample of 100 customers and found that 40% prefer vanilla. Based on this data, estimate the population proportion, including the margin of error.

5. 180 out of 300 randomly selected students in a school say that they own a pet. If there are 1,800 students in the school, how many students would we expect to own a pet, using a 95% confidence interval?

16.6 Mean Difference [CC]

Key Terms and Concepts

In an experimental study, we often want to determine the difference between the means of the data from two groups to determine whether the treatment received by one group may have had a significant influence. The group that receives the treatment is called the **treatment group** and the group that does not is called the **control group**. The process of randomly assigning the subjects to the two groups is called **randomization**.

For example, suppose we randomly divide 100 math students into two groups of 50. Both groups are given the same exam in separate rooms, but group A (the treatment group) is allowed to look at their math notebooks during the exam and group B (the control group) is not. After grading their exams, we find that the mean score for group A is 75 and the mean score of group B is 69. The difference of their means is $75 - 69 = 6$.

At first glance, it seems that group A has benefited by the use of their notebooks, but how certain are we that this alone was the reason?

We can test our hypothesis by rerandomizing the grouping. To do so, we'll shuffle the test results and divide them into two equal groups, and find the mean difference between the two groups. Then, we'll repeat this process many times.

The graph below shows the mean differences of 1,000 rerandomizations. The average mean difference is -0.185. We can see that at least 100 of the 1,000 groupings, or at least 10%, resulted in a mean difference of 6 or more, so the difference is not unusual. Even though we observed an effect between the two groups, it may have been due to nothing more than the random grouping of the students.

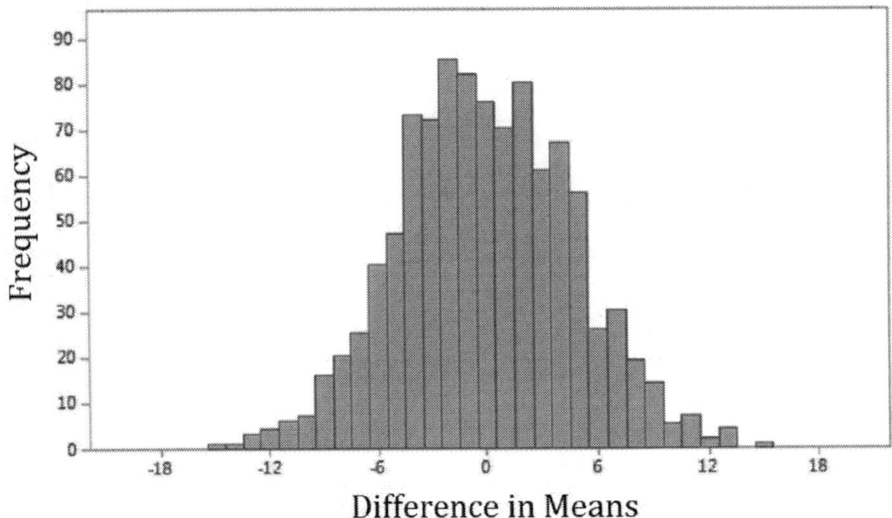

From the data collected above, we can calculate a confidence interval for the mean differences by using a margin of error that is 2 standard deviations from the overall mean.

If we don't use the rerandomizing method, we can still calculate a confidence interval using a formula for the standard error of the difference between means,

$$CI = (\overline{x_1} - \overline{x_2}) \pm 1.96 \cdot \sqrt{\frac{(s_1)^2}{n_1} + \frac{(s_2)^2}{n_2}}.$$

Once again, we'll use a constant critical value of 1.96 for our purposes.

Example: Suppose we collected data from 200 people divided into two groups.

Group	$\bar{x}$	s	n
1	97.25	3.65	80
2	87.25	9.60	120

The difference in means for our sample is $\bar{x}_1 - \bar{x}_2 = 97.25 - 87.25 = 10.0$. Our confidence interval would be

$$CI = (97.25 - 87.25) \pm 1.96 \cdot \sqrt{\frac{(3.65)^2}{80} + \frac{(9.60)^2}{120}} \approx 10 \pm 1.9$$

So, the confidence interval for estimating the population mean difference would be from 8.1 to 11.9.

CALCULATOR TIP

To find the confidence interval for estimating the population mean difference on your calculator:

7. Press STAT TESTS 0 for 2-SampTInt.
8. For Inpt, select Stats. For group 1, enter the sample mean, $\bar{x}$, the standard deviation, *Sx*, and the sample size, *n*. Do the same for group 2. Use a C-Level of 0.95, "No" for Pooled, and press ENTER to calculate.
9. The next screen shows the confidence interval on the first line.

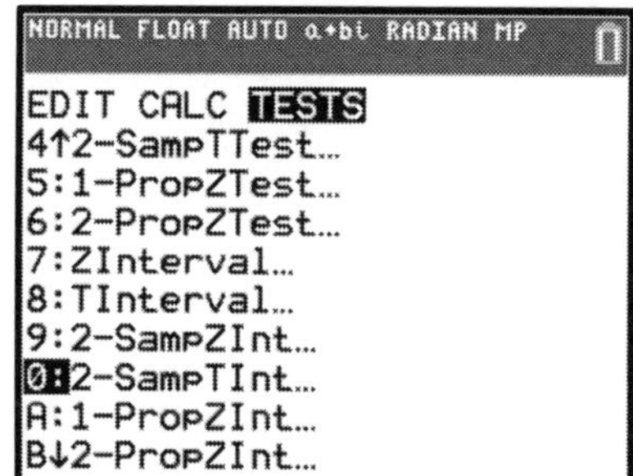

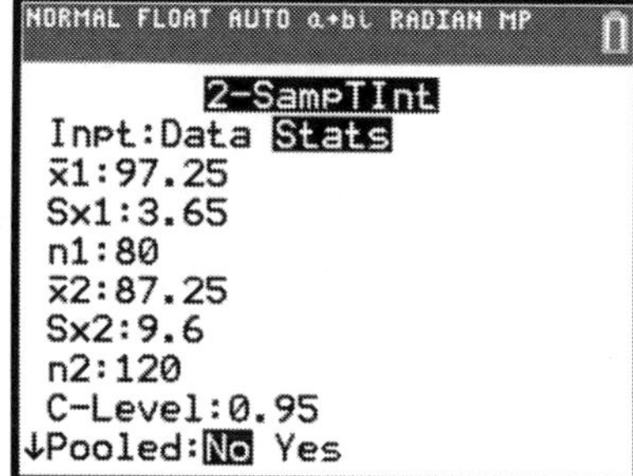

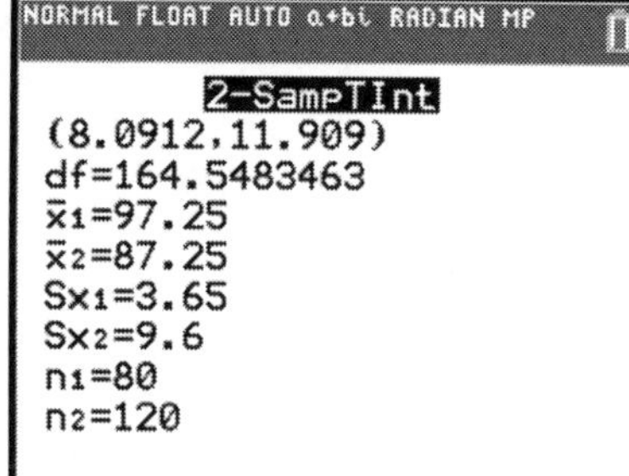

Just remember that the calculator's results may vary due to the use of a different critical value. This difference will be more significant when smaller sample sizes are used.

MODEL PROBLEM

To calculate the difference in mean systolic blood pressures between men and women, the following sample data is recorded. Determine the 95% confidence interval for estimating the population mean difference.

Group	$\bar{x}$	s	n
Men	128.2	17.5	1623
Women	126.5	20.1	1911

Solution:

$$CI = (128.2 - 126.5) \pm 1.96 \cdot \sqrt{\frac{(17.5)^2}{1623} + \frac{(20.1)^2}{1911}} \approx 1.7 \pm 1.24$$

$$CI \approx (0.46, 2.94)$$

Explanation of steps:

Enter the appropriate values into the confidence interval formula and evaluate.

[The formula is $CI = (\overline{x_1} - \overline{x_2}) \pm 1.96 \cdot \sqrt{\frac{(s_1)^2}{n_1} + \frac{(s_2)^2}{n_2}}$ *]*

Alternately, use the calculator's 2-SampTInt formula.

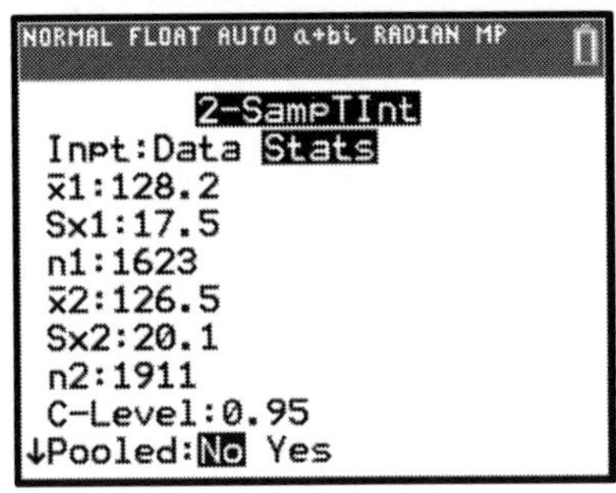

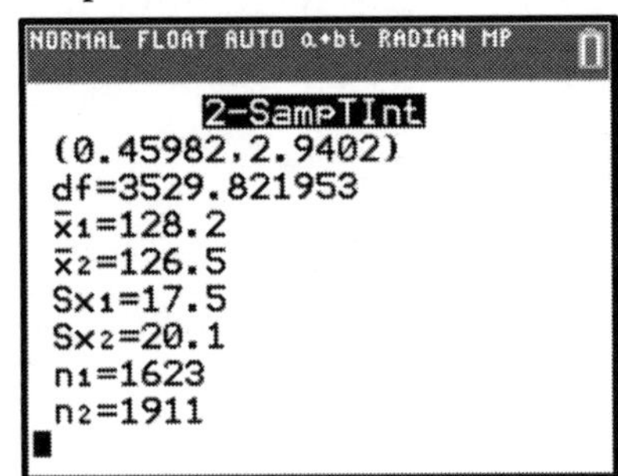

PRACTICE PROBLEMS

1. In an experimental study, the 50 subjects in group A were given an experimental hair replacement drug while the 50 subjects in group B were given a placebo. After one week, group A showed a mean of 3.2 inches in hair growth while group B showed a mean of 2.5 inches.

   The study then rerandomized the grouping 100 times, with the difference in the means calculated each time and the results were graphed in the histogram below.

   a) Calculate the difference of the means between groups A and B.

   b) Based on the histogram, is there strong evidence of the positive effects of the drug on hair growth?

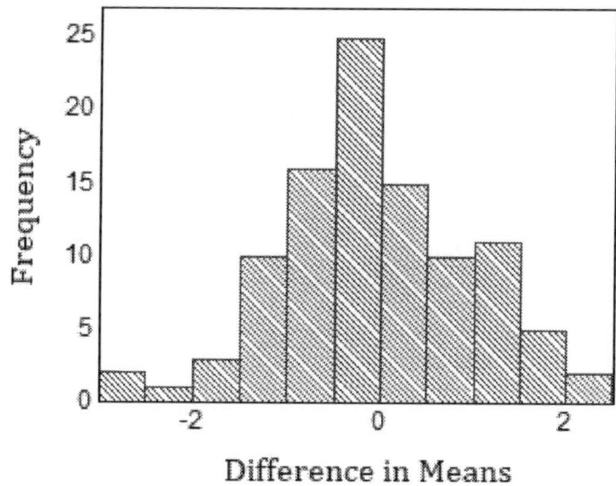

2. In a controlled experiment, groups 1 and 2 of the study were measured to have average heart rates at rest of 71.5 and 82.6 beats per minute, as shown with their respective standard deviations in the table below. There are 25 subjects in group 1 and 32 subjects in group 2.

Group	$\bar{x}$	s	n
1	71.5	11.3	25
2	82.6	9.1	32

Determine the 95% confidence interval for estimating the population mean difference from this sample.

3. Sample data for a medical study are given below.

	Men			Women		
	$\bar{x}$	s	n	$\bar{x}$	s	n
Height	68.9	2.7	1545	63.4	2.5	1781
Weight	194.0	33.8	1612	157.7	34.6	1894
BMI	28.8	4.6	1545	27.6	5.9	1781
TC	192.4	35.2	1544	207.1	36.7	1766

From these samples, determine the 95% confidence interval for estimating the population mean difference between men and women for each parameter:

a) height

b) weight

c) body mass index (BMI)

d) total cholesterol (TC)

Appendix I. Index

R

S

T

U

Made in the USA
Middletown, DE
07 March 2021